Lecture Notes in Civil Engineering

Volume 297

Lecture Notes in Civil Engineering (LNCE) publishes the latest developments in Civil Engineering—quickly, informally and in top quality. Though original research reported in proceedings and post-proceedings represents the core of LNCE, edited volumes of exceptionally high quality and interest may also be considered for publication. Volumes published in LNCE embrace all aspects and subfields of, as well as new challenges in, Civil Engineering. Topics in the series include:

- Construction and Structural Mechanics
- Building Materials
- Concrete, Steel and Timber Structures
- Geotechnical Engineering
- Earthquake Engineering
- Coastal Engineering
- Ocean and Offshore Engineering; Ships and Floating Structures
- Hydraulics, Hydrology and Water Resources Engineering
- Environmental Engineering and Sustainability
- Structural Health and Monitoring
- Surveying and Geographical Information Systems
- Indoor Environments
- Transportation and Traffic
- Risk Analysis
- Safety and Security

To submit a proposal or request further information, please contact the appropriate Springer Editor:

- Pierpaolo Riva at pierpaolo.riva@springer.com (Europe and Americas);
- Swati Meherishi at swati.meherishi@springer.com (Asia—except China, Australia, and New Zealand);
- Wayne Hu at wayne.hu@springer.com (China).

All books in the series now indexed by Scopus and EI Compendex database!

Kasinathan Muthukkumaran ·
Rajesh Sathiyamoorthy · Arif Ali Baig Moghal ·
S. P. Jeyapriya
Editors

Ground Improvement Techniques

Proceedings of the Indian Geotechnical Conference 2021 Volume 3

Springer

Editors
Kasinathan Muthukkumaran
Department of Civil Engineering
National Institute of Technology Tiruchirappalli
Tiruchirappalli, Tamil Nadu, India

Rajesh Sathiyamoorthy
Department of Civil Engineering
Indian Institute of Technology Kanpur
Kanpur, Uttar Pradesh, India

Arif Ali Baig Moghal
Department of Civil Engineering
National Institute of Technology Warangal
Warangal, Telangana, India

S. P. Jeyapriya
Department of Civil Engineering
Government College of Technology
Coimbatore, Tamil Nadu, India

ISSN 2366-2557 ISSN 2366-2565 (electronic)
Lecture Notes in Civil Engineering
ISBN 978-981-19-6726-9 ISBN 978-981-19-6727-6 (eBook)
https://doi.org/10.1007/978-981-19-6727-6

This Springer imprint is published by the registered company Springer Nature Singapore Pte Ltd.
The registered company address is: 152 Beach Road, #21-01/04 Gateway East, Singapore 189721, Singapore

Preface

The Indian Geotechnical Society, Trichy (IGS-Trichy) Chapter, and National Institute of Technology (NIT) Tiruchirappalli, India, organized the Indian Geotechnical Conference (IGC-2021) at Trichy during 16–18 December 2021. The main theme of the conference was **"GEO-INDIA"—GEO**TECHNICS FOR **IN**FRASTRUCTURE **D**EVELOPMENT AND **I**NNOVATIVE **A**PPLICATIONS.

The sub-themes of the conference included:

1. Soil Behaviour and Characterization of Geomaterials
2. Geotechnical, Geological and Geophysical Investigation
3. Foundation Engineering
4. Ground Improvement Techniques
5. Geo-environmental Engineering
6. Soil Dynamics and Earthquake Geotechnical Engineering
7. Earth Retaining Structures, Dams and Embankments
8. Slope Stability and Landslides
9. Transportation Geotechnics
10. Geosynthetics Application
11. Computational, Analytical and Numerical Modelling
12. Rock Engineering, Tunnelling, Deep Excavations and Underground Constructions
13. Forensic Geotechnical Engineering and Case Studies
14. Others: Behaviour of Unsaturated Soils, Offshore & Marine Geotechnics, Remote Sensing & GIS, Instrumentation & Monitoring, Retrofitting of Geotechnical Structures, Reliability in Geotechnical Engineering, Geotechnical Education, Codes & Standards, & any other relevant topic.

The proceedings of this conference consists of selected papers presented at the conference. The proceedings is divided into six volumes. A special issue on IGC-2021 keynote and theme lecture presentations were published by Indian Geotechnical Journal.

We sincerely thank all the authors who have contributed their papers to the conference proceedings. We also thank all the theme editors and reviewers who have been instrumental in giving their valuable inputs for improving the quality of the final papers. We greatly appreciate and thank all the student volunteers for their unwavering support that was instrumental in preparation of this proceedings. Finally, thanks to Springer team for their support and full cooperation for publishing six volumes of this IGC-2021 proceedings.

Trichy, India
2021

Kasinathan Muthukkumaran
Chairman, IGC

Contents

About the Editors

Dr. Kasinathan Muthukkumaran is currently Professor in Civil Engineering at National Institute of Technology, Tiruchirappalli, India. He obtained Ph.D. in Soil-Structure Interaction and Marine Geotechnical Engineering from Indian Institute Technology Madras. He has published more than 150 papers in international and national journals and conferences. He has completed 5 R&D (including ISRO—**Chandrayaan-2 Mission project**) and 70 major consultancy projects in Geotechnical Engineering and published two patent including **"Moon Soil"** (A Method for Manufacture of Highland Lunar Soil Simulant). He has guided 10 Ph.D. scholars and six more are in progress, five MS (by research) and more than 40 M.Tech. students in Geotechnical and allied research areas. He is the **Founder Chairman** of Indian Geotechnical Society (IGS-Trichy) Trichy Chapter. He is a member of Technical Committee (TC-301 on **"Preservation of Historic Sites"**) of International Society for Soil Mechanics and Geotechnical Engineering. Prof. Muthukkumaran area of research is in geotechnical engineering, which includes pile foundation, soil-structure interaction, marine geotechnics and foundations, field instrumentation, geotechnical physical modeling, ground improvement and forensic geotechnical engineering. He has received DST Young Scientist Award, IGS- Smt. Indra Joshi Biennial Award and Keynote Paper Award—GEOMATE Conference 2015 at Osaka, Japan. He is an Associate Editor of *Australian Journal of Civil Engineering and Serving* editorial board member of several journals. Prof. Muthukkumaran has significant administrative contribution as Estate Officer, Associate Dean (Planning and Development), Member of Buildings and Works Committee and Member of **Board of Governors (BoG)** of National Institute of Technology, Tiruchirappalli and Member of Buildings and Works Committee, IIM Trichy. He has received **NIT Trichy Achiever Awards** for research publications, research projects, maximum citation and consultancy projects.

Dr. Rajesh Sathiyamoorthy is Associate Professor of the Department of Civil Engineering at Indian Institute of Technology (IIT) Kanpur, India. He holds Doctor of Philosophy (2010) in Geotechnical Engineering from IIT Bombay and Master of Technology (2001) from College of Engineering Guindy, Anna University. He was at University of Joseph Fourier, LTHE, Grenoble, France as a Postdoctoral Fellow (2010–2011); Tongji University, Shanghai, China, as a Talented Young Scientist (2019–2020); University of Applied Sciences, Darmstradt, Germany, as Young Researcher (2004–2005). His research studies focus on hydro-mechanical behaviour of geomaterials, numerical and physical modelling of geostructures, application of geosynthetics and engineered cover systems and has published more than 95 technical papers in journals and conferences and delivered more than 25 keynote/invited lectures at various conferences and other events. He is a recipient of Institution of Engineers (India) Young Engineers Award-2013 (Civil Engineering), Talented Young Scientist Award-2018 from Ministry of Science and Technology China and Distinguished Alumni Award-2019 (Academic and Research) from VIT university. Dr. Rajesh is also a recipient of Prof. G. A. Leonards Biannual Award-2011, IGS—Prof. A. V. Shroff Biannual Award-2013, IIT Kanpur Best Instructor Award (seven times) and few best paper awards (IGS-Bangalore Chapter YGE Award-2015 and 2019, IGS-Prof. C. S. Desai Biennial Award-2020, IGC-2020). He is ISSMGE, TC-106 (Unsaturated Soils) committee member representing India.

Dr. Arif Ali Baig Moghal is Distinguished Researcher and an Associate Professor of the Department of Civil Engineering at the National Institute of Technology (NIT), Warangal. Dr. Arif received his Ph.D. in Civil Engineering from the Indian Institute of Science (IISc), Bangalore, India. Prior to joining NIT, he had worked at Bugshan Research Chair in Expansive Soils at King Saud University, Riyadh as professor. He has over 16 years of research, teaching and consulting experience within the broad fields of civil, geotechnical, geoenvironmental engineering addressing the nexus between sustainability and the environment. His research was funded by KACST (King Abdulaziz City for Science and Technology, Saudi Arabia) and Department of Science and Technology (DST), India. Dr. Moghal is the author of 53 journal papers, one edited book, 11 book chapters, 33 ASCE Geotechnical Special Publications and 25 full conference papers. He has given 23 invited presentations in India and Saudi Arabia. He is Fellow of the Indian Geotechnical Society (FIGS), Fellow of the Institute of Engineers, India (FIE), and a member of the American Society of Civil Engineers (MASCE).

Dr. S. P. Jeyapriya is currently working as Professor, Department of Civil Engineering (Geotechnical), Government College of Technology, Coimbatore, Tamilnadu, India. She has over 20 years of experience in teaching and research. She has published around 30 research articles in international and national journals, 50 papers in international and national conferences. She has coordinated around 15 programs that include faculty development programs, workshops, conferences and seminars. She has delivered invited lectures in several academic institutions. She is a reviewer of research articles submitted to journals and conferences and had done peer reviewing of e-course developed in the domain environmental engineering under NAIP. She has received two best paper awards. She is serving as State Government Nominee to the Board of Governors of few institutes. She is a life member of Indian Geotechnical Society (IGS), New Delhi, India, and Honorary Secretary of IGS-Coimbatore Chapter and has coordinated in establishing more than 20 student chapters at various institutes. Her areas of interests include soil mechanics, foundation engineering, ground improvement techniques and environmental geotechnology. She has been doing consultancy services to various government, quasi government and private organisations.

Chapter 1
Enhancement of Soil Properties by Using Red Mud and Lime

Sujit Kumar Rout, Rupashree Ragini Sahoo, Soumya Ranjan Satapathy, and Barada Prasad Sethy

Introduction

The term red mud (RM) is used synonym of bauxite tailing. Annually over 150 million tons of RM produced internationally out of which India is producing 9 million tons per year (Ministry of Mines, Government of India, 2019). Red mud is produced from the refining process of bauxite in to alumina followed by Bayer's process, which is not disposed satisfactorily. According to Yang and Xiao [1] the RM is highly alkaline in nature and stored either in a red mud pond with low solid content about 15–40% by volume or in dry form more than 65% [2]; either way, it uses large amount of land. The properties of RM like chemical composition typically depend on the extraction of alumina from bauxite which influences the overall properties of RM. The primary composition of RM is Fe_2O_3 (48–54%), Al_2O_3 (17–20%), SiO_2 (4–6%), Na_2O (3–5%), TiO_2 (3–4%), and CaO (1–2%) [3]. Various methods have been adopted to find out the properties and its utilization [4–6] but, due to very less utilization rate, huge quantity of RM leftovers in ponds. Numerous methods have been implemented by the different organizations to utilize the RM in an effective way. Due to its cohesive property, the RM is considered as an effective material to improve the engineering properties of the soil. Many methods have been implemented by the different organizations to dispose the waste material which will cause benefits to the society. Due to the cohesive property, RM is the efficient material for enhancing the engineering properties of soil. Parekh [7, 8], studied the behavior of RM and suggested that it is highly alkaline in nature and have clay fraction (20–30%). Also,

S. K. Rout · B. P. Sethy (✉)
NIST (Autonomous), Berhampur, India
e-mail: barada.jeetu@gmail.com

R. R. Sahoo
VSSUT, Burla, India

S. R. Satapathy
Nalanda Institute of Technology, Bhubaneswar, India

K. Muthukkumaran et al. (eds.), *Ground Improvement Techniques*, Lecture Notes in Civil Engineering 297, https://doi.org/10.1007/978-981-19-6727-6_1

he reported that major particles are silt (CaCo3, goethite, hematite, sodalite, and gibbsite. Vick [9] found out its low plasticity property and found out LL to be 45% and PL to be 10% with high specific gravity ranging from 2.8 to 3.3. Kalkan [10] had explained stabilization of expansive clay with red mud and cement-RM. He observed that strength is increasing and permeability and swelling pressure are decreasing. Sundaram [11] studied the behavior of RM and suggested that it can be used as foundation material in in situ condition itself. He also found that the RM is highly alkaline in nature ranging from 9.3 to 10.2. The Atterberg limits range from 39 to 45%, 27 to 29%, and 19 to 22% as liquid limit, plastic limit, and shrinkage limit, respectively. Rout et al. [12] used RM for the road embankment design based on its geotechnical properties. They observed that specific gravity, MDD, and soil friction angle values are more compared to soil without addition of RM. Satyanarayana et al. [13] studied the characterization of lime stabilized red mud mix for feasibility in road construction. RM was stabilized with 2–12% of lime at an increase rate 2%, and tests like UCS, split tensile strength, and CBR were conducted at 1, 3, 7, and 28 days curing periods, respectively. It has been detected that 10% of lime shows higher values when compared with other percentage. Singh et al. [14] stabilized the RM using cement kiln dust (CKD). The percentage of CKD varies from 2 to 12% with an increment of 2%. The strength criteria like MDD and UCS values have been checked. The optimum percentage of CKD is 8%. After the optimum value, further addition of CKD has no effect on strength criteria. Deelwal et al. [15] did the characterization of RM (both index and engineering properties). They concluded that RM is suitable for base and subbase course of road where the traffic is less. They also suggested that it is suitable for some geotechnical work. Pandey and Jawaid [16] did the stabilization of disturbed soil by adding fly ash and RM. The optimum value of RM is found to be 30% with 3% of fly ash which shows the higher CBR values. Lakshmi et al. [17] had stabilized red mud with cement in different proportions and obtained increase in OMC and UCS and decrease in MDD value with increased dosage of cement. Cement and RM mixture changed the dispersed structure of RM to flocculated structure; hence, it increased the strength of cement-RM mixture. Aswathy et al. [18] studied the behavior of clay by adding RM and found that addition of 20% RM with 2% of lime gives more strength compared to only addition of RM (15%).

This study involves the utilization RM, which can be used for stabilization of weak clay soil which has been collected from NALCO, Odisha. The strength criteria (i.e., compaction, UCS, direct shear tests, CBR) of have been studied for the maximum utilization of RM. The RM and soil mix are then treated with lime to obtain the optimum utilization of RM with lime treatment. All the tests have been executed confirming to IS 2720.

Table 1.1 Different properties of RM and soil

Properties		Soil	Red mud
Maximum dry density (g/cc)		1.66	1.73
Optimum moisture content (%)		19	24
Specific gravity		2.36	3.27
Grain size Distribution (%)	(a) Sand	5	12
	(b) Silt	72	36
	(c) Clay	23	52
Liquid limit (%)		48.72	33
Plastic limit (%)		26.12	24
Plasticity index		22.6	9
Classification		CI	ML
Cohesion (kN/m^2)		29.42	20.59
Soil friction angle (in degree)		19	36.5

Materials and Research Methodology

Soil

The soil is collected from Godavaga village, Sambalpur. To make the soil free from vegetation, pebbles, gravel, etc., a depth of 0.3 m from ground level is chosen for sample collection. After collection, the soil lumps are broken into small pieces and passed through 4.75 mm IS sieve. Based on the index properties, the soil is called CI conferring to IS classification system. The geotechnical properties are mentioned in Table 1.1.

Red Mud

Red mud is collected from NALCO which is located in the district of Odisha called. RM is a multifaceted material that varies due to the dissimilar types of bauxites rummage sale and has diverse parameters. Test results are shown in Table 1.1; it shows that RM is subjugated by silt elements and also high plasticity features. The basic geotechnical characteristics of RM are stated in the Table 1.1. Table 1.2 shows the chemical elements. The chemical compositions and metal content of the red mud were determined using energy dispersive spectroscopy (EDS) in Fig. 1.1.

Table 1.2 Chemical elements of RM and lime

Red mud		Lime	
Element	Weight (%)	Element	Weight (%)
$CaCO_3$	0.72	$CaCO_3$	3.12
SiO_2	31.95	SiO_2	36.12
Al_2O_3	7.10	Al_2O_3	0.12
Fe	45.38	Fe	0.26
Ti	2.33	Ca	52.52
Na	6.24	MgO	0.71
Ca	1.34	YbF_3	6.21
P	0.17	–	–
K	0.13	–	–

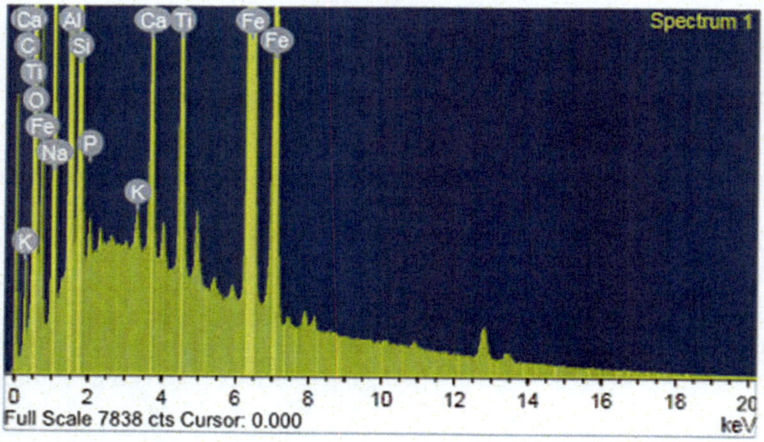

Fig. 1.1 EDS breakdown of RM

Lime

Lime is collected from nearby market in Sambalpur. The chemical composition (elements) is mentioned in Table 1.2, and chemical compounds present in lime are shown in Fig. 1.2 using EDS test.

Sample Preparation

The soil used in this study is clay. It is oven dried at 105 °C approximately and grounded before use to get uniform mixture. First, the desired proportion of clayey

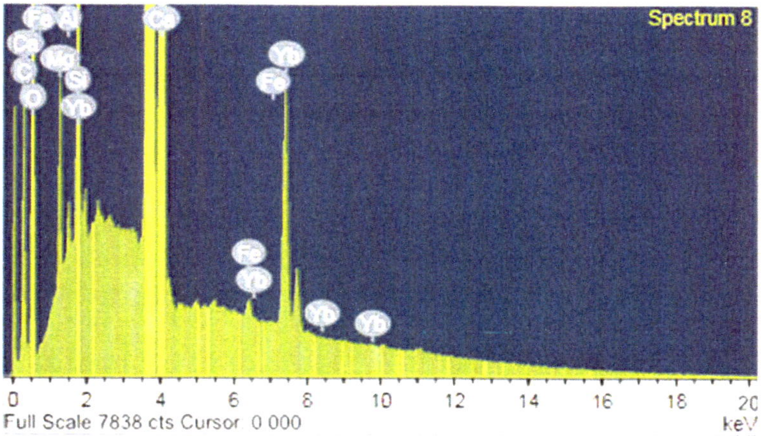

Fig. 1.2 EDS breakdown of lime

Table 1.3 Experimental program for different proportions of soil with dissimilar proportion of RM and lime

Sl. No.	Symbol	Soil (%)	Red mud (%)	Lime (%)
1	R1	100	0	0
2	R2	70	30	0
3	R3	60	40	0
4	R4	50	50	0
5	RL1	60	40	2
6	RL2	60	40	3
7	RL3	60	40	4
8	RL4	60	40	5

soil, red mud, and lime has been blended together under dry condition. The percentages of red mud are 30, 40, and 50% of the total weight of soil. Lime was taken as 2, 3, 4, and 5% by weight of total volume of mixture (soil and RM) (Table 1.3).

Results and Discussion

Compaction Characteristics

Proctor test (standard) has been conducted to determine the compaction characteristics of the red mud and soil and RM stabilized soil through lime with reference IS:

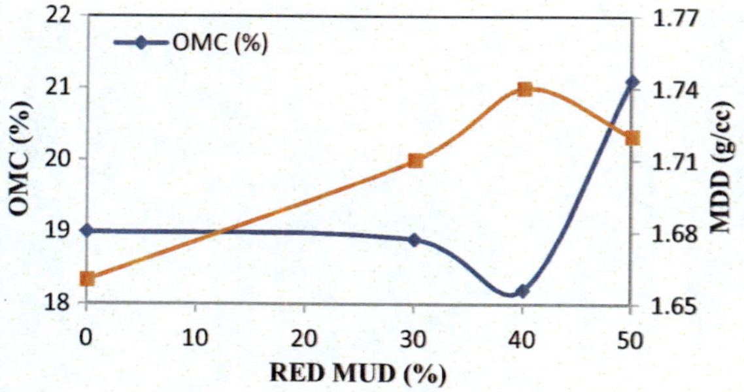

Fig. 1.3 OMC and MDD values for soil with red mud

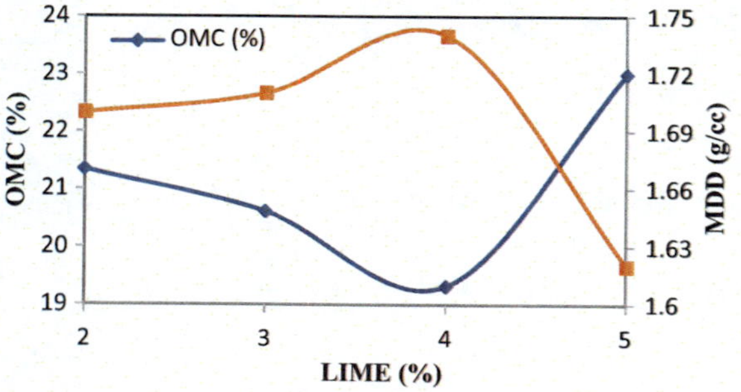

Fig. 1.4 OMC and MDD of RM stabilized soil with lime

2720 (Part 7). The variations of diverse combinations of OMC and MDD are plotted in Figs. 1.3 and 1.4.

The test result in Fig. 1.3 shows that, OMC of soil reduces significantly with the intensification of red mud percentage up to 40 after which the OMC value increases with increase in RM content. Simultaneously, the MDD of red mud mix soil increases significantly with red mud content (increase) up to a certain percentage after which there is decrease in MDD value. Originally, the OMC value and MDD value of virgin soil were 19% and 1.66 g/cc, respectively. But after adding red mud in different proportions, the optimal value of MDD and OMC was 18.2% and 1.74 g/cc, respectively.

Figure 1.4 shows that the optimum red mud mixed soil was again treated with different proportions of lime in order to get the best result. It has been observed that, the addition of $Ca(OH)_2$, the OMC decreases about 4% then it is increasing and the vice versa pattern is observed in case of MDD. The OMC and MDD values obtained in addition with 4% lime were 19.32% and 1.74 g/cc.

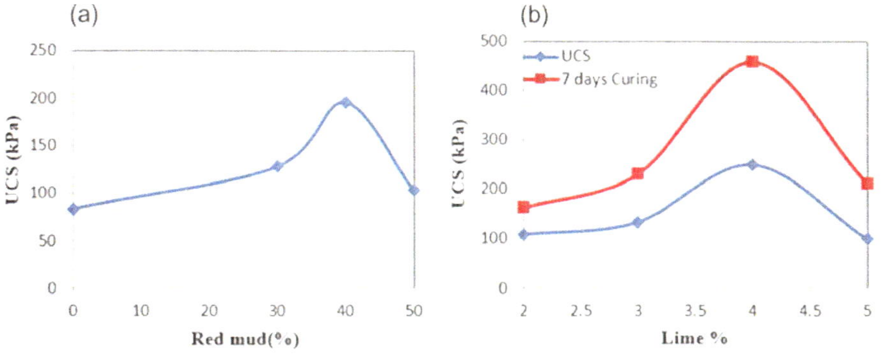

Fig. 1.5 Variation of UCS **a** with red mud, **b** with lime

Uniaxial Compression Test

The uniaxial compression test was conducted with strain rate of 1.25 mm/min as per IS: 2720 (Part 10), and static compaction method has been used to reach MDD and OMC. Figure 1.5 shows that the UCS is improved by 1.5 times in comparison with the soil without reinforcement and occurred at RM content of 40%. In addition, the lime has considerable effect on UCS when stabilized with RM. The UCS of soil increased to 249.58 kN/m^2 from 195.83 kN/m^2 when lime was added to RM stabilized clayey soil. The accumulation of lime more than 4% decreases the value (UCS) to 99.63 kN/m^2. The UCS value hits the peak with addition of 4% lime and 40%, or RM attains the maximum value when the lime is 4%. The UCS values of the virgin soil increases three times by the combined effect of RM and Ca $(OH)_2$. The cause of this consequence is the pozzolanic reactions of Ca $(OH)_2$ with soil and red mud. After addition of 5% lime, the strength decreases because of the availability of extra lime.

California Bearing Ratio (CBR)

The CBR test is used to evaluate the strength of sub-grade of road embankment. This test has been conducted as per IS: 2720 (Part 16).

The CBR value of RM mixed soil rises significantly through increasing the RM content up to 40% after which there is a decreasing trend. The CBR of RM stabilized soil increases to 4% from 2.8% of virgin soil in unsoaked condition and 2.69% from 1.7% of virgin soil in soaked condition.

By adding different percentages of lime show significant properties on the CBR (soaked and unsoaked) of the RM alleviated soil. Adding different percentage of lime, the unsoaked CBR values of RM stabilized soil increased to 19.43% from 6.32%, at 4% lime, by an increasing factor of 1.75; further the addition of lime decreases

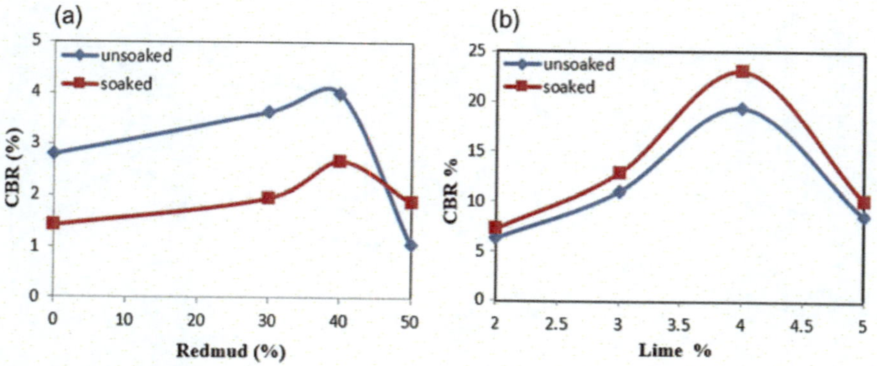

Fig. 1.6 Variation of CBR **a** with red mud, **b** with lime

the CBR value (soaked) of soil. Likewise, the CBR value (soaked) of RM stabilized soil increased to 23.07% from 7.27%, when 4% lime was added, by an increasing factor of 1.78 and further decreases (Fig. 1.6).

Shear Strength Parameters

Soil shear strength properties include the cohesion (c) and the soil friction angle. The test is conducted by putting the sample at OMC and MDD inside the shear box. The sample has been compacted in the box (shear) of (60 × 60 × 60) by tamping at MDD to obtain the specimens. The samplings were tested at stresses of 50 kN/m^2, 100 kN/m^2, and 150 kN/m^2 in UU conditions confirming to IS code 2720 (Part 13) 1986. The load is applied at a strain rate 0.002 mm/s. The readings were noted down at a fixed interval of horizontal dial gauge readings to study the displacement performance of soil RM mix and soil–RM–lime mix.

The observed shear parameters are c and φ in the Fig. 1.7 specifies that the stabilized soil shows an increase in the cohesion (c) and the soil friction angle (φ) up to 40% of RM content then decreases. The soil friction angle is increased considerably from 19° to 30°, and the cohesion increases from 29.42 to 49.03 KN/m2.

Figure 1.8 illustrates that there is a rise in the cohesion (c) and soil friction angle (φ) up to 4% of lime content. The soil friction angle increased significantly from 19° to 40° at 4% of lime and then decreased. Similarly, cohesion increased from 33.33 to 58.82 kN/m2 and after that it decreased.

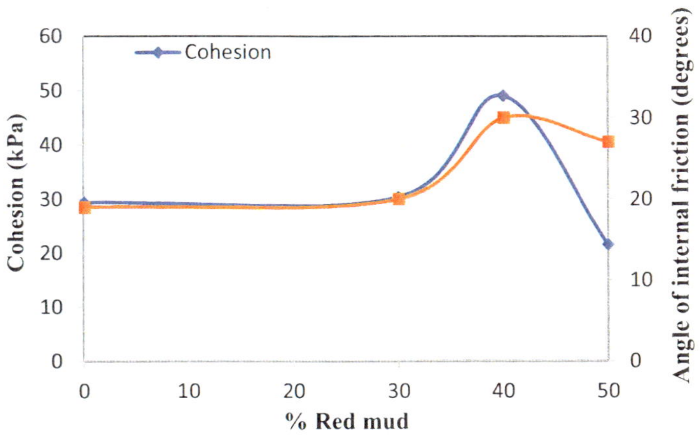

Fig. 1.7 Effect of red mud on shear strength parameters

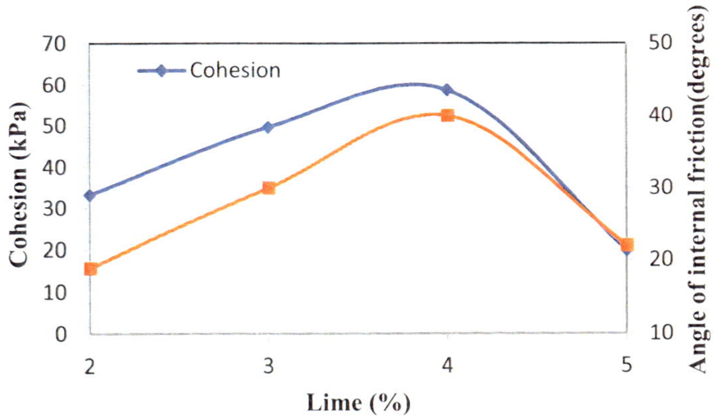

Fig. 1.8 Effect of lime on shear strength parameters

Analysis of EDS

The EDS analysis of particles is shown in Figs. 1.9 and 1.10 for RM stabilized soil and RM-soil mix with lime. The Ca content was increased with addition of lime that is 1.55%–5.78%. The iron content decreases from 45.38 to 29.0% in RM stabilized soil and 26.57% in lime stabilized red mud-soil mix.

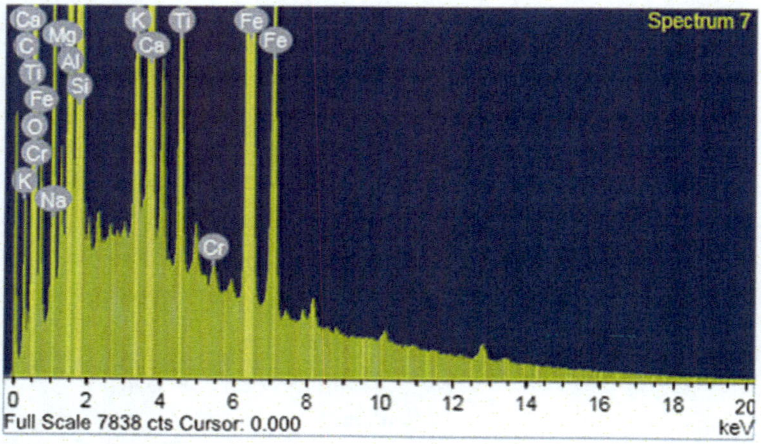

Fig. 1.9 EDS analysis of sample (red mud 60% + soil 40%)

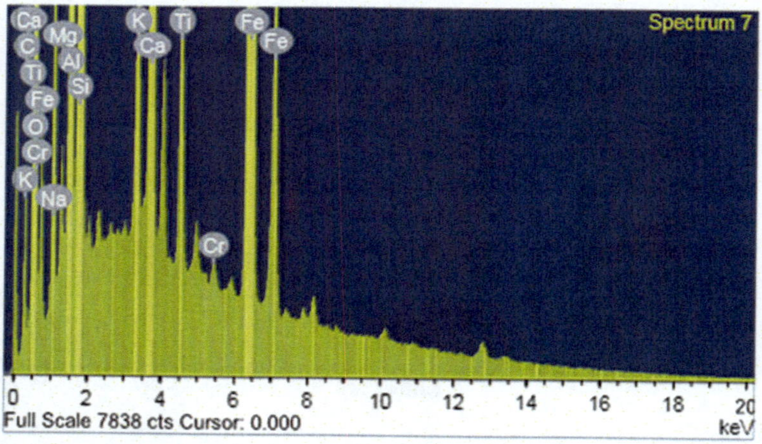

Fig. 1.10 EDS analysis of sample (red mud 60% + soil 40% + lime 4%)

Conclusion

Based on shear strength parameter, UCS results, and CBR results, the optimum proportion of soil to RM was 60:40 by weight. Addition of lime (up to 4%) to the stabilized clayey soil with optimal ratio of RM, the OMC value decreased with intensification in MDD. Further addition of lime results in decreasing in MDD and increasing in OMC. The soaked CBR of RM stabilized soil added with lime increased by a factor of 1.78 at 4% lime and further decreased. In unsoaked condition, the CBR value of RM stabilized soil added with lime increased by a factor of 1.75 at 4% lime and then decreased. The UCS achieves the maximum value when 4% of lime is added with RM stabilized soil. The UCS value of the virgin soil increases 3 times by the combined effect of RM and lime. After curing, the maximum value of UCS was found at 4% Lime. So, the optimum percentage of Soil: Red mud: Lime was found to be 60:40:4. The red mud is utilized with lime to enhance the behavior of soil strata for better results. Hence, it may be concluded from this study that the RM may be used in soil enhancement of poor clayey soil in its place of simply being predisposed on the land.

References

1. Yang J, Xiao B (2008) Development of unsintered construction materials from red mud wastes produced in the sintering alumina process. Constr Build Mater 22(12):2299–2307
2. Power G, Grafe M, Klauber C (2011) Bauxite residue issues. I: Current management, disposal and storage practices. Hydrometallurgy 108(1–2):33–45
3. Chaddha MJ, Rai SB, Goyal RN (2007) National seminar on environmental concern and remedies in Alumina Industry at NALCO, Damanjodi, India, Characteristics of red mud of Indian alumina plants and their possible utilization pp 41–44
4. Wang S, Boyjoo Y, Choueib A, Zhu ZH (2005) Removal of dyes from aqueous solution using fly ash and red mud. Water Res 39(1):129–138
5. Rubinos DA, Spagnoli G, Barral MT (2016) Chemical and environmental compatibility of red mud as liners for hazardous waste containment. Int J Environ Sci Technol 13(3):773–792
6. Alam S, Das SK, Rao BH (2017) Characterization of coarse fraction of red mud as a civil engineering construction material. J Clean Prod 168:679–691
7. Parekh B, Goldberger W (1976) An assessment of technology for possible utilization of Bayer process muds. US EPA, EPA-600/2-76-30
8. Somogyi F, Gray D (1977) Engineering properties affecting disposal of red mud. In: Proceedings in conference on geotechnical practice for disposal of solid waste materials, ASCE, pp 1–22
9. Vick SG (1990) Planning, design, and analysis of tailings dams
10. Kalkan E (2006) Utilization of red mud as a stabilization material for the preparation of clay liners. Eng Geol 87(3–4):220–229
11. Sundaram R, Gupta S (2010) Constructing foundations on red mud. In: 6th international congress on environmental geotechnics, New Delhi, India, pp 1172–1175
12. Rout S, Sahoo T, Das S (2012) Utility of red mud as an embankment material. Int J Earth Sci Eng (5):1645–1651
13. Satyanarayana PVV, Ganapati Naidu P, Adiseshu S, Hanumanth Rao CHV (2012) Characterization of lime stabilized red mud mix for feasibility in road construction. Int J Eng Res Develop 3(7):20–26
14. Singh K, Pandey RK, Mishra CS, Rai AK, Bind YK (2014) Analysis on utilization of cement kiln dust stabilized red mud for road construction. Int J Civ Eng Technol 5(8)
15. Deelwal K, Dharavath K, Kulshreshtha M (2014) Evaluation of characteristic properties of red mud for possible use as a geotechnical material in civil construction. Int J Adv Eng Technol 7(3):1053
16. Pandey PK, Jawaid A (2015) Soil improvement using red mud and fly ash. Glob J Eng Sci Res 1(12):7–9
17. Lakshmi TDV, Prasad DSV, Kumar MA, Raju GP (2015) Stabilization of industrial waste red-mud with cement. Int J Res Innov Earth Sci 2(1)
18. Aswathy M, Salini U, Gayathri VG (2019) Utility of lime and red mud in clay soil stabilization. In: Geotechnical characterization and geo-environmental engineering. Springer, Singapore, pp 19–26

19. IS 2720-Part 7 (1987) Methods of test for soils—part 8: determination of water content and dry density relation using light compaction, Bureau of Indian Standards, New Delhi
20. IS: 2720-Part 10 (1973) Methods of test for soils—determination of unconfined compressive strength, Bureau of Indian Standards, New Delhi
21. IS: 2720-Part 13 (1986) Methods of test for soils—direct shear test, Bureau of Indian Standards, New Delhi
22. IS 2720-Part 16 (1987) Methods of test for soil—laboratory determination of CBR

Chapter 2
Ground Improvement for Open Foundation on Soft Clays Using Stone Columns and PVD Drains for Retaining Walls and Approaches of a Cable Stayed ROB

Anurag Goyal and Anjali Gupta

Introduction

During the design of any embankment, the designer mainly emphasizes on checking of the bearing pressure, settlement, and overall stability. In case of embankment over soft soil, it is necessary to improve the sub-soil to increase the required bearing capacity and other related parameters. There are many ground improvement methods available nowadays, however, the engineering properties of the soft sub-soil can be improved considerably using preloading with prefabricated vertical drains (PVD) [1].

For design of retaining walls ground, improvement by provision of granular piles/stone columns is an effective method to strengthen the soft clays and control settlements within acceptable limits. Therefore, to optimize the cost and find a technically viable solution, combination of two ground improvement techniques has been proposed using stone columns beneath the retaining wall footings and using PVDs beneath the intermediate soil fill. The paper presents the case study of ground improvement carried out for the approaches to a cable stayed bridge—an ROB at Chandmari, West Bengal. Figures 2.1 and 2.2 show the typical cross section of approach ramps and elevation of ROB including approach ramps.

A. Goyal (✉)
Geotechnical Consultant, Liniva Consultants, New Delhi 110029, India
e-mail: anurgoyal@yahoo.com

A. Gupta
Department of Civil Engineering, Manav Rachna International Institute of Research and Studies, Faridabad 121003, India

K. Muthukkumaran et al. (eds.), *Ground Improvement Techniques*, Lecture Notes in Civil Engineering 297, https://doi.org/10.1007/978-981-19-6727-6_2

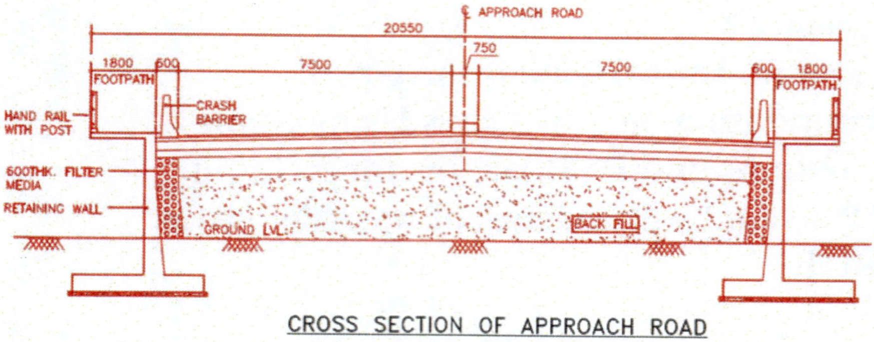

Fig. 2.1 Typical cross section of approach ramp

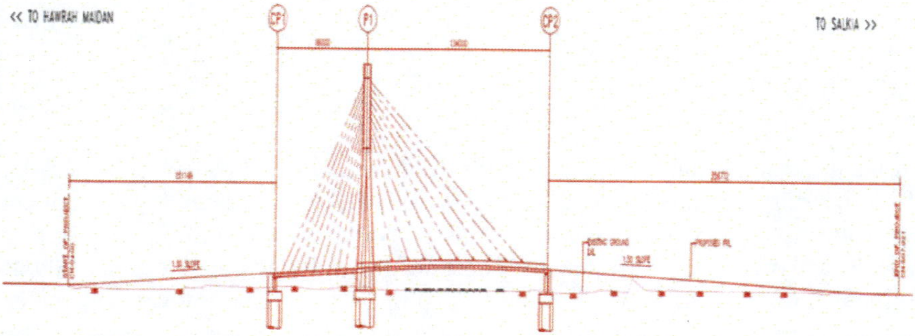

Fig. 2.2 Elevation of ROB and approaches

Sub-Soil Characteristics

Soil investigation report indicated that the top soil was observed to be filled up with overburden loose soil underlain by clayey soil. The clayey soil was found to be soft to medium till a depth of around 15.0 m. However, below this soil, a layer of stiff to very stiff clayey soil was also observed till the 30.0 m depth. This layer was followed by a hard clay layer till 50.0 m depth and further underlain by very dense sand.

Based on soil investigation report, the soil properties reported in Table 2.1 are recommended by considering total four layers and were utilized for the calculations of safe bearing capacity in shear and settlement. It is essential to note that in all three boreholes, slight variation in the depth of layer is observed. The depth of filled up soil was varying from 1.2 m to 5.7 m. Similarly, the depth of different layers was also found to be variable to some extent.

Table 2.1 Sub-soil strength parameters

Layer No.	Depth. M	Strength parameters	
		C (kPa)	$\emptyset$ (Degrees)
1	00.00–15.00	35	0
2	15.00–30.45	70	0
3	30.45–50.00	200	0
4	50.00–75.00	34	34

Recommendations Within Soil Investigation Report

As per soil investigation report, well foundation was suggested for the main span of ROB. For the approaches and retaining walls, the optimum solution had to be worked out. Hence, in order to optimize, it was essential to identify the alternative solution which can be adopted for the approach portion and side retaining walls. Hence, looking at the present condition, this exercise was undertaken where the existing capacity of ground was evaluated, and the same was compared with the required capacity of ground.

Safe Bearing Capacity of Ground

Looking at the ground model suggested in the soil investigation report, the ground has clayey stratum from the depth 0.0–15.0 m. Considering the approach requirements, a raft of width 6.0 m was assumed for calculating ultimate bearing capacity and allowable bearing pressure.

Ultimate bearing capacity for clayey soil, $q_{\text{ult}} = C \times N_c = 35 \times 5.14 = 179.9$ kPa.

Hence, allowable bearing pressure in shear criteria, q_{ult}/FOS = 179.9/2.5 = 71.96 kPa = 7.2 T/sqm.

Consolidation settlement at 71.96 kPa for a 12 m depth of clay layer (2 × B),

Using $C_c = 0.405$ (As per Bowels, $C_c = 0.009 \times (w_l - 10)$).

$e_o = 1.167$ (As per soil report).

$H = 12.0$ m (2 × B).

and equation [2],

$$\Delta H = \frac{C_c H}{1 + e_o} \log \frac{p'_o + \Delta p}{p_o} \tag{2.1}$$

Settlement = 300 mm (approximately).

Therefore, for a 7.2 T/m^2 allowable bearing pressure, a settlement of approximately 300 mm is expected which exceeds the permissible value of 100 mm for a plastic clay.

Expected Loading and Expected Pressure on the Ground

Looking at the approaches of ROB, the maximum height of retention above ground is approximately 8.0–9.0 m. This would require approximately $9 \times 2 + 2.4 = 20.4$ T/m^2 including live load surcharge of 2.4 T/sqm. Moreover, for retaining walls, it is essential to restrict the settlements within permissible limits.

Hence, on comparison of existing ground capacity with the expected pressure on the ground, it can be concluded that the ground is not suitable for the open foundation. Hence, the other option was to provide deep foundation, i.e., pile foundation or well foundation. However, the cost and time required for completion of these solutions are very high. This necessitates considering alternative solutions such as ground improvement techniques which can help in increasing the allowable bearing capacity by controlling shear failure and reducing settlement.

Ground Improvement Techniques

With the aim of improving bearing capacity and reducing settlement, mainly two ground improvement schemes are adopted for this project as stated below:

(I) Ground improvement using stone columns for side retaining walls.
(II) Ground improvement using PVDs for approach embankments.

Design of Ground Improvement Using Stone Columns for Retaining Walls

Stone columns are considered as one of the most versatile and cost-effective techniques and extensively used over the past few decades in numerous ground improvement and foundation projects. It has gained acceptance due to applicability to an array of soil conditions and soil strengths. The overall performance of stone columns is controlled by the lateral support provided by the surrounding soils, which typically increases with depth. Stone column is laid by drilling holes in the soft soil (clay in present case) and subsequently filling it up with granular material compacted in stages so as to improve the strength and consolidation properties of soft soil (clays).

Typically, stone column helps in improving safe load carrying capacity due to three mechanisms:

(a) By increasing load bearing capacity as the stones have higher modulus which results in higher load carrying capacity within the unit cell (area where stone column provides resistance).
(b) By densifying the surrounding soil which results in increased lateral confinement and improvement of shearing strength of soil.

Table 2.2 Soil parameters for the design of stone columns

Material	Cohesion, c (T/m^2)	Angle of internal friction, ϕ (degrees)	Unit weight, γ (T/m^2)
Clay	3.5	0	1.7
Gravel (for stone column)	0	40	1.9

(c) By applying surcharge effect which contributes to increase in load carrying capacity when stone columns are installed.

Soil parameters used for the design of stone columns. Selection of soil parameter is one of the critical aspects for the design of stone columns, and its design entirely depends upon the selected parameters. Shear strength parameters of existing ground and column material (typically aggregates/gravels) are essential parameters for the design of stone columns. Soil properties of clayey soil are considered as the same that recommended in the geotechnical investigation report. For aggregates to be filled within stone columns, $\phi = 400°$ in compacted/rammed condition is assumed for the design, which is reasonable and generally considered for the design of stone columns. The same is required to be ensured at site. The summary of shear strength parameters for the design of stone columns is mentioned in Table 2.2.

Geometrical parameters used for the design of stone columns. Diameter, spacing, length, and pattern are critical parameters for the design of stone columns. Any change in these parameters can result in significant variation in capacities of stone columns.

(a) Diameter of stone columns: Looking at the ground conditions, stone columns of 0.60 m diameter are selected.
(b) Spacing of stone columns: Spacing of stone column is generally directly proportionate to the load carrying capacity of stone column system. Hence, the spacing is determined based on the expected loading conditions. As per the required load carrying capacity, different spacing of stone columns is used, and summary of the same is shown in the design section.
(c) Length of stone columns: Length of stone columns is not directly used for the estimation of load carrying capacity. However, the length of stone columns plays a critical role for transferring the load to the ground. As per the present condition, the length of stone column is suggested such that the stone column is placed though soft, firm, medium, and rested on firm to very firm stratum. In order to ensure this, a benchmark of $N = 15$ is considered to decide length of stone columns. As per present ground condition, $N = 15$ is available at a depth of approximately 15.0 m. Considering the columns is to be rested at $N = 15$ and assuming a foundation depth of around 2.5 m and drainage blanket of 0.5 m, total length of stone column proposed is 12.0 m.
(d) Pattern of stone columns: Generally, triangular and square arrangements of stone columns are used throughout in India and abroad. Looking at industry trends,

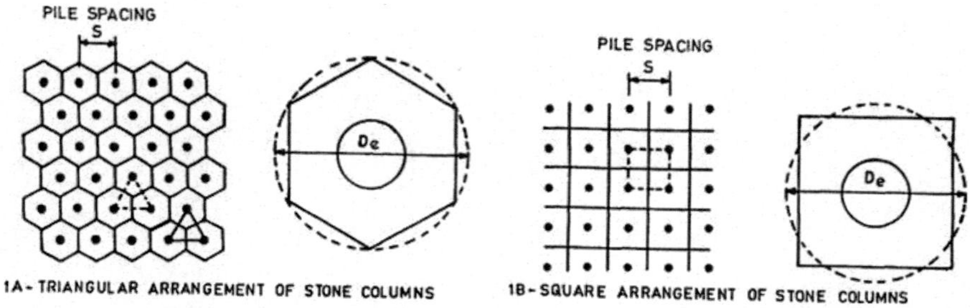

Fig. 2.3 Various patterns of stone columns

triangular pattern is most popular as it provides denser configuration. Hence, for this project, triangular pattern is adopted (Fig. 2.3).

Design of stone columns. The design of stone columns is done as per IS-15284 part-1 [3], and the load carrying capacities of ground where stone columns are provided are evaluated by adding the contribution of each of the following components:

(a) Capacity of column resulting from the resistance offered by the surrounding soil against its lateral deformation (bulging) under axial load.
(b) Capacity of stone column resulting from increase in resistance offered by the surrounding soil due to surcharge over it.
(c) Bearing support provided by the intervening soil between the columns. In general, for selected parameters such as shear strength of stone column and clay and selected diameter and spacing, the capacity offered by unit cell (area surrounding a single stone where resistance is offered by column itself and surrounding ground in load sharing mechanism) is evaluated. This capacity is checked against the required capacity. In order to optimize and achieve the required capacity, iterative process is adopted by keeping all parameters constant except spacing of stone column. By varying the spacing, different capacities can be obtained.

The design summary of stone columns is presented in Table 2.3.

Field performance of stone columns. The design capacities achieved using the calculations are based on theoretical approach. However, field performances are the key for the success of ground improvement projects. Hence, field trials and testing shall be done for single column and group column prior to actual ground improvement installation works in accordance with IS15284 (Part-1): 2004.

Also, frequent routine column load tests (single column and group column) shall be performed to assess effectiveness of ground improvement. Moreover, post-installation of stone columns, boreholes shall be driven within the soil to check the performance of ground improvement and cross verified with the parameters used in the design of stone columns.

Table 2.3 Design summary for stone columns

S. No.	Required safe load carrying capacity (T/m^2)	Spacing for 0.6 m diameter stone columns (m)
1	5.5 + 3.8 = 9.3	2.00
2	6.8 + 3.8 = 10.6	2.00
3	8.2 + 3.8 = 12.0	1.88
4	9.9 + 3.8 = 13.7	1.60
5	11.4 + 3.8 = 15.2	1.45
6	12.9 + 3.8 = 16.7	1.33
7	14.2 + 3.8 = 18.0	1.25
8	15.4 + 3.8 = 19.2	1.19

Design of Ground Improvement Using PVDS

Due to construction of a new embankment or structure on soft ground following problems may occur:

a. Increase in the pore water pressure (excess hydrostatic pressures) in the sub-soil.
b. Increase in effective stress due to decrease in pore water pressure with lapse of time which further lead to the increase in shear strength of the sub-soil.
c. Reduction in the volume of the soil causing settlements.

In order to handle these problems, prefabricated vertical drains (PVDs) are recommended in infrastructure projects to increase rate of consolidation [4]. PVDs also known as band drains due to its shape consist of a plastic core covered by geotextile filter with a width of about 100 mm and thickness around 4 mm. These are suitable for soft saturated and normally to slightly over-consolidated soils, prior to loading. Pattern used for laying PVDs is generally equilateral triangular with effective spacing between the band drains less than 1200 mm as per IRC: 75–2015 [5]. More details about PVDs can be referred from MORTH specifications for roads and bridge works clause 704.2.2.

Preloading is an effective and economical technique to accelerate rate of consolidation in comparison with other methods of ground improvement provided proper monitoring of settlements, development, and dissipation of pore water pressures are observed. In this approach, measurement of ground settlements and pore water pressure could support to estimate quantitatively the level of ground improvement concerning improvement in the shear strength and forecast its future behavior. But, the preloading techniques may not be alone sufficient because of long time duration required to obtain significant consolidation settlements following considerable attainment in strength to support the embankment loads. Installation of vertical drains along with preloading would not only reduce the preload time but also shorten the drainage path which will support rapid consolidation.

Field instrumentation, such as piezometers, settlement plate, and inclinometers, is used to monitor performance and possibly control the rate of construction of

Table 2.4 Soil parameters used for the design of PVDs with preloading

Sr. No.	Parameter	In situ soil parameters
1	Depth	16 m
2	Saturated unit weight	17 kN/m^3
3	Cohesion	35 kPa
4	Initial void ration (e_o)	1.167
5	Liquid limit (%)	55
6	Plastic limit (%)	28
7	C_c (=0.3 (e_o—0.27))	0.269
8	C_v (Coefficient of consolidation—vertical)	3.41 × 10^{-4} cm/s
9	C_h (Coefficient of consolidation—horizontal)	5.12 × 10^{-4} cm/s

embankment and/or surcharge. Settlements measuring devices and deep settlement points are used to measure only the rate and total amount of consolidation.

Soil parameters used for the design of PVDs with preloading. Selection of soil parameter is one of the critical aspects for the design of PVDs with preloading. The designing of this type of ground improvement entirely depends upon the selected parameters. Consolidation parameters and shear strength parameters of existing ground are summarized below in Table 2.4.

Geometrical parameters used for the design of prefabricated vertical drains and preloading. Cross section, length, spacing, and pattern of PVDs are critical parameters for the design of prefabricated vertical drains. Any changes within these parameters can result in significant variation in time for the consolidation. Moreover, the geometry of preload is also very important to achieve the required degree of consolidation. The details are mentioned below:

(a) Cross section of PVDs: PVDs are manufactured in factories under controlled environment. The most common size of PVDs is 100 mm × 4 mm. Hence, the same size is adopted for this project also.
(b) Spacing of PVDs: Spacing of PVDs is generally directly proportionate to the time required to achieve consolidation. More the spacing, more will be the time required for consolidation. Hence, the spacing is determined based on available time duration for any project. Analyzes are performed for various spacing, and timelines are derived to achieve 90% consolidation.
(c) Length of PVDs: Length of PVDs is dependent on the two factors, i.e., the influence zone of loading and the extent of very soft to soft soil. It is general practice to install PVDs through very soft to soft stratum even if the load intensities are not expected to reach the very soft to soft clay layer. This helps in preventing punching shear failure during an unusual loading event. Hence, in present case, the length of PVDs is kept constant at 20.0 m depth beneath and surroundings of the retaining wall structure. The length is decided such that a minimum *N*

value of 15 is targeted at the termination depth of PVDs. Hence, all the soil with N value less than 15 will be treated using PVDs.

(d) Pattern of PVDs: PVDs in triangular and square patterns are most commonly used in India and abroad. However, triangular pattern is most popular as it provides better results based upon industry trends. Hence, for this project, triangular pattern has been selected.

(e) Geometric parameters of preload: The height of surcharge is selected such that at all the time the loading during consolidation process is in excess of expected pressure intensity post-ground improvement. The details of height are discussed in design section. Moreover, the width of preload is selected such that sufficient preload is applied during the ground improvement procedures.

Results of PVDs design with preloading. The design of prefabricated vertical drains along with preloading is performed as per the procedures described in IRC: 75-2015. The design process of PVDs requires evaluating time required for any given degree of consolidation (U), coefficient of consolidation—horizontal (C_h), equivalent area of band drain (d), and spacing/equivalent diameter of cylindrical column (D) using the following Hansbo's equation.

$$t = \frac{D^2}{8 \times C_h} \times \left[\frac{1}{1 - \left({}^{d}/_{D} \right)^2} \times Ln\left(\frac{D}{d} \right) - \frac{3}{4} + \frac{1}{4}\left(\frac{d}{D} \right)^2 \right] \times Ln\left(\frac{1}{1 - U} \right) \tag{2.2}$$

This process is repeated by varying the degree of consolidation. Hence, a curve representing time versus degree of consolidation is generated. Using this curve, the settlement behavior of field can be predicted. The same is subsequently used to design the preloading. This process is then repeated for variable. As an assessment to the PVDs designs with preloading, the height of preloading is decided for targeted loading intensities (the height of loading is kept as the pressure applied by preloading soil equal to the required bearing capacity). Different PVD spacings are adopted, and the time required for the consolidation is evaluated. The summary of the same is presented in Table 2.5 for 0.8 m spacing.

Table 2.5 Summary of PVD design for 0.8 m spacing of PVDs

Required SBC (T/m^2)	Height of loading (m)	Applied stress at the end of preloading (T/m^2)	Waiting period in days
15.5	8.69	15.6	84
14.0	7.8	14.0	84
12.0	6.7	12.0	84
10.0	5.6	10.0	84
8.0	4.5	8.10	84
6.0	3.4	6.10	84

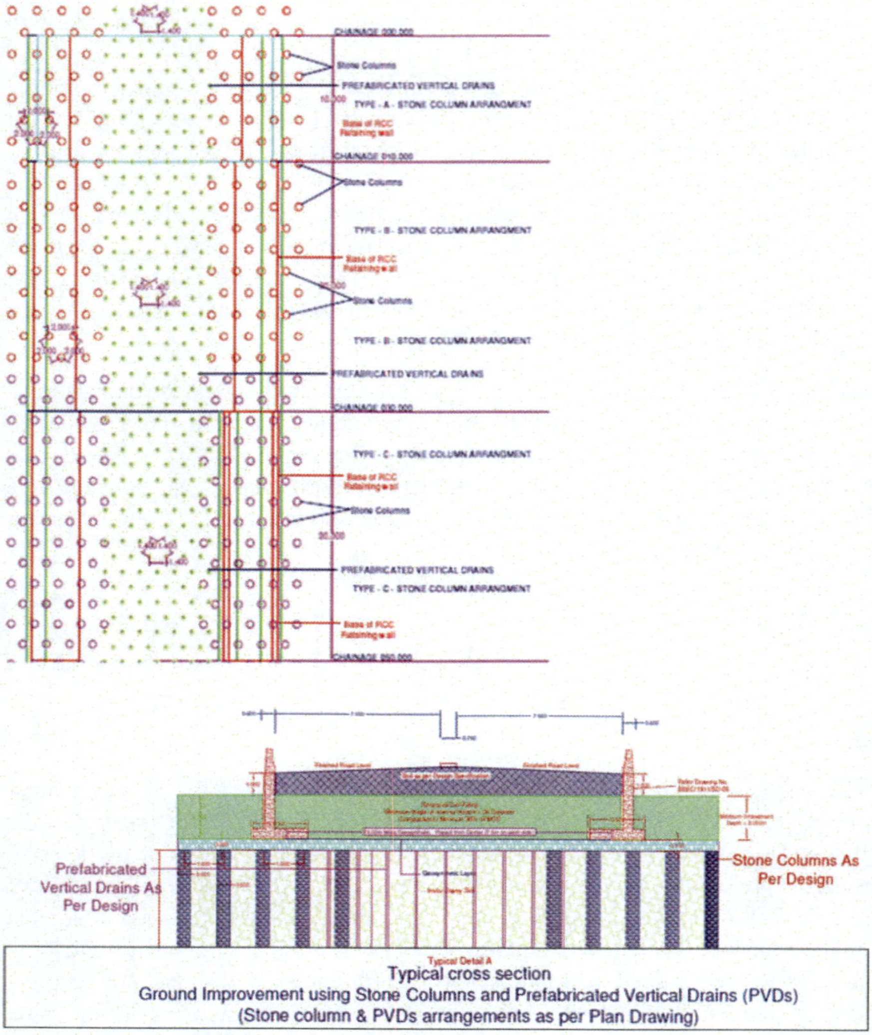

Fig. 2.4 Plan and section showing stone columns below retaining walls and PVDs below earthfill

Typical plan and section of the open foundation with ground improvement using stone columns and PVDs are depicted in Fig. 2.4.

Conclusion

Ground condition of the site consisted of clayey soil with very low SBC and large settlements is expected at site under the expected loading varying from 6 to

19.5 T/m^2. Stone columns are considered suitable beneath the retaining wall due to site constraints as the space required for preloading is not available. The diameter of 0.60 m is considered suitable for this project. The spacing of stone columns is decided based on the required allowable bearing pressures on the stone columns. Preloading along with PVDs is considered as a suitable ground improvement method for this project beneath embankment fill. The height of preloading is decided based on conservative estimation. These heights are such that it exerts approximately 1.5 T/m^2 excess pressure on ground as compared to expected loading intensity. The preloading embankments are checked for the embankment slope stability. The design of PVDs is performed for different spacings and different waiting periods are derived. As the applied preload is exerting higher stresses as compared to expected actual stresses, the 90% consolidation settlement values at preloading intensity are expected to be more around 85–90% settlement values at actual loading intensity. The estimated post-construction settlement values are expected to be less than 25 mm for all loading intensities post-ground improvement. The estimated shearing capacity of the soil is expected to give a factor of safety in excess of 2.0 against the recommended factor of safety of 1.5 while using PVDs as ground improvement.

References

1. IS 13094: Selection of ground improvement techniques for foundations in weak soils-guidelines (1992, Reaffirmed 2011)
2. Terzaghi K, Peck RB (1967) Soil mechanics in engineering practice. Wiley, New York
3. IS 15284-Part 1 (2004) Design and construction for ground improvement guidelines, stone columns, Bureau of Indian Standards, New Delhi
4. IS 15284-Part 2 (2004) Design and construction for ground improvement guidelines, preconsolidation using vertical drains, Bureau of Indian Standards, New Delhi
5. IRC75 (2015) Guidelines for design of high embankments, Bureau of Indian Standards, New Delhi

Chapter 3
State of the Art on the Extent of Smear Zone and Variation of Permeability During the Installation of Drain in Clayey Soil

R. P. Aparna, R. G. Robinson, and S. R. Gandhi

Introduction

The challenge to the geotechnical engineers is to develop technically viable and economically feasible solutions to improve unfavorable grounds to construct civil engineering structures. Adopting deep foundation is a possible solution as it can transfer the load to the competent layer. But this method is generally uneconomical, as the thickness of weak deposits is often very large in most of the regions. Ground improvement technique is a better alternative, and the selection of a specific method depends on several aspects like the geological formation of the soil to be treated, soil type, effectiveness, cost, time, experience, etc. Every ground improvement method should lead to the enhancement in properties like shear strength, bearing capacity, stiffness, resistance to liquefaction, compressibility, etc.

Preloading is commonly used in order to attain the required consolidation and appropriate gain in shear strength. In this technique, a surcharge pressure (preload) is placed to the ground prior to the construction of superstructure. The applied preload should be equal to or greater than the contact pressure of the superstructure. The application of preload will result in the consolidation of soft soil, and it will bring about a rise in shear strength of soil and decrease in post-construction deformation. By installing vertical drains, the slow consolidation rate of preloaded soft soil is enhanced. The increase in the consolidation rate is due to the provision of horizontal drainage in addition to the standard vertical drainage. Vertical drains (VD) are upright

R. P. Aparna (✉)
Department of Civil Engineering, Indian Institute of Technology Madras, Chennai, India
e-mail: draparnaiitm@gmail.com

R. G. Robinson
Department of Civil Engineering, Indian Institute of Technology, Madras 600036, India
e-mail: robinson@iitm.ac.in

S. R. Gandhi
National Institute of Technology, Surat, Gujarat 395007, India

K. Muthukkumaran et al. (eds.), *Ground Improvement Techniques*, Lecture Notes in Civil Engineering 297, https://doi.org/10.1007/978-981-19-6727-6_3

columns of permeable (sand or fibrous) material inserted in clayey soils. In order to fasten the rate of consolidation by shortening the drainage trail, the ground improvement with vertical drain is used with preloading, vacuum or surcharge method. The main factors that control the efficiency of the vertical drains which are considered in the design includes (1) drain spacing (S), (2) equivalent diameter (d_w), (3) well resistance and (4) effect of disturbance [1, 2]. In this paper, the smear effect and the theoretical, experimental analytical and field studies related to the extent of smear and permeability ratio are discussed.

Smear Effect

The PVDs are installed into the soil deposits with the help of a mandrel. The drains are inserted either by pushing (static method) or by means of driving/vibration (dynamic approach). The static method is preferred for driving the mandrel since dynamic method creates immediate excess pore water pressure leading to greater disturbance. The method for installing sand drain can also be classified as non-displacement and displacement method. In the displacement method, the closed-end mandrel is inserted into the soil causing movements in vertical and horizontal directions. In this method, the disturbance to the surrounding soil is more. It should not be installed in sensitive clays or in soil with highly developed microfabric. In the non-displacement method, a bore hole is drilled by power auger or water jets. The bore hole is then filled with sand. This will be more effective in limiting disturbance, but there will be a smear zone around borehole walls. It may give irregularly shaped drains of unknown size [3].

During the installation of PVD, the mandrel is progressed until it reaches sufficient depth. After reaching the installation depth, the mandrel is withdrawn, and the drain is placed in position by the anchor. This installation and withdrawal of the mandrel will result in remolding of soils around the drain and will result in a reduction of permeability in the remolded zone [4]. This area of remolding and reduced permeability is called the smear zone. The characteristics of soil within this smear zone is not similar to that of the intact zone. The variance in properties in the smear zone leads to the rise in problems when applying radial consolidation theory. It was often assumed in the earlier days that the insertion of a drain does not change the properties of nearby soil. In actual field condition, however, installation of drain disturbs the soil, and the effect of disturbance depends on the type of soil. Therefore, the property of the soil including smear effect that impacts the time rate of radial consolidation is very important.

Extensive studies have been done to examine the diameter of the smeared zone around the vertical drain and to calculate the decrease in permeability in the smeared zone [5–11]. The studies have been reported on the radius of smear and the variation of permeability due to smear effect by theoretical, experimental, numerical and also by field studies. A literature review of these works has been carried out and is briefly presented in the following sections.

Theoretical Studies on the Extent of Smear Zone and Permeability Ratio

Barron [12] stated that smearing affects drain performance and considered the smear zone around the vertical drains. For explaining smear zone generation, two main theories are proposed: (a) the soil remolding concept and (b) the reconsolidation theory. According to the first theory, the soil close to the drain is parted into two zones such as (i) remolded zone near the drain and (ii) undisturbed zone beyond the disturbed region. This is the two-zone hypothesis. Beyond the smear zone, existence of transition zone (three zone hypothesis) has also been reported by few researchers [13–17]. The permeability of the transition zone (k_{tr}) is not constant unlike the smear zone but rather slowly rises from smeared zone permeability, k_s to the undisturbed radial permeability value k_h as the horizontal distance from the drain increases. Also the extent of transition zone can even extend up to about ten times the mandrel diameter [4, 8, 18–21]

The variation in properties of soil caused by insertion of mandrel may be due to the reconsolidation by means of the decrease in the excess pore pressure (or) as a result of remolding of soil (or) due to the combination of both. Cavity expansion theory analyzed the soil reconsolidation due to the insertion of the mandrel [20, 22]. The effect of smearing after and before reconsolidation can also be investigated by electrical resistance probe and micro-cone [23].

Summary of theoretical studies on the extent of smear and permeability variation is given in Table 3.1. In the classical theories (Table 3.1), the impact of remolding is analyzed with a two-zone model, i.e., an outer intact area with undisturbed permeability and an inner smeared region near to the drain with reduced permeability (k_s). The precision of calculations using conventional theories is governed by the exact assessment of radial permeability and the radius of smeared zone. Several studies reported in the literature have attempted to determine smear zone extent along with the permeability variations in the zone of intense remolding.

In all these studies, the flow of water to the drain is considered as axisymmetric. While solving the radial consolidation equation, Barron [12] developed the axisymmetric analysis of vertical drain by assuming that the extent of the smeared zone is the

Table 3.1 Theoretical studies on the extent of smear and variation of permeability in the smeared and undisturbed zones

References	d_s[a]	k_h/k_s
Barron [12]	(1–1.5) d_m	3
Holtz and Holm [24]	2 dm	–
Hansbo [18, 25–27]	(2–3) d_w	2–4 ($k_s = k_v$)
Jamiolkowski et al. [28]	(2.5–3) d_m	3
Rixner et al. [29, 30]	(2.5–3) d_m	–

[a] d_s—smear diameter, d_m—equivalent diameter of mandrel, d_w—equivalent diameter of drain

same as that of the drain area which was later extended by Hansbo [18] and Onoue [31]. All these solutions considered axisymmetric conditions of a unit cell. Holtz and Holm [24] and Hansbo [18, 25] developed a model, based on equal strain condition, wherein the diameter of smear zone (d_s) is two to three times the equivalent diameter of the drain (d_w).

Due to the smearing of soil around the drain during installation, the anisotropy in terms of permeability is nullified resulting in reduced horizontal permeability in comparison with the undisturbed soil. The previous studies [8, 18, 32] have reported that the vertical permeability of the soil (k_v) is the same in both smeared and undisturbed zones. Hansbo [33], Bergado et al. [6] postulated that the anisotropy in permeability (k_h/k_v) in the smeared region is close to unity. Jamiolkowski et al. [28] stated that the drainage trail is generally condensed from the soil thickness to half the drain spacing, and the value of (k_h/k_v) can range from 1 to 15.

Rixner [29, 30] suggested that installation of drain will induce deformation of soil around the PVD and shear strain. The shearing will increase the pore water pressure and the total stress. The area of the mandrel will be greater than that of the drain, and hence, after the removal of mandrel, an annular space is developed around the drain. The installation results in remolding to the soil surrounding the drain, and its radial extent ranges from 2.5 to 3 times the mandrel radius. Rixner [29, 30] suggested that the disturbance mainly depends on:

(1) Mandrel size and shape: the cross section should be close to the drain, and the tip of mandrel and anchor has to be tapered as possible.
(2) Soil macrofabric/Soil layering: the k_h/k_v ratio can be very high (>10) for soils with pronounced macrofabric. But within the smeared zone, the beneficial effects of soil stratification will be eliminated. The smearing of the pervious layer with less pervious soil can impede the lateral seepage of water from the permeable layer into the drain, thereby reducing k_h/k_v.
(3) Installation procedure: static pushing is ideal with respect to driving/vibrating the mandrel. The stiffness of the mandrel and rate of penetration should be selected to limit buckling.

It was suggested by some researchers [6, 24, 28] that for the design purpose, complete remolding within the disturbed zone can be assumed, and this zone has one value of hydraulic conductivity (k_s).

Experimental Studies on the Extent of Smear Zone and Permeability Ratio

Extensive laboratory studies were performed by various researchers on a wide variety of clay with different index properties to find the extent of remolded zone and permeability ratio. The summary of experimental studies on the diameter of smear and permeability variation is given in Table 3.2.

Table 3.2 Experimental studies on the extent of smear and variation of permeability in the smeared and undisturbed zones

References	Soil (type)	Index properties (%)		d_s	k_h/k_s
		w_L	I_P		
Bergado [6]	Bangkok clay (RM)	75–100	40–60	2 dm	1.5–2 $k_s = k_v$
Indraratna and Redana [8]	Alluvial clay (RM)	70	40	4–5 dm	$k_s = k_v$
Hird and Mosely [10]	Kaolin (RM)	70	30	2–4 dm	1.5
Sharma and Xiao [21]	Kaolin (RM)	70	30	4 d_w	1.3
Indraratna and Rujikiakamjorn [38]	Alluvial clay (RM)	42	25	3 d_w	–
Fang and Yin [39]	Hongkong marine clay (RM)	51.1	25	2 dm	2
Sathananthan and Indraratna [34]	Mourya clay (RM)	42	25	2.5 dm	1.5
Walker and Indraratna [40]	Alluvial clay (RM)	70	40	8.4 d_w	–
Shin et al. [23]	Busan clay (RM)	46.4	24.1	5.3 dm	–
Saowapakiboon et al. [41]	Bangkok clay (RM)	102.2	62.7	2 dm	2.7
Tran-Nguyen and Edil [42]	Hydrite R Kaolinite (RM)	49	25	3–4 dm	1–1.5
Rujikiakamjorn et al. [43]	Bulli clay (UD)	50	25	3.7 dm	1.3–2.9
Indraratna et al. [44]	Ballina clay (UD)	94.7	32	6.3 dm	2.7
Joseph et al. [45]	Cochin marine clay (RM)	156	122	5–6 dm	1.3–1.4
Pajouh et al. [46]	Kaolinite, bentonite(RM)	67–87	40–43	3 d_w	2
Choudhary et al. [47]	Ballina clay (RM)	98	66	2.5 dm	1.3
Sengul et al. [37]	Craney Island Dredgings (RM)	51	21	2.3–3.3 dm	2.9–3.1
Barral et al. [48]	Ballina clay (UD)	94–102	58–74	5 dm	–

Note RM—Remolded, UD—Undisturbed

A common procedure was adopted for all these laboratory studies. The lab test commences typically by the preparation of soil sample. It is difficult and costly to make undisturbed soil samples in a large scale. Hence, the degree of disturbance (d_s/d_m or d_s/d_w) and the permeability variation (k_h/k_s) of the soil were normally measured in the laboratory (Table 3.2) using reconstituted (or) remolded samples. The reconstituted clayey samples were prepared by mixing at water content approximately equal to 1.1–2 times the liquid limit (w_l). The sample was then filled into the large-scale consolidation tank in three or four layers, extracting trapped air.

The preconsolidation pressure (ranging from 10–100 kPa) was then applied to the samples. After preconsolidation, scaled down PVD/sand drain was installed to the soil sample using specially fabricated mandrel.

The rate of installation ranges from 0.5 to 20 mm/s which is very less compared to the field installation rate. In the field, it is around 1–1.5 m/s. This stage is followed by coring of specimens in the vertical and radial directions. The extraction of specimens from large-scale consolidometer can be done in different stages: (1) soonafter the decrease of excess pore water pressure produced during insertion of mandrel [21], (2) once the primary consolidation of soil with drain is completed [8, 34, 35] or (3) by directly performing the permeability test on soil stabilized using PVD with the help of special instrumentation [10, 36, 37].

Analytical and Numerical Studies on the Extent of Smear Zone and Permeability Ratio

Various studies have been performed by the finite element method (FEM) and finite difference method (FDM) to study the influence of parameters that characterize the smear zone. The theory of radial consolidation under axisymmetric condition was described by Hansbo [18]. Several equations were developed for transforming axisymmetric to plane strain permeability in the intact as well as in the smear zone, by various researchers [46, 49–52] summarized and compared these equations in their study and suggested that the formula developed by Hird et al. [51] lead to a precise prediction in the primary stage of consolidation. The summary of analytical or numerical studies that investigates the extent of smear and permeability ratio is given in Table 3.3.

Field Studies on the Extent of Smear Zone and Permeability Ratio

Extensive case studies were reported in the literature to find the degree of smear and permeability ratio. The smear zone characteristics are generally assessed by the back calculation method. The estimation was performed based on the observed time-settlement data that is fitted in the equation or processed with a numerical program. The summary of case studies on the extent of smear and variation of permeability is given in Table 3.4.

Table 3.3 Numerical studies on the extent of smear and variation of permeability in the smeared and undisturbed zones

References	Analytical/Numerical method	d_s	k_h/k_s
Indraratna and Redana [53]	FEM (CRISP92)	2.6 dm	25
Arulrajah et al. [54]	FEM (PLAXIS)	–	7.6
Sathananthan et al. [22]	CET, FEM (PLAXIS)	2.5 dm	k_s varies from 61 to 92% of k_h
Lin and Chang [55]	FDM (FLAC)	2	3.2
Tarefder et al. [56]	FEM (SAGE CRISP)	2.5 d_w	10
Ghandeharioon et al. [20]	CET	3.1 dm	–
Pajouh et al. [4]	FDM (FLAC 2D)	3 dm	2
Lam et al. [57]	FEM (ABAQUS)	2 dm	5.9–10
Liu and Row [58]	FEM (ABAQUS)	–	3
Yildiz and Uysal [52]	FEM (PLAXIS)	5 dm	20
Chen et al. [59]	FEM (PLAXIS 3D)	1.1–3 dm	–

Table 3.4 Field studies on the extent of smear and variation of permeability in the smeared and undisturbed zones

Reference	d_s	k_h/k_s
Casagrande and Poulos [5]	1 dm	–
Bergado et al. [6, 7]	2–2.5 dm	5–10 ($k_s = k_v$)
Almeida et al. [60]	2.5–3 dm	3–6
Mesri et al. [61]	2–4 dm	–
Chai and Miura [9]	2–3 dm	$k_s = k_v$ 10
Erikson et al. [62]	2 dm	6
Chai et al. [1]	6.7 d_w	13.8
Bo et al. [63]	4–7 d_w	10
Saowapakpiboon et al. [41]	2 dm	6.6–7.2
Hiep and Chung [64]	2 dm	2

Conclusion

All the studies reviewed above related to smearing of clay due to PVD installation have concluded that smearing affects the performance of the drain. The smearing is primarily controlled by the remolded zone near the drain. Hence, neglecting the effects of anisotropy/smearing will result in inaccurate predictions of soft soil response. It was also reported that without considering the smear effect, i.e., perfect drain, may overestimate the settlement. It can be noted from Tables 3.1, 3.2, 3.3 and 3.4 that the smear ratio (s) reported in the studies ranges from 1 to 7. The laboratory studies reporting the results on the effects of smearing, during drain installation, have generally used reconstituted soil samples tested in large tanks. But the behavior of soil in the field is different, owing to its structure. When the reconstituted sample is prepared, its structure is entirely destroyed. It can be perceived that the value of k_h/k_s for lab studies is in the range of 1–5. However, in field studies, the value of k_h/k_s actually varies from 2 to 14. The rate of installation of PVD in the laboratory studies is much lesser compared to the field studies. In the field, it can be up to 1–1.5 m/s. The studies reviewed have clearly brought out two parameters such as the radial extent of the smear around the drain and the consolidation features of the remolded soil in the disturbed zone that affects the rate of consolidation.

References

1. Chai JC, Shen SL, Miura N, Bergado DT (2001) Investigation of factors affecting vertical drain behavior. J Geotech Geoenvironmental Eng 127(11):965–972
2. Lee NK, Chung S (2010) Reevaluation of the factors influencing the consolidation of ground by incorporating prefabricated vertical drains. KSCE J Civ Eng 14(2):155–164
3. Barends FBJ, Lindenberg J, Lunger HJ, Quelerij L.de, Verruijt A (1999) Geotechnical engineering for transportation infrastructure. In: Proceedings of the 12th European conference on soil mechanics and geotechnical engineering, Amsterdam, Netherlands, 7–10 June 1999
4. Pajouh AP, Fatahi H, Khabbaz B, Vincent P (2014) Analysing consolidation data to predict smear zone characteristics induced by the vertical drain. Geomech Eng 7(1):105–131
5. Casagrande L, Poulos S (1969) On the effectiveness of sand drains. Can Geotech J 6(3):287–326
6. Bergado DT, Asakami H, Alfaro MC, Balasubramaniam AS (1991) Smear effects of vertical drains on soft Bangkok clay. J Geotech Eng 117(10):1509–1530
7. Bergado DT, Alfaro MC, Balasubramaniam AS (1993) Improvement of soft Bangkok clay using vertical drains. Geotext Geomembr 12:615–663
8. Indraratna B, Redana IW (1998) Laboratory determination of smear zone due to vertical drain installation. J Geotech Eng ASCE 124(2):180–184
9. Chai JC, Miura N (1999) Investigation of factors affecting vertical drain behavior. J Geotech Geoenvironmental Eng 125(2–3):216–226
10. Hird CC, Moseley VJ (2000) Model study of seepage in smear zones around vertical drains in layered soil. Geotechnique 50(1):89–97
11. Chu J, Bo MW, Choa V (2004) Practical considerations for using vertical drains in soil improvement projects. Geotext Geomembr 22(1–2):101–117
12. Barron RA (1948) Consolidation of fine graded soils by drain wells transactions of ASCE 2346(113):718–724

13. Onoue A, Ting NH, Germaine JT, Whitman RV (1991) Permeability of the disturbed zone around vertical drains. In: Proceedings of the ASCE geotechnical engineering congress, Colorado, American Society of Civil Engineers, Reston, VA, USA, pp 879–890
14. Madhav MR, Park YM, Miura N (1993) Modelling and study of smear zones around band-shaped drain. Soils Found 33(4):135–147
15. Hawlader BC, Goro I, Balasingam M (2002) Numerical study of the factors affecting the consolidation of clay with vertical drains. Geotext Geomembr 20, pp 213–239
16. Basu D, Basu P, Prezzi M (2006) Analytical solutions for consolidation aided by vertical drains. Geomech Geoengineering Int J 1(1):63–71
17. Azari B, Fatahi B, Khabbaz H (2016) Assesment of the elastic-viscoplastic behavior of soft soil improved with vertical drains capturing reduced shear strength of a disturbed zone. Int J Geomech ASCE 16(1):1–15
18. Hansbo S (1981) Consolidation of fine-grained soils by prefabricated drains. In: Balkema AA Proceedings of the 10th international conference on soil mechanics and foundation engineering, 3rd edn, pp 677–682
19. Bo MW, Bawajee R, Chu J, Choa V (2000) Investigation of smear zone around the vertical drain. In: Proceedings of the third international conference on ground improvement techniques, Singapore, pp 109–114
20. Ghandeharioon A, Indraratna B, Rujikiatkamjorn C (2010) Analysis of soil disturbance associated with mandrel-driven prefabricated vertical drains using an elliptical cavity expansion theory. Int J Geomech 10(2):53–64
21. Sharma JS, Xiao D (2000) Characteristics of a smear zone around vertical drains by large-scale laboratory tests. Can Geotech J 37(6):1265–1271
22. Sathananthan I, Indraratna B, Rujikiatkamjorn C (2008) Evaluation of smear zone extent surrounding mandrel driven vertical drains using the cavity expansion theory. Int J Geomech 8(6):355–365
23. Shin DH, Lee C, Lee JS, Lee W (2009) Detection of smear zone using micro-cone and electrical resistance probe. Can Geotech J 46:719–726
24. Holtz RD, Holm BG (1973) Excavation and sampling around some sand drains at Ska-Edeby, Sweden. In: Proceedings of the 6th Scandinavian geotechnical meeting, Trondheim, Norway, Norwegian Geotechnical Institute, pp 79–85
25. Hansbo S (1979) Consolidation of clay by band-shaped prefabricated drains. Ground Eng 12(5):16–25
26. Hansbo S (1994) Foundation engineering. Elsevier, Developments in geotechnical engineering, 75
27. Hansbo S (1997) Practical aspects of vertical drain design. In: Proceedings of the 14th international conference on soil mechanics and foundation engineering, Hamburg, 3, pp 1749–1752
28. Jamiolkowski M, Lancellotta R, Wolski W (1983) Precompression and speeding up consolidation. In: Proceedings of 8th European conference on soil mechanics and foundation engineering, Helsinki 3(6):1201–1226
29. Rixner JJ, Kraemer SR, Smith AD (1986) Prefabricated vertical drains, vol. I, Engineering guidelines, Federal highway administration, USA (1986a)
30. Rixner JJ, Kraemer SR, Smith AD (1986) Prefabricated vertical drains, vol. II, summary of research effort, Federal highway administration, USA (1986b)
31. Onoue A (1988) Consolidation by vertical drains taking well resistance and smear into consideration. Soil Found 28(4):165–174
32. Saye SR (2002) Assessment of soil disturbance by the installation of displacement sand drains and prefabricated vertical drains. Soil Behav Soft Ground Constr GSP 119:325–362
33. Hansbo S (1987) Design aspects of vertical drains and lime column installation. In: Proceedings 9th Southeast Asian geotechnical conference 2(8):1–12
34. Sathananthan I, Indraratna B (2006) Laboratory evaluation of smear zone and correlation between permeability and moisture content. J Geotech Geoenvironmental Eng 132(7):942–945

35. Ghandeharioon A, Indraratna B, Rujikiatkamjorn C (2012) Laboratory and finite-element investigation of soil disturbance associated with the installation of mandrel-driven prefabricated vertical drains. J Geotech Geoenvironmental Eng 138(3):295–308
36. Tran-Nguyen HH, Edil TB (2011) The characteristic of PVD smear zone. Geo-Frontier ASCE:748–757
37. Sengul T, Edil T, Ozaydin K (2017) Laboratory determination of smear and transition zones due to prefabricated vertical drain installation. Mar Georesour Geotechnol 35:895–904
38. Indraratna B, Rujikiatkamjorn C (2004) Laboratory determination of efficiency of prefabricated vertical drains incorporating vacuum preloading. In: The 15th Southeast Asian geotechnical conference. Bangkok, Thailand, 1, pp 453–456
39. Fang Z, Yin JH (2006) Physical modelling of consolidation of Hong Kong marine clay with prefabricated vertical drains. Can Geotech J 43:638–652
40. Walker R, Indraratna B (2006) Vertical drain consolidation with parabolic distribution of permeability in smear zone. J Geotech Geoenvironmental Eng 132(7):937–941
41. Saowapakpiboon J, Bergado DT, Youwai S, Chai JC, Wanthong P, Voottipruex P (2010) Measured and predicted performance of prefabricated vertical drains (PVDs) with and without vacuum preloading. Geotext Geomembr 28(1):1–11
42. Tran-Nguyen HH, Edil TB (2011) The characteristic of PVD smear zone. Geo-Frontier 2011 ASCE:748–757
43. Rujikiatkamjorn C, Ardana M, Indraratna B, Leroueil S (2013) Conceptual model describing smear zone caused by mandrel action. Géotechnique 63(16):1377–1388
44. Indraratna B, Zhong R, Fox PJ, Rujikiatkamjorn C (2015) Large-strain vacuum-assisted consolidation with non-Darcian radial flow incorporating varying permeability and compressibility. ASCE J Geotech Geoenvironmental Eng
45. Joseph, Chandrakaran S, Sankar N, Jose BT (2015) Laboratory evaluation of extent of smear zone due to columnar intrusion for Cochin marine clays. In 50th Indian geotechnical conference, Pune, India
46. Pajouh AP, Fatahi H, Khabbaz B (2015) Experimental and numerical investigations to evaluate two-dimensional modelling of vertical drain–assisted preloading. Int J Geomechancis:1–14
47. Choudhary K, Indraratna B, Rujikiatkamjorn C (2016) Pore pressure based method to quantify smear around a vertical drain. Géotechnique Lett:211–215
48. Baral P, Rujikiatkamjorn C, Indraratna B, Kelly R (2018) Radial consolidation characteristics of soft undisturbed clay based on large specimens. J Rock Mech Geotech Eng:1–9
49. Indraratna B, Rujikiatkamjorn C, Sathananthan I (2005) Radial consolidation of clay using compressibility indices and varying horizontal permeability. Can Geotech J 42:1330–1341
50. Lin DG, Kim HK, Balasubramaniam (2000) Numerical modeling of prefabricated vertical drain. Geotech Eng J 31(2):109–125
51. Hird CC, Pyrah IC, Russel D (1992) Finite element modeling of vertical drains beneath embankments on soft ground. Geotechnique 42(1):499–511
52. Yildiz A, Uysal F (2015) Numerical modelling of Haarajoki test embankment on soft clays with and without PVDs. Geomechanics Eng 8(5):707–726
53. Indraratna B, Redana IW (2000) Numerical modeling of vertical drains withsmear and well resistance installed in soft clay. Can Geotech J 37:132–145
54. Arulrajah A, Nikraz H, Bo MW (2005) Factors affecting field instrumentation assessment of marine clay treated with prefabricated vertical drains. Geotext Geomembr 22(5):415–437
55. Lin DG, Chang KT (2009) Three-dimensional numerical modelling of soft ground improved by prefabricated vertical drains. Geosynth Int 16(5):339–353
56. Tarefder RA, Zaman MM, Lin DG, Bergado DT (2009) Finite element modeling of soft ground with PVD under vacuum and embankment preloading. Int J Geotech Eng 3(2):233–249
57. Lam LG, Bergado DT, Hino T (2015) PVD improvement of soft Bangkok clay with and without vacuum preloading using analytical and numerical analyses. Geotext Geomembr 43(6):547–557
58. Liu KW, Rowe RK (2015) Numerical modelling of prefabricated vertical drains and surcharge on reinforced floating column-supported embankment behaviour. Geotext Geomembr 43(6):493–505

59. Chen J, Shen SL, Yin ZY, Xu YS, Horpibulsuk S (2016) Evaluation of effective depth of PVD improvement in soft clay deposit: a field case study. Mar Georesour Geotechnol 34(5):420–430
60. Almeida MSS, Ferreira CAM (1993) Field, in situ and laboratory consolidation parameters of very soft clay. In: Predictive soil mechanics, proceedings of the worth memorial symposium, Thomas Telford, London, pp 73–93
61. Mesri G, Lo DOK, Feng TW (1994) Settlement of embankments on soft clays. In: Vertical and horizontal deformations of foundations and embankments: proceedings of Settlement'94. American Society of Civil Engineers, New York, 1, pp 8–56
62. Eriksson U, Hansbo S, Torestensson BA (2000) Soil improvement at Stockholm-Arlanda airport. Ground Improv 4(5):73–80
63. Bo MW, Chu J, Low BK, Choa V (2003) Soil improvement: prefabricated vertical drain technique. ISBN 981-243-044-X, Thomson Learning, Singapore, p 341
64. Hiep H, Chung SG (2018) Back-analysis of geotechnical parameters on PVD-improved ground in the Mekong Delta. Geotext Geomembr 46:402–413

Chapter 4
State-Of-The-Art Review on Improvement of Strength Characteristics of Soil Using Nano Silica

Jayanti Munda and Supriya Mohanty

Introduction

In recent times owing to fastest increase in population leads to scarcity of adequate lands for the construction of various civil engineering projects. The biggest challenges are the presence of problematic soils at the construction site because of its weak shear strength tends to differential settlements. In such cases, ground improvement can be one of the appropriate solutions to counteract various problems related to problematic soils. Soil stabilization remarkably strengthens the properties of the weak soil. With the development in Nanotechnology instead of using some traditional stabilizing materials, nanomaterials can be used to strengthen the soil. The importance of adding nanomaterials for soil modification is because of their large specific surface area (SSA), tremendous cation exchange capacity (CEC) and its active bonding characteristics with other soil particles. Even very little amount of nanoparticles can significantly change the geotechnical properties of soil.

Nanotechnology is a new advance technology in the field of civil engineering and many researchers are investigating its feasibility in the development of the soil properties. Nanomaterials in the field of ground improvement are still not accepted thoroughly. Nanotechnology is an expansion of sciences and technologies concentrated on forming nanometer sized particles and the particles sized between 1 and 100 nm. The whole idea of nanotechnology was initially suggested by Richard Feynman's talk in 1959 [1]. The present paper reviews the applications of Nano silica in the field of geotechnical engineering.

J. Munda (✉) · S. Mohanty
Department of Civil Engineering, Indian Institute of Technology (BHU), Varanasi, Uttar Pradesh 221005, India
e-mail: jayantimunda.rs.civ19@itbhu.ac.in

K. Muthukkumaran et al. (eds.), *Ground Improvement Techniques*, Lecture Notes in Civil Engineering 297, https://doi.org/10.1007/978-981-19-6727-6_4

Plasticity Characteristics of Composite Materials

Bahmani et al. [2] reported the performance of SiO_2 nanoparticles of different sizes 15 and 80 nm (0.2, 0.4, 0.6, 0.8, and 1%) along with cement (4, 6 and 8%) mixed in residual soil. It was found that plasticity index of stabilized soil with 15 nm nano silica (NS) was minimum as compared to soil treated with 80 nm nano silica at cement contents of 4 and 6%. Furthermore, with the cement content of 8%, the plasticity index of soil treated with 80 nm nano silica was lesser compared to soil samples treated with 15 nm silica nanoparticles. The lowest plasticity index obtained with 0.4% NS at 8% cement content. Thomas and Rangaswamy [3] studied the effect of addition of cement (1%) treated soft soil mixed with nano silica particles (0.25, 0.5, 0.75, 1, and 2%) on the plasticity index. The results concluded that there was decrement in liquid limit (LL) with the increase in NS contents but no significant change in plastic limit (PL) was noticed. Ghasabkolaei et al. [4] observed that plasticity index decreases in the cement treated clayey soil. Further, the inclusion of nano silica to the cement stabilized clayey soil reasonably increases the plasticity index. Changizi and Haddad [5] observed that inclusion of silica particles to the clayey soil, the plasticity index decreased to about 52%. Kalhor et al. [6], Malik et al. [7] explored that addition of varied proportions of nano silica to the clayey soil resulting increase in PL and LL. Lv et al. [8] observed that addition of 10 and 29 nm nano silica to the loess soil, consistency limits increases. However, with the addition of 100 nm nano silica causes slight reduction in consistency limits of the loess soil.

Compaction Characteristics of Composite Materials

Changizi and Haddad) [5, 9]studied the effects of nano silica (i.e., 0.5, 0.7, and 1.0%) on clayey soil and found that with each increment of nano silica content, both optimum moisture content (OMC) and maximum dry density (MDD) of the stabilized soil increases. This is due to the absorption of moisture and filled up the pore spaces with nano silica in the soil. Bahmani et al. [2] presented the effect of nano silica on the residual soil and concluded that there is increment in OMC and a slight reduction in MDD. The maximum value of the MDD was observed with diameter of 80 nm nano silica than that of 15 nm nano silica at different cement contents. Malik et al. [7] reported that when nano silica (i.e., 5, 10, 15, and 20%) mixed with clayey soil, the OMC increases and the MDD decreases. Samala and Mir [10] noticed that the OMC increases and MDD decreases with the inclusion of Nano-SiO_2 (i.e., 0.5, 1.0, 1.5, and 2.0%) on soft soil. Sarli et al. [11] investigated the performance of recycled polyester (i.e., 0.5, 1, and 1.5%) and Nano-SiO_2 (i.e., 2, 4, and 6%) of loess. It was noticed that with each increment of both fiber and Nano-SiO_2 there is decrease in MDD and increase in OMC of loess. Tomar et al. [12] examined the various proportions of polypropylene fiber (i.e., 0.1, 0.4, 0.7,1, and 1.3%) and nano silica (i.e., 1, 3, 5, and 7%) on clayey soil resulting decrement in MDD and increment in OMC. The results in reduction of MDD are owing to the lower density of PPF compared to that of clayey soil. Nezhad et al. [13] reported the influence of prepared clayey soil to the gas oil (i.e., 0, 3, 6, and 9% by dry weight of soil) using nano silica and hydrated lime (i.e., 0, 1, 2, and 3%). The results showed that after inclusion of nano silica-lime mixture to the contaminated soil, the MDD of stabilized soil samples decreased owing to lower specific gravity of nano silica and lime than that of soil while OMC increased because of higher surface area of nano silica and shorter reaction between lime and soil.

Strength Characteristics of Composite Materials

Changizi and Haddad [9] explored the effect of nano silica (i.e., 0.5, 0.7, and 1.0%) to the clayey soil and found that with each increment of nano silica doses results increase in both shear strength parameters (c & φ). The maximum increase in shear strength parameters noticed at 1.0% of nano silica and the increase of treated soil observed by factors 2.1 and 1.23. The highest unconfined compressive strength value of stabilized soil observed at 0.7% of nano silica content by a factor of 1.56. Moayed and Rahmani [14] studied the enhancement of strength characteristics of kaolinite using nano silica (i.e., 1, 3, 4, and 5%) and found that nano silica content can increase the unconfined compressive strength value of soil up to 1.43 times that of the parent soil. Bahmani et al. [2] observed that addition of nano silica (i.e., 0.2, 0.4, 0.8, and 1.0%) and cement (i.e., 4, 6, and 8%) can enhance the strength characteristics of residual soil. It was noticed that 15 nm nanoparticles can more effectively strengthen the soil samples compared to 80 nm nanoparticles. The maximum strength obtained was 85% to the optimum content of 0.4% nano silica and 8% cement. Samala and Mir [10] noticed that UCS value of two different locations (site-1 and site-2) treated with nano silica (i.e., 0.5, 1, 1.5, and 2.0%) was increase by about 2.7 and 2.5 times, respectively. The curing period of 7 days and 14 days can increase the UCS value by 4.5 and 5.5 times and 2.5 and 3.5 times for the site-1 and site-2, respectively. The optimum amount of nanoparticles obtained was 1.5% and 1.0%. Sarli et al. [11] assessed the performance of loess soil treated with recycled polyester fiber (i.e., 0.5, 1 and 1.5%) and nano silica (i.e., 2, 4, and 6%). They concluded that highest shear strength parameters obtained with the optimum amount of fiber and nano silica was 4% of loess in proportions of 33% and 50%, respectively. Tomar et al. [12] reported that addition of nano silica (i.e., 1, 3, 5, and 7%) in combination with polypropylene fiber (i.e., 0.1, 0.4, 0.7, 1, and 1.3%) to the clayey soil increases the UCS value. The highest UCS value achieved at 7% of nano silica and 0.7% polypropylene fiber. Thomas and Rangaswamy [3] noticed that even very small amount of nano silica can effectively enhance the UCS value of the cement treated soft clayey soil with the optimum value of 1% cement. Nezhad et al. [13] investigated on treatment of prepared gas oil contaminated soil with nano silica. They found that there is significant gain in the UCS value by the addition of 3% nano silica in combination with hydrated lime and with the increase in curing period. Choobbasti et al. [15] revealed that when cement treated sand mixed with nano silica, the mechanical properties was modified. The maximum increase in UCS value observed at 5% cement with 10% nano silica. Choobbasti et al. [16] continued investigation on the influence of nano silica on the cemented sandy soil. The test results revealed that the highest value of UCS was observed at 8% nano silica with the cement treated soil. Further, Choobbasti et al. [17] extended the research on the improvement of cemented sand mixed with nano silica by conducting static and cyclic triaxial tests. Based on test results, the authors noticed that the maximum value of cyclic modulus of elasticity was observed at 10% nano silica and 5% cement content and confining pressure of 200 kPa. Zomorodian et al. [18] explored on ground modification with various combinations of nanoparticles. The effect of lean and kerosene contaminated soil was studied using nano clay combined with nano silica. They found that maximum improvements in strength and stiffness were noticed with nano clay and nano silica content of 1% and 1.5%, respectively. Cui et al. [19] noticed that the shear strength parameters can improve remarkably with the inclusion of carbon fiber in combination with nano silica to the silty soil. The maximum shear strength was observed with the optimum amount of 2wt% carbon fiber and 3wt% nano silica value at all normal pressures levels. Kulanthaivel et al. [20] reported that the addition of nano silica combined with white cement was intended for the improvement of UCS value 7.01 times larger than the parent clay soil. Furthermore, the combination has positive development in CBR value, i.e., increased up to 31% in comparison to untreated clayey soil.

Hydraulic Conductivity of Composite Materials

The hydraulic conductivity is one of the vital parameter for most of the embankment/dam structures. Kulanthaivel et al. [20] found that when nano silica mixed in combination with white cement to the clayey soil, there was considerable amount of reduction in permeability, i.e., 45% to the parent soil. The optimum content of nano silica and white cement was observed as 2% and 3%, respectively. Bahmani et al. [2] found that the addition of two different sizes of nano silica (15 and 80 nm) have significant effect on hydraulic conductivity. The authors concluded that the reduction in hydraulic conductivity of 15 nm nanoparticles was larger compared to the 80 nm nanoparticles and the minimum conductivity was noted at 0.4% of nano silica.

Interaction Mechanism of Nano Silica in Soil

Changizi and Haddad [9], Cui et al. [19] reported that in clay mixed nano silica composite, the mechanism of strength increase is due to the strong bonding developed between soil particles and the formation of viscous gel and nano silica that creates shorter gap between the soil particles, which leads to the strong contact between the soil particles. Kulanthaivel et al. [20] reported that nano silica has higher specific surface area and evolution of silicate cementitious materials that leads to the immediate strong reaction between the clay particles, which enhances the soil strength. Bahmani et al. [2] found that due to the chemical reactions between cement and nano silica there is formation of secondary C-S-H gel, which improves the soil strength. Tomar et al. [12] observed the bridge effect between polypropylene fiber and nano silica that minimize the cracks developed in soil and results the strength improvement of soil. Many authors [3, 4, 16] suggested that increment of UCS value of cement stabilized sand mixed with nano silica achieved due to the formation of calcium hydroxide (CH) crystals which formed during the hydration process at initial stage of curing periods and transforms all the unhydrated crystals into CSH gel. Nezhad et al. [13] noticed that the pozzolanic reaction between lime and nano silica in the soil leads to the improvement in UCS value of the gas oil contaminated clayey soil.

Conclusions

This paper presents the state of the art review on improvement of strength characteristics of soil using nano silica. The test results achieved from laboratory experiments and physiochemical interaction between the nanoparticles and soil were inconsistent still undoubtedly inclusion of very little amount of nanoparticles can drastically enhance the soil properties. This is owing to the presence of nanoparticles larger surface area and high reactive surface charges. The results revealed that the inclusion of nano silica in the soil can improve the geotechnical properties including shear strength parameters, unconfined compressive strength, plasticity and compaction characteristics and reduces the hydraulic conductivity. Therefore, it is recommended that nano silica can be used in the enhancement of strength properties of various structures including foundations, dams, embankments, and landfills.

References

1. Feynman RP (1960) There's plenty of room at the bottom. Eng Sci 23(1):22–36
2. Bahmani SH, Huat BBK, Asadi A, Farzadnia N (2014) Stabilization of residual soil using SiO_2 nanoparticles and cement. Constr Build Mater 64:350–359
3. Thomas G, Rangaswamy K (2020) Strengthening of cement blended soft clay with nano-silica particles. Geomechanics Eng 20(6):505–516
4. Ghasabkolaei N, Janalizadeh A, Jahanshahi M, Roshan N,Ghasemi SE (2016) Physical and geotechnical properties of cement-treated clayey soil using silica nanoparticles: an experimental study. Eur Phys J Plus:131–134
5. Changizi F, Haddad A (2017) Improving the geotechnical properties of soft clay with nano silica particles. Proc Inst Civ Eng
6. Kalhor A, Ghazavi M, Roustaei M, Mirhosseini SM (2019) Influence of Nano-SiO_2 on geotechnical properties of fine soils subjected to freeze-thaw cycles. Cold Reg Sci Technol 161:129–136
7. Malik A, Puri SO, Singla N, Naval S (2019) Strength characteristics of clayey soil stabilized with nano-silica. In: Recycled waste materials, p 11–17

8. Lv Q, Chang C, Zhao B, Ma B (2018) Loess soil stabilization by means of SiO_2 nanoparticles. Soil Mech Found Eng 54(6):409–413
9. Changizi F, Haddad A (2015) Effect of Nano-SiO_2 on the geotechnical properties of cohesive soil. Geotech Geol Eng 34(2):725–733
10. Samala HR, Mir BA (2020) Some studies on microstructural behaviour and unconfined compressive strength of soft soil treated with SiO_2 nanoparticles. Innov Infrastruct Solutions 5(1)
11. Sarli JM, Hadadi F, Bagheri RA (2019) Stabilizing geotechnical properties of loess soil by mixing recycled polyester fiber and Nano-SiO_2. Geotech Geol Eng 38:1151–1163
12. Tomar A, Sharma T, Singh S (2020) Strength properties and durability of clay soil treated with mixture of nano silica and polypropylene fiber. Mater Today Proc
13. Nezhad RS, Nasehi SA, Uromeihy A, Nikudel MR (2021) Utilization of nanosilica and hydrated lime to improve the unconfined compressive strength (UCS) of gas oil contaminated clay. Geotech Geol Eng 39:2633–2651
14. Moayed RZ, Rahmani H (2017) Effect of Nano SiO_2 solution on the strength characteristics of kaolinite. Int J Geotech Geo Eng 11(1)
15. Choobbasti AJ, Vafaei A, Kutanaei SS (2015) Mechanical properties of sandy soil improved with cement and nanosilica. Open Eng 5(1):111–116
16. Choobbasti A, Kutanaei SS (2017) Microstructure characteristics of cement-stabilized sandy soil using nano silica. J Rock Mech Geotech Eng 9:981–988
17. Choobbasti AJ, Vafaei A, Kutanaei SS (2018) Static and cyclic triaxial behaviour of cemented sand with nano silica. J Mater Civ Eng 30(10)
18. Zomorodian SMA, Shabnam M, Armina S, O'Kelly BC (2017) Strength enhancement of clean and kerosene-contaminated sandy lean clay using nanoclay and nanosilica as additives. ApplClay Sci 140:140–147
19. Cui H, Jin Z, Bao X, Tang W, Dong B (2018) Effect of carbon fiber and nanosilica on shear properties of silty soil and the mechanisms. Constr Build Mater 189:286–295
20. Kulanthaivel P, Soundara B, Velmurugan S, Naveenraj V (2020) Experimental investigation on stabilization of clay soil using nano-materials and white cement. Mater Today Proc

Chapter 5
Influence of Aquaculture Sludge on Volume Change Behavior of Expansive Clays

T. V. Nagaraju, B. M. Sunil, and Babloo Chaudhary

Introduction

Expansive clays and their properties are well known globally [1]. The volume behavior of expansive clays must be calculated; mainly pavement construction, canal lining, shallow foundations, excavation, and fills [1, 2]. The volume change behavior mainly depends on the moisture content, over burden pressure, and loading conditions. Pavements or highways are particularly vulnerable due to the seasonal variation of moisture content in expansive subgrades. Light weight residential buildings rests on expansive clays exhibits severe cracks due to low weight building or strength to resist the swell potential [1, 3]. Moreover, few cases, due to leachate form solid waste landfill reacts with clays shows significant increase in hydraulic conductivity and decrease in swell behavior [4, 5]. Many authors have been investigated on the swell-shrink behavior of expansive clays. Effect of chemical additives such as cement, lime, calcium chloride, sodium silicate, fly ash, rice husk ash and ground granulated blast furnace slag on properties of expansive clays are well documented globally [6, 7]. According to their findings, blending of chemical additives with clays may cause flocculation or agglomeration or formation of cementitious compounds [7]. They may also allow cation exchange between clay particles and the chemicals such as Al^{3+}, Ca^{2+}, Mg^{2+}, K^{+}, Na^{+}, NH^{4+}, and Li^{+} [8, 9].

In a landfill, clay barrier, due to exposure of leachate with clays affect the mineralogical units of clays, distance between clay platelets, osmotic pressure between layers, and repulsive forces between layers [10]. In microscopic and mesoscopic scale, distance between the clay platelets (volume change) depends on the valance of cations and ionic strength of the leachate. The effect of leachate on the volume change behavior of clays has been reported by many authors [11]. The landfill

T. V. Nagaraju (✉) · B. M. Sunil · B. Chaudhary
Department of Civil Engineering, National Institute of Technology Karnataka, Surathkal 575025, India
e-mail: varshith.varma@gmail.com

K. Muthukkumaran et al. (eds.), *Ground Improvement Techniques*, Lecture Notes in Civil Engineering 297, https://doi.org/10.1007/978-981-19-6727-6_5

leachate (NH^{4+}) interaction with smectite clays cause exchange of the inter layer cations, decrease in double diffusion layer, and improvement in hydraulic conductivity. Further, long term exposure of leachate (NH^{4+}) may cause hazardous consequences of environment [11, 12]. Landfill leachate due to high alkalinity, particularly presence of ammonia (NH^{4+}) cations reacts with clays and form colloidal content, due to mineral disintegration and faulted structures. In long term, clays losses its plasticity and improvement in double diffusion layer, allows clays to behave as silts [12]. Ammonia ions (NH^{4+}) are one of the major cations which affect the clays including bentonite [13].

At present, aquaculture ponds sludge leachate is most concern because of high alkalinity (NH^{4+}) concentration [14]. Moreover, no research had been conducted on the effect of aquaculture leachate on the volume change behavior of expansive clays. In India, coastal region of Andhra Pradesh, intensive aquaculture farming is taking place and stands first in production in India [15]. However, to date, no data from journal publications on the effect of aquaculture leachate on ground water, especially regarding aquaculture leachate clay interactions. Intensive aquaculture farming demands high density seed, feed, antibiotics, disinfectants, and chemical allows negative impact on environment due to effluents generation. Currently, especially in India, no aquaculture effluent treatments were in practice [16]. The aquaculture effluents were diverted into nearby irrigation canals without any treatment, leads to eutrophication in the canals [17]. Other hand, continuous farming or crops, age of aquaculture sludge increases and reacts with subsoil clay and may pollute ground waterbodies.

In this paper, to study the volume change behavior of aquaculture leachate exposed expansive clays, a series of one-dimensional swell-consolidation tests were conducted. This paper also explores the effect of aquaculture leachate on the environment.

Materials and Experimental Procedure

Expansive clay with high compressibility behavior (CH) was used in this study. The expansive clay was collected from the delta region of Andhra Pradesh, India. Index properties and engineering properties of the clay sample was summarized in Table 5.1.

Table 5.1 Properties of clay samples

Properties	Liquid limit (%)	Plastic limit (%)	Soil Classification (IS)	Clay content (%)	Heave (mm)	Swell potential (%)	Free swell index (%)
Value	85	24	CH	94	1.74	8.7	170

Table 5.2 Aquaculture sludge leachate characteristics

Properties	pH	Electrical conductivity (μs/cm)	Total dissolved solids (ppm)	Calcium (mEq/100 g)	Sodium (mEq/100 g)	Ammonia (mEq/100 g)	Potassium (mEq/100 g)
Value	8.6	1020	2080	38.4	16.21	46.24	14.12

Aquaculture sludge was taken from the bottom of the aquaculture pond having age of 7 years in Bhimavaram, Andhra Pradesh, India. Leachate was collected from the aquaculture sludge using the electro-magnetic stirring process. The leachate characteristics are illustrated in Table 5.2.

To understand the effect of aquaculture leachate on the volume change behavior of expansive clays, a series of one-dimension swell-consolidation tests were conducted on both virgin clay and clays exposed with aquaculture waste sludge. To prepare the oedometer samples, clay was pulverized and passed through 425 μm IS sieve. Then the clay powder was compacted in the oedometer ring in three layers with an arbitrary density as 1.24 g/cc. This is followed by oedometer testing by allowing initial pressure of 5 kPa on clay sample with an exposure of aquaculture leachate.

Other hand, scanning electron microscopy (SEM) and snapshots before and after expose of aquaculture leachate was observed to know the texture and structure.

Results and Discussion

Effect of Aquaculture Waste Sludge on Swell-Shrink Behavior of Clays

In Table 5.3 shows the entire test data of swell-shrink before of expansive clays treated with aquaculture dry sludge as 10, 20, and 30% of the dry weight of expansive clay.

Figure 5.1 shows the variation in rate of heave with increase in aquaculture waste sludge content in the blended clay samples. The rate of heave of blended clay samples

Table 5.3 Effect of expansive clay before and after aquaculture sludge leachate exposure

Properties	Aquaculture waste sludge content (%)			
	0%	10%	20%	30%
Heave (mm)	1.74	1.56	1.21	0.84
Swell potential (%)	8.7	7.8	6.05	4.2
Swelling pressure (kPa)	115	95	86	71
Linear shrinkage (%)	12	7	5	2
Rebound (mm)	0.58	0.41	0.22	0.08

are decreased with increase in aquaculture sludge content. This can be attributed due to the ion exchange between ammonia and clay particles. Test data shows the swell potential and swelling pressure of the blended clay samples are gradually decreased with increase in aquaculture sludge content. This is due to the decrease in plastic behavior.

Ion exchange process between the clay particles and aquaculture leachate reduces the plasticity and subsequently reduction in the linear shrinkage values of blended clay samples is obtained (vide Fig. 5.2). Ion exchange is a key factor in limiting volume change behavior of clays exposed with leachate having ammonia presence [13]. The Fig. 5.2 indicates that rebound values decrease with increasing aquaculture sludge content. The decrease in rebound values indicates the reduction in plastic behavior that result in less rebound potential (Fig. 5.2).

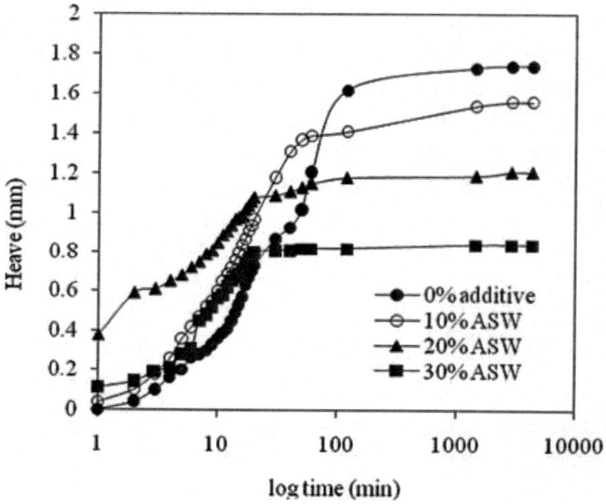

Fig. 5.1 Rate of heave of aquaculture sludge blended clays

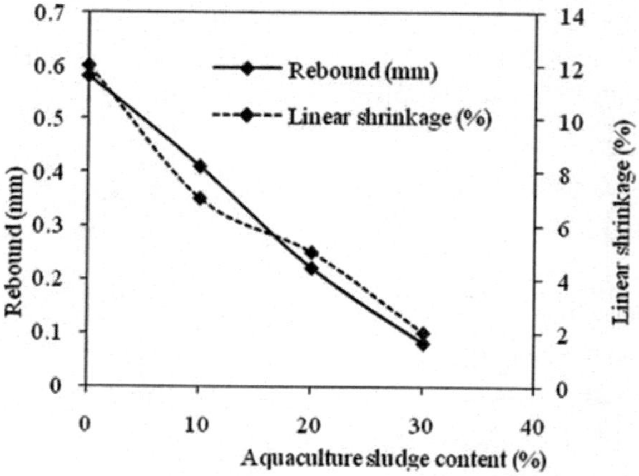

Fig. 5.2 Effect of aquaculture sludge content on linear shrinkage and rebound

Fig. 5.3 SEM and snapshot images of untreated expansive clay

Effect of Aquaculture Waste Sludge on Clay Structure

It was clearly evident that the leachate exposure samples were flocculated and size of the particle increased, but also linear shrinkage of the samples were decreased (vide Fig. 5.4). Further, particles become powdered like substance without any intact due to chemical reactions. In addition to this, cracks or linear shrinkage were apparent in the untreated expansive clay (montmorillonite). Similar trend was seen in the landfill leachate exposure with the semectite clays [11, 13]. In general expansive clays will have high potential double diffusion layer, it can be helpful to volume change behavior in the clays. So, when the clays exposed with aquaculture leachate having high ammonia (NH^{4+}) concentration contributes the cation exchange and may be broken the double diffusion layer. In this context, the clay mineral transformation occurred by means of cationic exchange and clay minerals yielded to lose plasticity behavior.

Fig. 5.4 SEM and snapshot images of expansive clay exposed with aquaculture leachate

Conclusions

The following principal conclusion may be drawn from the study:

The addition of aquaculture sludge to the expansive clay decreased the rate of heave and swell potential. It is found the addition of aquaculture sludge causes cation exchange between the clay particles and aquaculture sludge.

Rebound and linear shrinkage of soil was found to decrease with increasing aquaculture sludge content. SEM and snapshots of the treated and untreated expansive clays confirm the cation exchange between the clay particles and aquaculture sludge, which contributed to reduction in plasticity.

Moreover, the high ammonia concentration presence in the aquaculture sludge causes cation exchange and may break the double diffusion layer, and hydraulic conductivity may increase.

Determination of the volume change of behavior of expansive clays exposed with aquaculture leachate is an important means of assessment of aquaculture leachate on ground water.

References

1. Basma AA, Al-Homoud AS, Malkawi AIH, Al-Bashabsheh MA (1996) Swelling-shrinkage behavior of natural expansive clays. Appl Clay Sci 11(2–4):211–227
2. Nagaraju TV, Satyanarayana PVV (2019) Geotechnical aspects of various constructions along the canal embankment using rice husk ash as stabilizer. In: Ground improvement techniques and geosynthetics. Springer, Singapore, pp 143–150
3. Bell FG, Maud RR (1995) Expansive clays and construction, especially of low-rise structures: a viewpoint from Natal South Africa. Environ Eng Geosci 1(1):41–59
4. Hettiaratchi JPA, Hrudey SE, Smith DW, Sego DCC (1988) A procedure for evaluating municipal solid waste leachate components capable of causing volume shrinkage in compacted clay soils. Environ Technol 9(1):23–34
5. Khodary SM, Negm AM, Tawfik A (2018) Geotechnical properties of the soils contaminated with oils, landfill leachate, and fertilizers. Arab J Geosci 11(2):1–17
6. Phanikumar BR, Nagaraju TV (2018) Engineering behaviour of expansive clays blended with cement and GGBS. Proc Inst Civ Eng Ground Improv 171(3):167–173
7. Goodarzi AR, Akbari HR, Salimi M (2016) Enhanced stabilization of highly expansive clays by mixing cement and silica fume. Appl Clay Sci 132:675–684
8. Prusinski JR, Bhattacharja S (1999) Effectiveness of Portland cement and lime in stabilizing clay soils. Transp Res Rec 1652(1):215–227
9. Tertre E, Dazas B, Asaad A, Ferrage E, Grégoire B, Hubert F, Delay F (2021) Connecting molecular simulations and laboratory experiments for the study of time-resolved cation-exchange process in the interlayer of swelling clay minerals. Appl Clay Sci 200:105913
10. Karpiński B, Szkodo M (2015) Clay minerals–mineralogy and phenomenon of clay swelling in oil & gas industry. Adv Mater Sci 15(1):37–55
11. Oztoprak S, Pisirici B (2011) Effects of micro structure changes on the macro behaviour of Istanbul (Turkey) clays exposed to landfill leachate. Eng Geol 121(3–4):110–122
12. Oyediran IA, Olalusi DA (2019) Leachate effects on some index properties of clays. In: IAEG/AEG annual meeting proceedings, San Francisco, California, 2018, vol 6. Springer, Cham, pp 159–164

13. Gautier M, Muller F, Le Forestier L, Beny JM, Guégan R (2010) NH4-smectite: characterization, hydration properties and hydro mechanical behaviour. Appl Clay Sci 49(3):247–254
14. Summerfelt RC (2000) Water quality considerations for aquaculture. Department of animal ecology, pp 2–7
15. Belton B, Padiyar A, Ravibabu G, Rao KG (2017) Boom and bust in Andhra Pradesh: development and transformation in India's domestic aquaculture value chain. Aquaculture 470:196–206
16. Jayanthi M, Ravisankar T, Nagaraj G, Thirumurthy S, Muralidhar M, Saraswathy R (2019) Is aquaculture abandonment a threat to sustainable coastal resource use ?—a case study of Andhra Pradesh, India, with options for reuse. Land Use Policy 86:54–66
17. WU DS, Xiong Q, Du JY (2009) Study on the eutrophication by aquaculture. Jiangxi Science, 4

Chapter 6
Densification of Fly Ash Deposits Equipped with Rammed Stone Column—A Case Study

Shadab Gadhiya and Maunank Modi

Introduction

Utilization of fly ash is not likely to exceed beyond 10–15%, and therefore, various thermal power plants in our country have no other option but to create ash ponds. Such ash pond requires large land area and as most ash pond is filled to its design capacities; fresh lands in thousands of hectares are being acquired for thermal power plant extension/renovation.

The entire project site, under consideration, consists of fly ash deposits; hence, ground improvement in form of stone column is proposed to satisfy load requirements. Stone columns intended to increase the bearing capacity of soil and to reduce settlement for open foundation resting over fly ash. In addition, stone column will help in increasing lateral capacity of pile and to mitigate liquefaction potential during any seismic event.

Stone column with different diameters is proposed to suit the project requirements. 900-mm-diameter stone column with triangular pattern will be used for open foundation whereas 600-mm-diameter stone column will be used along with pile foundation.

S. Gadhiya · M. Modi (✉)
L&T—Sargent and Lundy, Vadodara 390019, India
e-mail: Maunank.Modi@Lntsnl.com

S. Gadhiya
e-mail: Shadab.Gadhiya@Lntsnl.com

K. Muthukkumaran et al. (eds.), *Ground Improvement Techniques*, Lecture Notes in Civil Engineering 297, https://doi.org/10.1007/978-981-19-6727-6_6

Geotechnical Conditions

Based upon detailed geotechnical investigation report, total 44 nos. of boreholes were drilled. Huge variation with respect to different facilities was noticed. Generalized soil profile is as below.

Stratum I—overburden soils—fly ash is met from 7.5 to 18.0 m depth. The field SPT N values range from 4 to 22 to about 18.0 m depth. Some layers of sandy silt/silty sand are encountered between fly ash and soil-rock interface. Refusal was met on soil-rock interface.

Stratum II—granitic gneiss (Rock)—below stratum I, very weak to moderately strong granitic gneiss encountered to the final explored depth of 30 m. Core recoveries are generally nil up to about 30.0 m depth.

Foundation Consideration Based upon Field Test Results

Bearing capacity for open foundation shall be restricted to 10 t/m^2 as per tender requirements. Hence, looking to the site conditions and findings from geotechnical investigation report, all open foundations requiring less than 10 t/m^2 bearing capacity shall be supported on 900-mm-diameter stone column in equilateral triangular pattern at 1.8 m c/c (cross-section as shown in Fig. 6.1). For load greater than 10 t/m^2, stone column of 600 mm diameter along with piles shall be provided (cross-section as shown in Fig. 6.2). Sand blanket of 500 mm over stone column is proposed as drainage layer. Extension of 1000 mm on the edges outside the stone column is provided to cater bulging due to foundation load. Design of stone column is made as per Indian standard 15,284, Part-1 code. Depth of fly ash deposit varies for entire project, hence optimized design suiting to nearby facilities are made.

Field Load Test on 900-mm Stone Column for Open Foundation

Densification of fly ash deposit due to provision of stone column will improve shear strength, bearing capacity, and reduces settlement. For validation of parameter enhancement, pre- and post-SPT's along with SCPT's were conducted followed by full-scale field load test on individual and group of stone column. Initial stone column test on single column (ISCT–1 & ISCT–3) and on group of stone columns (ISCT–2 & ISCT–4) for open foundation are conducted at different location as presented in Fig. 6.3.

Results of pre- and post-SPT along with pre- and post-CPT near stone columns are presented in Figs. 6.4 and 6.5. From pre- and post-SPT and SCPT test, it was

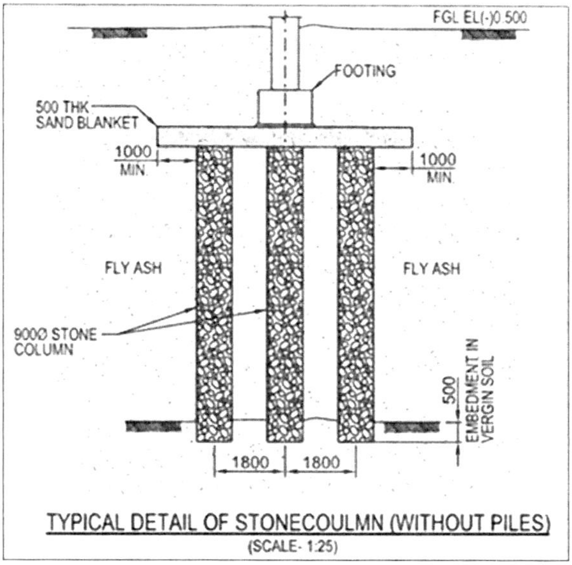

Fig. 6.1 Typical detail of stone column for open foundation

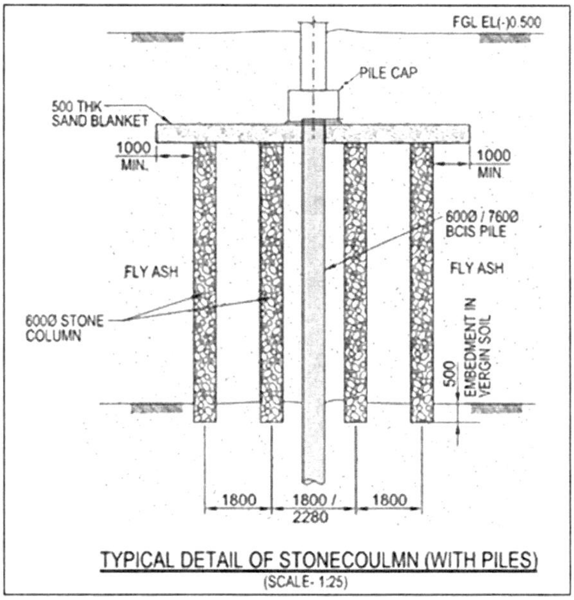

Fig. 6.2 Typical detail of stone column with piles

concluded that gain in shear strength happened significantly, and fly ash deposit got densified after addition of stone aggregates.

Field load test on single stone column of 900 mm dia. is conducted with maximum load of 43.5 t which is 1.5 times of design load of 29 t and on group of three stone

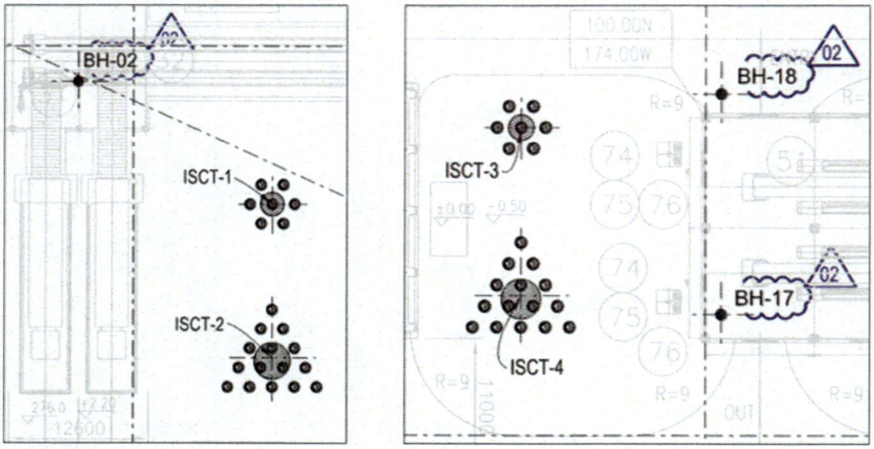

Fig. 6.3 Initial stone column test layout for individual and group of stone column

Fig. 6.4 SPT versus depth for ISCT—2

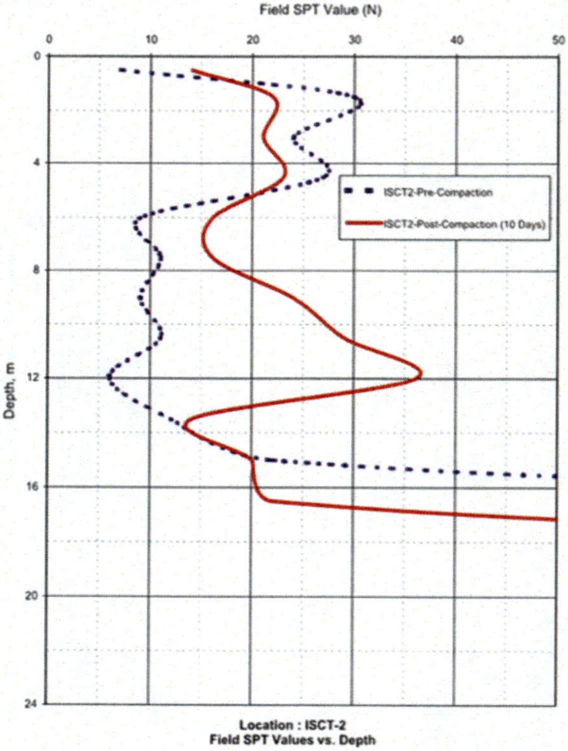

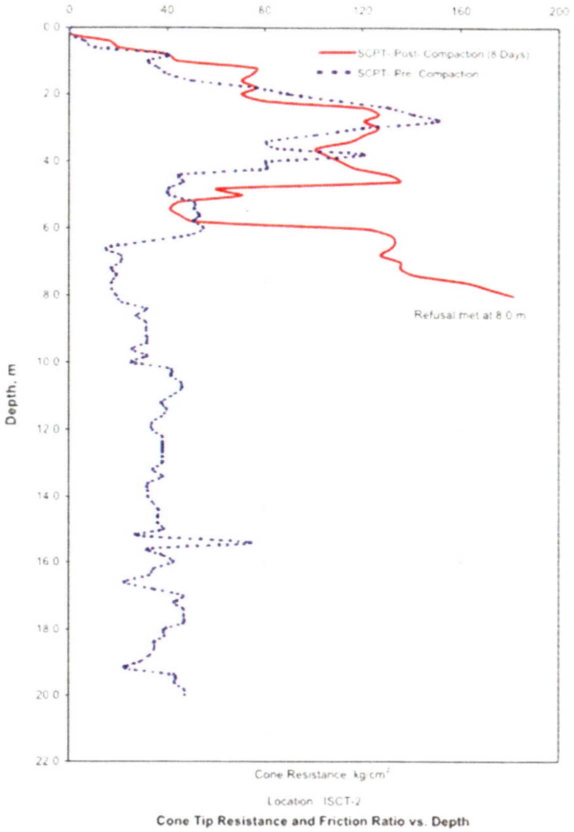

Fig. 6.5 Cone tip resistance and friction ratio versus depth for ISCT—2

column with maximum load of 127.5 t which is 1.5 times of design load of 83 t (Fig. 6.6).

As per IS 15284 (Part-1), clause 13.7, acceptable settlement criteria for single stone column load test shall be 12 mm, whereas total settlement achieved from field test is 4.90 mm which confirms safe load of 29 t on single stone column. Similarly, for group of stone column, acceptable settlement as per IS 15284 (Part-1), clause 13.7, shall be 30 mm, whereas achieved total settlement from field load test is 17.03 mm. Hence, test result confirms safe load of 83 t on group of three stone column. Results also prove actual load carrying capacity of stone column is much higher than design load; however, as per technical specification requirements, bearing capacity of 10 t/m^2 is limited for entire project.

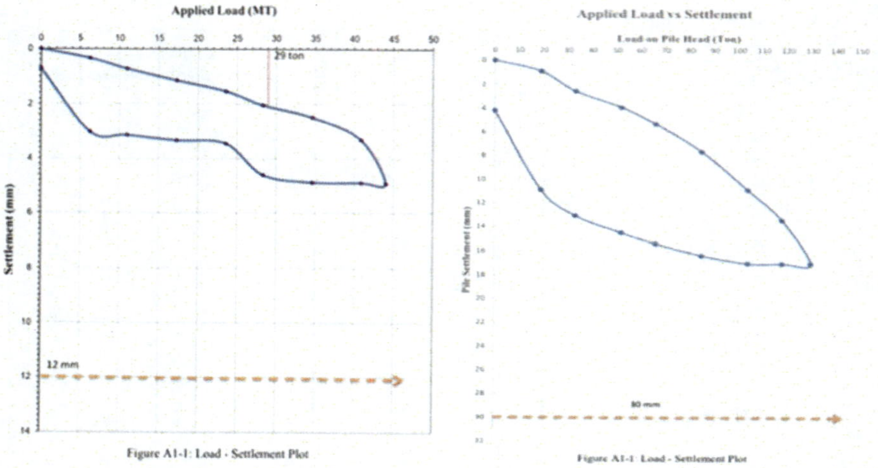

Fig. 6.6 Load versus settlement for individual and group of stone column

Field Load Test on 600-mm-dia. Stone Column for Pile Foundation

To suit to the project requirements, 600-mm-dia. and 760-mm-dia. bored cast in situ piles were designed to carry superstructure load. However, single type of stone column with 600 mm diameter is selected to suit both pile dia. configuration. Typical layout of stone column with different dia. piles is shown in Fig. 6.7. Total six pile groups of both dia. and different length were planned to be tested with stone column configuration as shown in Fig. 6.8.

Due to presence of fly ash to maximum 18.0 m depth, lateral pile capacity for entire project was very low. In addition, site was susceptible to liquefaction. Hence, rammed stone column in triangular pattern is provided around piles at spacing of three times dia. To verify lateral pile capacity, full-scale field load test is conducted.

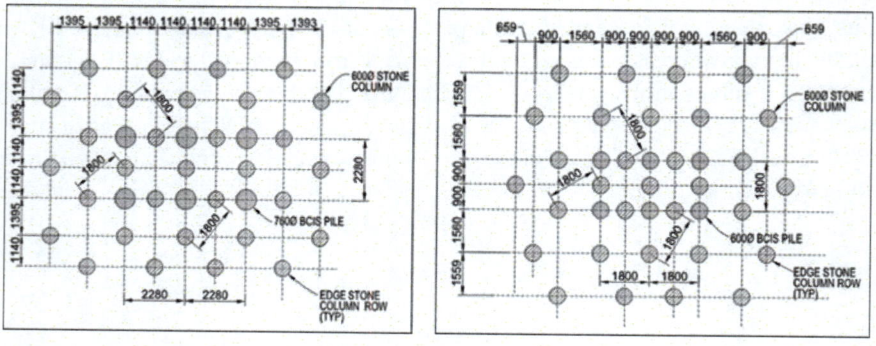

Fig. 6.7 Stone column configuration for different dia. piles

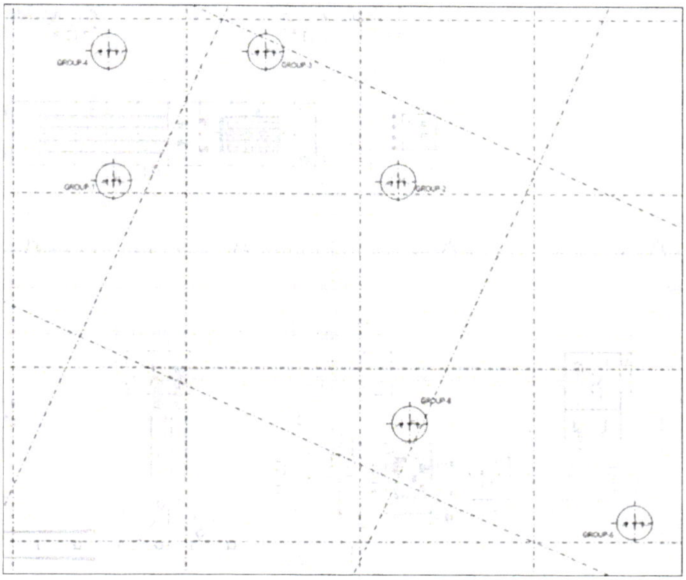

Fig. 6.8 Layout of pile load test with stone column configuration

For 600-mm-dia. pile, design load is 4.30 t which is loaded to three times of design load of 12.9 t. Similarly, for 760-mm-dia. pile, design load is 7.0 t which is loaded to three times of design load of 21.0 t. Typical load settlement graph of lateral pile load test for 600-mm-dia. pile from group 2 and 760-mm-dia. pile from group 6 is shown in Fig. 6.9.

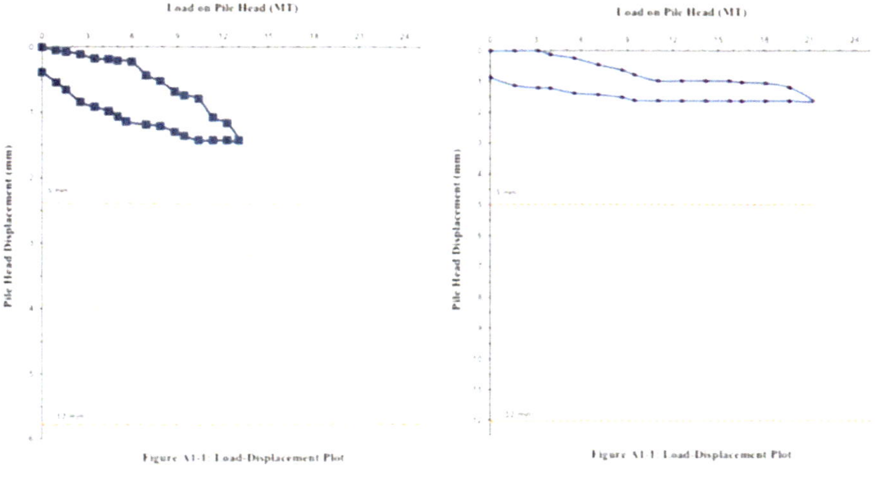

Fig. 6.9 Load versus settlement of 600-mm-dia. pile and 900-mm-dia. pile

Maximum lateral displacement for 600-mm-dia. pile has reached 1.64 mm after loading to three times of design load, whereas acceptable limit as per IS 2911 (Part 4) is 5 mm. Similarly, for 900-mm-dia. pile, maximum lateral displacement reached is 1.64 mm after loading to three times of design load. Hence, it proved that actual lateral load carrying capacity after ground improvement with stone column is much higher than designed value. Summary of pile load test results for compression, lateral, and tension test on all six pile group is presented in Table 6.1.

Table 6.1 Summary of pile load test

Group	Type of load test	Dia. Of pile (mm)	Length of pile (m)	Design load (t)	Test load (t)
1	Compression (TPV1)	600	28	85	255
	Tension (TPP1)			26	78
	Lateral (TPH1)			4.3	12.9
2	Compression (TPV2)	600	28	85	255
	Tension (TPP2)			26	78
	Lateral (TPH2)			4.3	12.9
3	Compression (TPV3)	760	25	250	750
	Tension (TPP3)			75	225
	Lateral (TPH3)			13	37.5
4	Compression (TPV4)	760	25	250	750
	Tension (TPP4)			75	225
	Lateral (TPH4)			12.5	37.5
5	Compression (TPV5)	760	28	139	417
	Tension (TPP5)			42	126
	Lateral (TPH5)			7	21
6	Compression (TPV6)	760	28	139	417
	Tension (TPP6)			42	126
	Lateral (TPH6)			7	21

Conclusion

a. Theoretical approach for use of rammed stone column to densify fly ash deposit is validated through number of different field test. Pre- and post-SPT and SCPT have shown substantial gain in densification of fly ash layers. The percentage increase in in situ properties after installing stone column varies from 130 to 450.
b. Field load test on 900-mm-dia. stone column for laying open foundation has shown significant lower values of displacement/deflection which proves that bearing capacity of soil can be taken more than 10 t/m^2; however, limiting to specification requirement, it is still considered as 10 t/m^2.
c. Pile load test with single row of stone column around piles and two rows of stone column around piles were conducted. Results show considerable amount of gain in lateral capacities with two rows of stone column.
d. Full-scale pile load test on 600-mm-dia. pile and 760-mm-dia. pile demonstrates much higher capacities in all types of test.
e. Lateral deflection has reduced to a greater extent. In addition, liquefaction susceptibility also mitigated efficiently.

Chapter 7
Application of Prefabricated Vertical Drains (PVDs) for Improvement of Soft Clays—A Case Study

Abhijeet Kanungo and V. Jaya Pragash

Introduction

Project Details

The current case study is concerning a facility located at Navi Mumbai. The project consists of warehouses, office buildings, container yard, internal roads, apron, pavers, and other ancillary buildings. The total site area of the following case study was around 1,79,000 m^2 (Approx. = 44 Acres). The site of the concerned case study is nearby sea and the approximate distance from the sea was 6.0–7.0 km. The soil profile of the site was the real challenge in the project, and this case study is about how the design team overcomes this problem related to geotechnical engineering. The soil strata of the following case study consist of filled up soil with boulders varying from a depth of 3.0–4.0 m followed by very soft marine clay up to a depth of 12.0–13.0 m from Existing Ground Level (EGL), i.e., the soft clay layer thickness was around 9.5 m. Soft clay was followed by hard/stiff clay up to the weathered/hard rock top. The rock top was varying from a depth of 14.0–26.0 m throughout the site. The soil was subjected to various loads consisting of structural loads (dead load, live load, wind, and earthquake loads), racking loads from grade slab in warehouses, vehicular loads on external sections, and container stacking load in container yards. Detailed study of strata was done through the soil investigation, preliminary surveys, and site visits. Based on the study, it was found that the soft clay layer was highly compressible in nature and highly susceptible to settlements on loading. If the construction takes place without ground improvement on such kinds of soils, it can cause excessive

A. Kanungo (✉) · V. J. Pragash
B&F IC—EDRC—Geotechnical Division, L&T Construction, Chennai 600089, India
e-mail: abhijeetkanungo@lntecc.com

V. J. Pragash
e-mail: vjayapragash@lntecc.com

K. Muthukkumaran et al. (eds.), *Ground Improvement Techniques*, Lecture Notes in Civil Engineering 297, https://doi.org/10.1007/978-981-19-6727-6_7

settlements which may lead to damage to the structure and affect the serviceability of the project and the structures.

Hence, to deal with the problem mentioned in the above paragraph, ground improvement was proposed. Based on the detailed study and analysis, Prefabricated Vertical Drains (PVDs) with surcharge were adopted as the most suitable option for improving the ground in this project and after the successful completion of the ground improvement process, and PVD with surcharge proved to be a most suitable and feasible solution. This case study will depict the application of PVD with surcharge technique of ground improvement at the concerned site.

Soil Investigation

Before design and execution of any ground improvement activity, we need to have detailed soil investigation parameters and comprehensive knowledge of soil profile of the site. Also, before adopting any ground improvement, we investigate the feasibility of the technique and other practical and economical aspects of the ground improvement technique. Apart from 10 boreholes in the tender stage, to acquire detailed parameters of soil, 27 boreholes and 15 trial pits were proposed in the concerned project. The depth of boreholes was up to the rock layer and the depth of trial pits was up to the top of the soft clay layer. The depth of boreholes and trial pit was decided such that we can estimate the approximate depth of rock top throughout the site. Field tests such as standard penetration test and vane shear test were conducted at the site, and samples were collected for laboratory tests. Oedometer test for consolidation parameters of soil was conducted in the laboratory. The groundwater table at the site was around a depth of 1.0–2.0 m from the Existing Ground Level (EGL). Design parameters adopted are given in Table 7.1. Also, the proposed boreholes and trial pits are shown in Fig. 7.1.

Table 7.1 Design parameters derived from soil testing

Layer	Description	Depth of layer from EGL	Layer thickness	Cohesion (KN/m^2)	Angle of friction (Degree)	Bulk unit weight (KN/m^3)	Average field SPT N values
Layer 1	Filled up soil with boulders	0–3.5 m	3.5 m	–	28	19	15–20
Layer 2	Very soft to soft clay	3.5–12.5 m	9.0 m	10	–	15.5	0–3
Layer 3	Hard stiff clay	12.5–18 m	5.0 m	100	–	18	30–40
Layer 4	Weather/hard rock	>18.0 m	–	–	40	20	Refusal

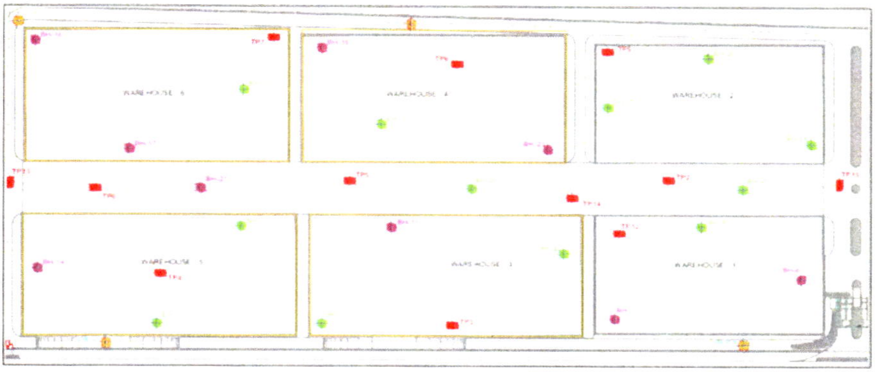

Fig. 7.1 Borehole and trial pit layout for warehouse area of the site

Ground Improvement Features

Selection of Technique

Based on the preliminary study of the project and soil investigation, it was observed that the consolidation of the soft clay layer was required to arrest the excessive settlement of structures in the future after construction. Another option was to transfer the load directly to hard strata, and the second option may be suitable for structural loads through pile foundation, but grade slab and pavement loads are not feasible practically to transfer to hard stratum. Hence, the first option was considered, i.e., consolidation of soft clay.

Consolidation: When cohesive soils which are saturated in nature, either fully saturated or partially saturated, are subjected to applied stress, the volume of saturated soils decreases and results in the reduction of compressibility of soil. In this process, removal of water takes place from the voids of soil and generally, and it is observed in clayey soils only.

There are various methods for consolidation of clayey soil. In this project, number of methods were analyzed such as stone columns, rigid inclusions, preloading, PVDs with vacuum consolidation, and drains/PVD with surcharge. Based on several factors such as availability of literature, requirement of skilled labor, feasibility at the site, equipment and vendor availability, time consumption, and cost, Prefabricated Vertical Drains (PVDs) with surcharge were adopted as ground improvement method for the project.

PVD will provide a shorter drainage path to water, to come out from the impervious strata, i.e., clay and surcharge will be generating the stress in the clayey soil which will ultimately force the water to come out of the soft clay.

Component of Ground Improvement

The ground improvement technique of PVD with surcharge consists of number of components which are described in this section.

Prefabricated Vertical Drains (PVDs): PVDs consist of a central core, which will act as a free drainage channel, and a non-woven filter jacket, which prevents the soil surrounding the drain from entering the central core but allows water to flow in. Vertical drains are installed under a surcharge load to accelerate the drainage of impervious soils and thus speed up consolidation. These drains provide a shorter path for the water to flow through to get away from the soil.

The use of PVDs is applicable for soils which are moderate to highly compressible under static loading and compress very slowly under natural drainage conditions due to low soil permeability and relatively great distance between natural drainage boundaries. Soils with these characteristics are almost exclusively cohesive, fine grained soils, either organic or inorganic. In this project, PVD of 12 m depth was installed in triangular pattern, from the existing ground and terminated in stiff clay layer. For faster execution of project, total site was divided in two phases based on spacing of PVDs. In phase 1 of the project, 0.8 m spacing of PVD was adopted, and in Phase 2 of the project, 1.0 m of PVD spacing was adopted.

Surcharge: Soil was used to apply pressure to the soft clay layer which results in generation of pore water pressure in the clayey soils. A surcharge of height varying from 2.5 m to 5.5 m was adopted in the project. The surcharge height was depending upon the load requirements of various structures. The surcharge was divided into two layers, namely, permanent surcharge and temporary surcharge to fasten the execution process.

Drainage Layer: It consists of drainage material that allow dissipation of pore water pressure generated due to surcharge. A drainage layer of 0.5 m thickness was provided in the present case study.

Geotextile: A layer of geotextile was used to prevent the blockage of drainage layer during the process of consolidation.

Monitoring Instruments: Monitoring instruments such as settlement gauges, vibrating wire piezometers, and post vane shear test were used to monitor the ground improvement process and results.

Details related to the ground improvement scheme are provided in Table 7.2 and components of ground improvement are shown in Fig. 7.2.

Design Parameters

PVD + Surcharge method of consolidation is designed to get an optimized PVD spacing, PVD depth, and waiting period of the total system. For design of ground

Table 7.2 Ground improvement scheme

Parameters	Value
Soft clay layer thickness, H	9.5 m
Time for consolidation without ground improvement	95 years (considering single drainage)
Ground improvement method recommended	PVD + Surcharge
Pattern of PVD	Triangular
Spacing of PVD	0.8 m (Phase 1) and 1.0 m (Phase 2)
Length of PVD (excluding projections)	12.00 m
Total surcharge height (depending on load requirement)	5.50 m, 5.00 m, and 2.50 m
Drainage layer thickness	0.50 m
Time for 90% consolidation	90 days (Phase 1) and 150 days (Phase 2)
Number of stages for surcharge application	ONE

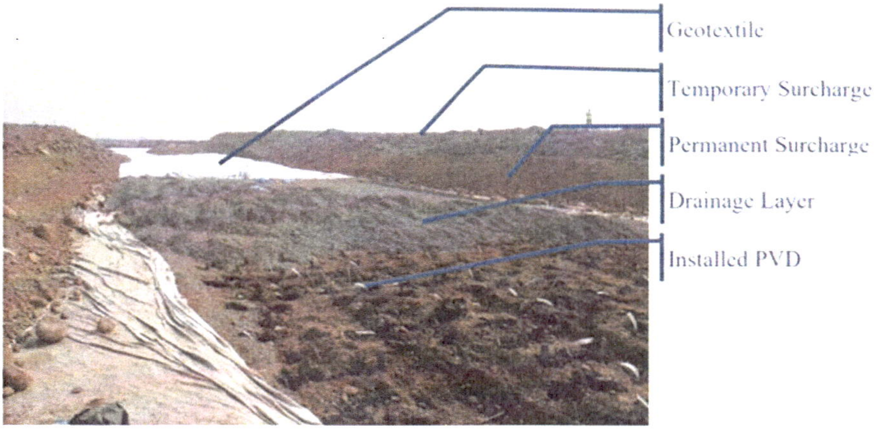

Fig. 7.2 Components of ground improvement

improvement system, various parameters were considered based on the soil investigation, codal provisions, and design engineer experience. Table 7.3 provides the detail of design parameters considered for design of the ground improvement system of the project. Ratio $Cc/(1 + e_0)$ is around 0.180 result in higher consolidation settlements and shows that soil is highly compressible in nature. Single drainage was considered for design as the soft clay layer was followed by an impervious layer of stiff to hard clay layer. Area treated by single PVD, A, and equivalent diameter of band drain, d,

Table 7.3 Design parameters for design of ground improvement system

Design parameters	Value
Soil property	
Liquid limit of soft clay	75%
Plastic limit of soft clay	40%
Plasticity index	35%
Specific gravity	2.65
Void ratio	2.25
Compression index (Cc)	0.585
Coefficient of vertical consolidation, Cv	0.8 m^2/Year
Coefficient of horizontal consolidation, Ch	1.6 m^2/Year
Drainage	Single
Drainage path (equal to thickness of soft clay layer)	9.5 m
PVD properties	
Width of band drain b	100 mm
Thickness of band drain t	4 mm
Area treated by single band drain, A	0.555 m^2
Equivalent diameter of cylindrical column, D	0.840 m
Equivalent diameter of band drain, d	0.067 m
Degree of radial consolidation, Ur	90%

depends on the pattern of the PVD, in which it is installed and the dimensions of the PVD. Degree of consolidation means the amount of consolidation designer wants to achieve through this system and by achieving the desired degree of settlement future settlements can be maintained within permissible limits. In this case, study degree of consolidation for which design has been done is 90%.

Analysis Steps

Consolidation Time Calculation

Consolidation of soft clay layer would have taken place without PVDs and surcharge as well, but it would have taken a lot more time as compared to that of present condition. Hence, during analysis, time for 90% consolidations was calculated, for the condition without PVD as well as, with PVD and surcharge. The time for without PVD condition was calculated using Eq. 7.1.

$$t = \frac{T_v * d2}{C_v} \tag{7.1}$$

where T_v is the time factor depending on the degree of consolidation, d is the drainage path, in the following case study it is taken as 9.5 m, which is equal to the clay layer thickness, and C_v is the coefficient of vertical consolidation C_v. The time without ground improvement was 95 years.

Now, to reduce the time of the consolidation, we have introduced the PVD into the clayey layer. Hence, to calculate the time for 90% of consolidation, Hansbo's Equation [2] was used. Equation 7.2 shows Hansbo's Equation

$$t = \frac{D^2}{8.Ch}.\left[\frac{1}{1-(d/D)^2}\cdot \ln(d/D) - \frac{3}{4} + \frac{1}{4}\left(\frac{d}{D}\right)^2\right].\ln\left(\frac{1}{1-U}\right) \quad (7.2)$$

Using above equation time for consolidation came out to be 84 days, hence 90 days were kept as a waiting period for PVD spacing of 0.8 m. Similarly, for phase 2, 150 days were kept as the waiting period with PVD spacing of 1.0 m.

Surcharge Height Calculation

As shown earlier in the article that the surcharge height was kept different for different loads and structures. The surcharge height depends on the estimated future load going to be exerted on soil after construction of structure. The same load will be applied to soil during the waiting period. Hence, the surcharge is also known as preloading. The load on the soil shall be the combination of dead load of section and live load of structure. For example, in the present case study, in warehouse area, total load was around 100 KN/m^2 and the density of surcharge was considered 18 KN/m^3, hence, the total surcharge height in the grade slab area was 5.50 m. Similarly, for yard area surcharge, height was 5.0 m, and for pavement area, it was around 2.50 m.

Increase in Cohesion Value

With removal of water, soil will be compressed and result in stiff and hard clay. Indirectly, this will result in an increase in the cohesion value of the soil. During the initial soil investigation, the cohesion value of the soft clay layer was around 15 KN/m^2, and after ground improvement, the cohesion value increased and the same was estimated using the equation from the IRC75-2015: GUIDELINES FOR THE DESIGN OF HIGH EMBANKMENTS. The equation used for computation of final cohesive value is shown in Eq. 7.3.

$$\Delta c = k \times U \times \Delta\sigma \quad (7.3)$$

where Δc is gain in cohesion after ground improvement, U is the degree of consolidation achieved after ground improvement, and $\Delta\sigma$ is the increase in the effective pressure at the center of the soft clay layer due to the surcharge loading.

Degree of consolidation can be calculated using the equation provided in clause 6.5.2 of IS15284-Part II: Design and Construction for Ground Improvement—Guidelines (preconsolidation using vertical drains).

The increased cohesion value for warehouses area was 37 KN/m^2. The design value calculated was validated using the field vane shear test.

Combined Degree of Consolidation

Generally, the consolidation will be taking will be radial, but the vertical consolidation will also take place. The vertical degree of consolidation can be calculated from Table 7.2 of IS15284-Part II: Design and Construction for Ground Improvement—Guidelines (preconsolidation using vertical drains). The combined degree of consolidation can be given by Eq. 7.4.

$$U = 1-(1 - U_z) * (1 - U_r) \tag{7.4}$$

The combined degree of consolidation for the concerned project was estimated as 91.57%.

Settlement Computations

Consolidation settlement was computed at the center of the soft clay layer using the dispersion of load from top to center as 1*H*: 2 *V*. Equation used for computation of settlement is shown below in Eq. 7.5.

$$\Delta H = \frac{c_c.H}{1 + e_0} \times \log\left(\frac{\sigma_0 + \Delta\sigma}{\sigma_0}\right) \tag{7.5}$$

The total settlement of clayey layer was 641 mm. As per Section "Combined Degree of Consolidation" of the present article, degree of consolidation achieved during the ground improvement is 91.57%, hence, future settlement that can take place is around 54 mm which is under the permissible limits. Hence, design is safe and showing the desired result. If the future settlement is not within the permissible limits, then we need to redesign by varying the spacing of PVD and degree of consolidations.

Execution of PVD and Surcharge

Ground improvement was done in the area of 1,76,170 m^2 (43.50 acres). Execution of PVD was done in two phases as discussed in previous sections. Based on the spacings, total site was divided into two phases. This was done to maintain a balance between cost and time.

PVD Installation: PVD was installed in a triangular pattern at designed spacing using PVD rigs. The depth of installation was fixed based on the thickness of soft clay layer and depth of stiff clay so that the anchor rod can be anchored in the hard strata. In the current case study, depth of installation was 12.0 m and a projection of approximately 150 mm on each side was left. PVD was embedded in drainage layer. PVD rig can install a PVD in 40–60 s. Around 2,50,000 PVDs were installed at the site, which means approx. 30,00,000 running meter of PVD was installed. A total of 6 PVD rigs were used for the timely completion of the project.

Surcharge: Laying of surcharge was done in two types so that the time for backfilling to reach desired levels can be saved. Surcharge in the current case study was done as the combination of permanent surcharge and temporary surcharge. Temporary surcharge was removed after waiting period, while the permanent surcharge was kept as it was required to reach desired Formed Ground Level (FGL) of the site. Total surcharge quantity of 8,00,000 Cum was used in the present ground improvement system. Surcharge was also including the drainage layer.

Above numbers are proof that the current case study is one of the biggest ground improvement case studies in India (Fig. 7.3).

Results and Observations

During and after the execution of surcharge, various monitoring instruments were installed at the project site to monitor the progress of ground improvement. The instruments used were including settlement gauges and vibrating wire piezometers. For checking the final field cohesion values, field vane shear tests were proposed and conducted after waiting period.

A total of 44 settlement gauges, 22 Piezometers were installed at site and 70 post ground improvement vane shear tests were proposed. All the monitoring instrumentations were done as per the specifications of Highway Research Board (HRB) 13 and 14 manuals. Full site was divided into segments so that the maintenance of record, execution, and removal of surcharge becomes easy and timesaving.

Settlement gauges used in the present articles were plate-type settlement gauges and these gauges were installed at the top of clay layer and readings were taken using the auto-level surveying instrument. Weighted average settlement of approx. 700 mm was observed throughout the site. A maximum settlement of 1.177 mm (1.177 m) was also observed in one of the segments.

Fig. 7.3 Execution images of PVD and surcharge

There was total of 44 settlement vs time curves were studied. Figure 7.4 shows some of the results of settlement vs time for various structures.

Piezometers were installed at the site before laying of surcharge. Piezometers were installed to measure the pore water pressure generation and observe the dissipation of pore water pressure. As the surcharge laying was under process, the pore water pressure was developing, reached its peak values, and as the waiting period started, the pore pressure start decreasing. It was observed that when the waiting period was about to over full pore, pressure was dissipated as the pore water pressure attains same value as the initial value or a value near to the initial value. This confirms the end of consolidation process and also validates design. If the pore pressure graph is not showing the correct curve as mentioned in the above discussion, waiting period can be increased based on the experience (Fig. 7.5).

Vane shear tests (70 Nos) were conducted after waiting period was completed. Vane shear test was done to check the increased cohesion value. During the time of soil investigation also vane shear tests were conducted. Cohesion value of both the vane shear test and design cohesion value was compared. In all the locations, the cohesion value was more than the cohesion value computed during the design stage. This shows that the soft clay which was present at the site was now consolidated

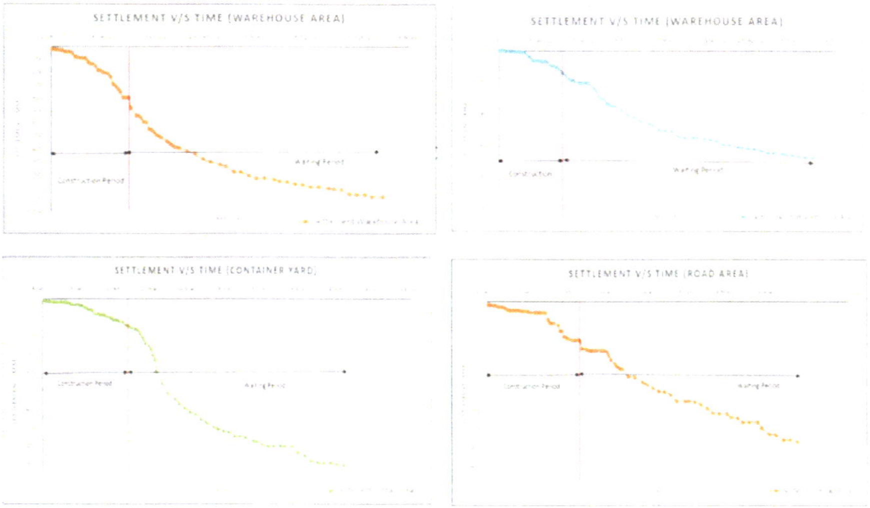

Fig. 7.4 Settlement versus time graph for various structures

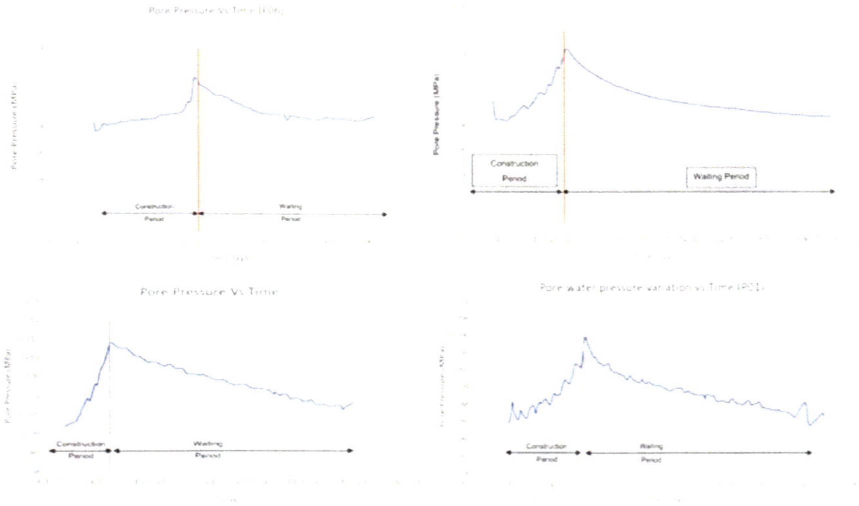

Fig. 7.5 Pore pressure versus time plots from piezometers

and the cohesion value was increased around 5–7 times of initial cohesion value (Fig. 7.6).

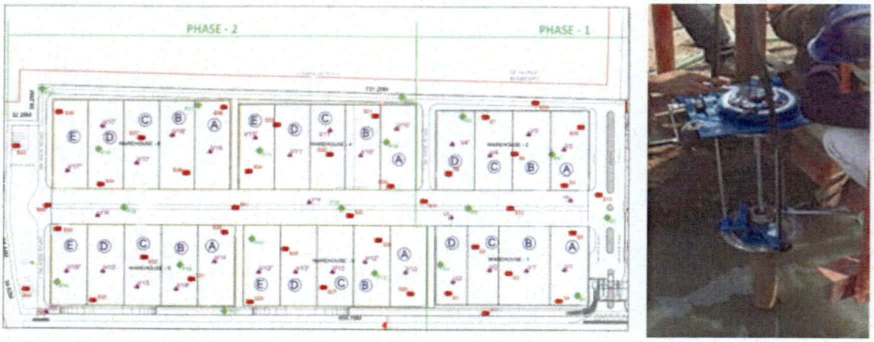

Fig. 7.6 Instrumentation layout and vane shear test

Conclusions

In the following case study, we have seen problems related to soft clay, solution of the highly compressible nature of the soil. In the article, design and execution of the ground improvement technique was discussed. Application and mechanism of the PVD + Surcharge were also shown. So, after completion of the ground improvement process, following are the observations.

1. Soil strata present in any site may have uncertainties; hence, detailed soil investigation is required for correct set of inputs, and which can lead to optimization of quantities and time.
2. PVD + Surcharge method is suitable for the consolidation of soil type described in the present case study. Also, it is an effective and economical method for soft clayey soils.
3. From the piezometer results, it can be incurred that the generation of pore pressure takes place during the surcharge laying and dissipation of pore pressures takes place in the waiting period. It is observed that the pore pressure generated in was not equal to the surcharge load as the some load would have dissipated in top fill-up soil of 3.0 m height. The curve generated for pore pressure vs time from piezometers was matching the curves cited in the references.
4. Settlement observed was non-uniform. Also, settlement readings observed were 100 to 200 more than that of predicted settlement, and this may be due the fact that design was done based on laboratory parameters, and in actual, the soil parameter would have vary in field.
5. Cohesion values observed for 0.8 m spacing of PVD were more than that of 1.0 m spacing of PVD, but theoretical computation of the increased cohesion as per IRC 75–2015[2] does not take spacing of PVD into account.
6. Phased wise approach, segmental surcharge laying, and two types of surcharges can be adopted depending upon the project requirement to optimize the cost and time of execution.

As India is having around 7517 km long coastline, which is having a higher probability of having marine soft clay along the coast, hence, there is major scope of Prefabricated Vertical Drains with surcharge in India. As a future scope or research, we can study the detailed affect of spacing of PVD on increased cohesion or smear effect in PVD for various mandrel size can be studied.

References

1. IS15284-Part II: design and construction for ground improvement—guidelines (Pre-consolidation using vertical drains)
2. IRC:75–2015: guidelines for the design of high embankments (First Revision)
3. HRB 13: State of the Art: high embankments on soft ground—part a—stage construction
4. HRB 14: State of the Art: high embankments on soft ground—part b—ground improvement
5. IS 4434–1978: code of practice for in-situ vane shear test for soils
6. FHWA/RD-86/168: prefabricated vertical drains (vol 1—engineering guidelines)

Chapter 8
Performance Evaluation of Earthen Embankment Underlain by Marine Clay Deposit with Ground Improvement Techniques—A Case Study of Mangaluru Region, Karnataka

Anand M. Hulagabali, R. Srujana, A. V. Rachana, and M. Y. Longkumer

Introduction

Background

Embankment constructed over soft ground encounters excessive settlements and large lateral deformation. Due to the development of new construction projects, embankment construction over soft clay is gaining lot of interest day by day. Properties and suitability of materials used as embankment fill are checked before being used for construction according to ASTM and USBR codes. Low bearing capacity and high compressibility characteristics of soft clay make it unsuitable and weak foundation soil for construction. Earliest method of embankment construction over soft clay was being carried out by using vertical drains [4]. It is necessary to adopt proper ground improvements like geogrid, geocell, encased columns, sand column, and stone columns to minimize the settlements to greater extent possible. It is essential to adopt suitable ground improvements to improve serviceability of road, for providing additional stability, essential shear strength, and to enhance bearing capacity of underlying sub-soil. The performance of improvement and their extent of mitigation depends upon nature of ground improvement technique adopted. This method sounds more accurate and cost-effective and is widely being used nowadays.

A. M. Hulagabali (✉) · R. Srujana · A. V. Rachana · M. Y. Longkumer
Department of Civil Engineering, The National Institute of Engineering, Mysuru 570008, India
e-mail: anandmh@nie.ac.in

K. Muthukkumaran et al. (eds.), *Ground Improvement Techniques*, Lecture Notes in Civil Engineering 297, https://doi.org/10.1007/978-981-19-6727-6_8

Brief History

Construction of embankment over soft ground is prevalent since civilizations. Lok Ma Chao Highway embankment was constructed over extremely wet lands and swamps. The foundation soil considered was of 10 m deep and had very low undrained shear strength of 3 kPa. It is known that drains are predominantly used since earlier ages for accelerating consolidation and hence to mitigate settlements. Prefabricated vertical drains and geotextile were soiled in this case for soil reinforcement instead of preloading and ground replacement techniques. Another example of WENT, landfill access road built on soft clay of 6 m thickness and strength of 10–25 kPa. Embankment is found to rest on alluvial plain and of 10 m height, and in order to increase the rate of consolidation, band drains were installed with a Geocell mattress. The quality of embankment is enhanced after increasing the stiffness of underlying subgrade after reinforcement. Above two examples witness stable embankment over soft ground is found in Hong Kong.

Statement of Problem

Construction of embankment on the soft soil deposit is challenging as they are prone to stability and settlement problems due to the poor engineering properties of subsoil which includes high plasticity, high compressibility, and low shear strength. PLAXIS 2D tool is used to analyze increased performance of sub-soil after installing stone columns, sand columns, and basal reinforcement layer. Marine clay which is characterized by soft clay properties is obtained from coastal Karnataka is used for this case study.

Scope of Study

Construction over soft clay is not stable due to unavoidable excessive settlements. Road embankment constructed on soft clay may lead to potential failure or collapse which would be dangerous for the road users. So, it is necessary to adopt proper ground improvements to mitigate the issue. Scope of the study includes.

Determination of the extent of settlement of embankment with and without ground improvement of soft clay sub-soil by using finite element method which serves as more accurate method.

To consider the practical critical condition of marine clay, this is abundantly found in coastal regions of Karnataka and to find the suitability and behavior of embankment if the embankment is constructed over it in the future.

Objectives

- To carryout stage-wise construction technique to deal with weak foundation problem in simple and economic way
- To carry out the stable construction of embankment on marine clay to avoid problems related to stability and settlement.
- To know the effectiveness of ground improvements in reducing settlements
- To graphically compare the improved embankment performance after installing stone column, sand column, and basal reinforcement layer.
- To consider the practical condition of Dakshina Kannada and estimate the settlements and its mitigation by various ground improvements.

Literature Review

Wu et al. [7] studied on experimental on geosynthetic-reinforced sand fill over marine clay with or without deep cement mixed soil columns under different loadings." In this paper, they presented an experimental study on a geosynthetic-reinforced sand fill over marine clay with or without DCM columns under different loadings. Two tests were conducted on the sand fill reinforced with fixed-end and free-end geosynthetics over marine clay under three-stage local loading to investigate the effects of the boundary conditions of geosynthetic reinforcement on reducing settlements. Another test was conducted on the fixed-end geosynthetic-reinforced sand fill over the marine clay improved by DCM columns under single-stage uniform loading. The results revealed that the stress concentration ratio increases with an increase in consolidation settlements, and the maximum tensile strain of the geosynthetic sheet occurs near the edge rather than at the center of the top surface of the DCM columns.

According to [9], soil reinforcement is carried out to deal with settlement problems which is commonly concerned in case of soft soils. Load bearing capacity and reduction in settlement can be improved by reinforcing cohesive soil with stone columns. Dissipation of excess pore water pressure helps in reduction of settlement at quick rate. High modulus and stiffness of stone column take the greater portion of applied vertical load and help in stress transfer to the underlying bearing layer. The conservative approach of settlement calculation assumes the load distribution between the soil and stone column as a ratio called stress concentration factor. However, the settlement reduction depends upon both area ratio and stress concentration factor. Length and spacing of the columns depend on settlement tolerances to the corresponding load. Sand columns are installed in the soil possessing low shear strength. By varying the diameter and spacing of the columns, the required relative density can be achieved. Greater confinement causes the greater denseness of the soil which decrease radially outward from the column. The load carrying capacity of weak subsoil is increased by the sand column by taking up the greater percentage of load and enhances settlement characteristics of the soil which is increased by accelerating the rate of consolidation settlement. The settlement of soil depends upon spacing and length of the column. The strain deformation can be effectively minimized with the application of geosynthetics. Polymers of high tensile strength are used for the soil reinforcement. Inadequate bearing capacity of the soft soil has to be improved and the soil has to be consolidated. Failure modes have to be analyzed to determine the required tensile strength of geogrid. The application of the geogrid helps in equitable stress distribution beneath the structure and prevents the stress concentration. The adherence of some geogrid with the soil, stability, high coefficient of friction, high tensile strength provides good reinforcing properties to the soil. The load distribution characteristics form the mechanical properties of geogrid which depend upon the material and structure of fabric. The performance depends upon the tensile stress strain characteristics of the soil and soil-geosynthetics interface friction.

Kamash et al. [8] studied the displacement of column supported embankment over soft clay after widening considering soil consolidation and column. It emphasizes the need of ground improvement by using vibro concrete columns, deep mixed columns, and others in reducing the settlement. The major objective of this study is to compare soil under existing environment and widened embankment for the settlements by various ground improvements. It is observed that the extra new embankment to the existing one would cause extra differential settlement and numerical analysis is carried out by using FLAC software. It analyzes the dependence of settlements on column location and spacing. Columns placed in the middle of foundation in the existing embankment condition reduce the total and differential settlement while column installation near the sides to reduce settlements in the widened embankment. Settlements are mitigated by reducing the column spacing near the widen portion decreases the settlement caused due to transverse gradient distortion because of widening.

Methodology

Two-dimensional finite element analysis is carried out for a plain strain model defined by 15 nodes of triangular elements. Numerical analysis initialized by creating a model of suitable dimension followed by assigning material properties, fixing up the boundary condition, calculation, and analysis. Plaxis 2D Version 8.2 is used for this purpose. Obtained results are graphically compared to analyze the performance of each ground improvements. Referring to relevant IRC standards [2], two lane 10 m wide highway embankment of 6 m height with 1.5 m paved shoulders is built over soft soil sub-grade. The sub-grade layer is non-homogeneous comprising 12 m deep marine clay and hard sand strata of 3 m depth below it. Side slope of 1:1.5 (V: H) is essential for the proper drainage and good geometry of embankment. The region witnessing material change due to change in soil properties has to be applied with interface element to develop proper relative displacement between soil structures [3]. A static uniformly distributed A-A system of loading is of magnitude 30kN/m^2 is imposed on the central portion of embankment. Maximum traffic volume is found to act only on central paved portion of embankment, hence, side offsets of

1.5 m are not loaded. After fine meshing of the model, it is required to set up initial conditions by placing ground water table at 13 m from the bottom of sub-grade to develop pore water pressure followed by initial stress generation. Poor undrained shear strength of marine clay restricts single-stage construction; hence, stage-wise construction technique is adopted by defining model in each stage. The shear stress developed due to vehicular loading is safely transferred to the sub-grade below it. Poor shear strength of marine clay fails to withstand the imposed load and failure is seen in the form of excessive settlement. Plastic calculation with increased number of steps to meet accuracy is performed. Long-term stability analysis is carried out by taking drained condition for embankment fill and short-term stability analysis by considering undrained condition for Marine clay sub-grade.

Locally available marine clay is obtained from Mangalore. Top layer of soil is removed and dug suitably to obtain dense mass of soil. Undisturbed soil sample free from debris is safely collected and checked for its engineering properties. Shear parameters are determined by carrying out unconfined compression test as it is the most suitable strength test for undrained cohesive soil samples. Shear strength parameters of the soil are obtained by conducting unconfined compression test adhering to IS: 2720-Part 10–1973 and IS: 4332-Part 5–1970. UCS test is carried out by enclosing the soil in split sample and axially loading it until it fails. From the transfer of shear plane and further numerical stimulation, c and Φ values are obtained. After determining density of the sub-grade and its respective dry density by adopting moisture content of 27%, the obtained values are transferred into Plaxis tool for further numerical analysis and obtained results are compared graphically.

Soon after creating geometry, single layer of basal reinforcement layer and suitable axial stiffness is provided in the region between embankment and sub-grade. For proper stimulation of interaction between basal reinforcement layer and soil, the interface element is provided both on top and bottom side of basal reinforcement layer. For the analysis of stone/sand column, end bearing stone and sand columns are used. They rest on sand bed of depth of 3 m which rests 12 m below the existing ground level. Group of columns of 1 m diameter are used. Constant 2 m spacing is maintained throughout. Interface element is provided all around the columns and the regions having change in soil properties. Columns are closely spaced throughout the bottom width of embankment.

Materials: Locally available marine clay acts as foundation soil and its properties are determined by conducting UCS test. Gravel having good stiffness forms the embankment fills and its properties are chosen adhering to ASTM codes; they are adopted from [1] (Fig. 8.1).

Properties of embankment fill, marine clay, and ground improvements are summarized in the table below (Fig. 8.2).

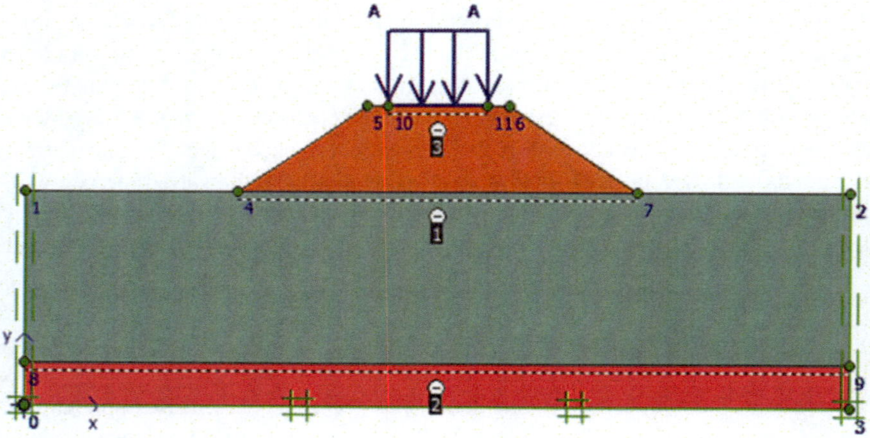

Fig. 8.1 Model considered for study

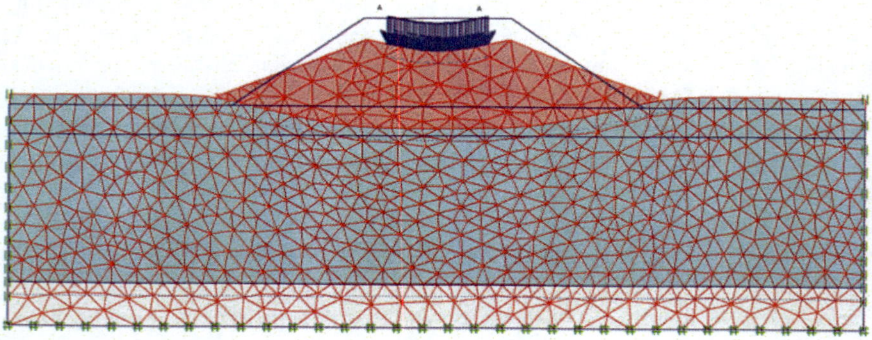

Fig. 8.2 Deformed mesh of virgin model

Soil	γ-unsaturated (kN/m^3)	γ-saturated (kN/m^3)	E (kN/m^2)	μ	C (kN/m^2)	Φ (degree)	Ψ (degree)
Embankment soil	17	18.9	70,000	0.35	1	32	4
Marine clay	14.17	17.8	1700	0.26	20	14	0
Bearing layer of sand	17	20	13,000	0.3	1	31	0
Stone column	15.38	20	30,000	0.33	0.1	45	10
Sand column	17	20	20,000	0.3	1	35	0

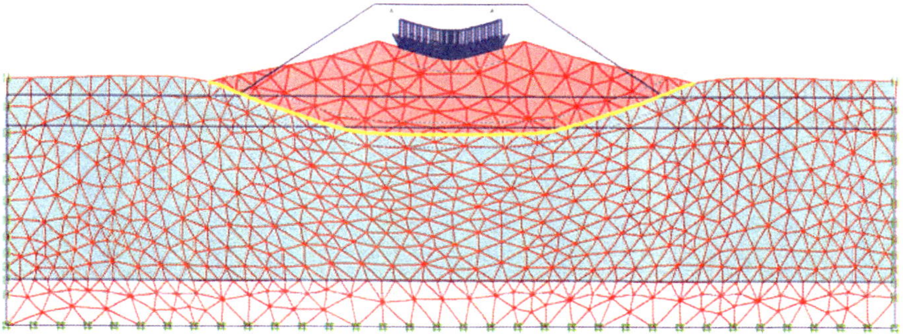

Fig. 8.3 Deformed mesh of model reinforced by basal reinforcement layer

Basal Reinforcement Layer

For enhancing the performance of earthen embankment and improve its stability, application of basal reinforcement layer is found to be cost-effective, simple, and time saving method and widely being used these days.

Basal reinforcement layer commonly known as geogrid is available in wide range of tensile strength. They spread the load equally on the ground and hence improve the service life of road. High tensile strength of these horizontal layers helps to withstand heavy embankment loading and safely holds the soil mass and hence prevents excessive settlements. Basal reinforcement layer of suitable tensile strength is chosen and placed at the base width of embankment. The axial strength of basal reinforcement layer of EA = 1180kN/m is fixed according to [5]. It is seen that both horizontal, vertical, and total deformation are minimized considerably by using basal reinforcement layer. Stresses get uniformly distributed below the embankment by use of basal reinforcement layer. There is no visible variation seen in the development of stresses. However, strain in both the Cartesian directions gets reduced as compared to virgin model (Fig. 8.3).

Sand Columns

Use of sand columns is more effective and commonly used solution to deal with settlement issues associated with soft clay. Sand with negligible cohesion and great friction reduces settlement enormously and enhance the performance of sub-grade by improving shear strength of the soil. The traffic load of embankment gets transferred to the hard sand strata below the foundation soil through the sand columns. It is seen that sand columns improve ultimate bearing capacity of foundation soil, in which they are installed. Sand columns are most effective in reducing deformation in both horizontal and vertical directions. Induced stresses developed in the soil increase in the interface section. However, stress distribution is not uniform unlike basal

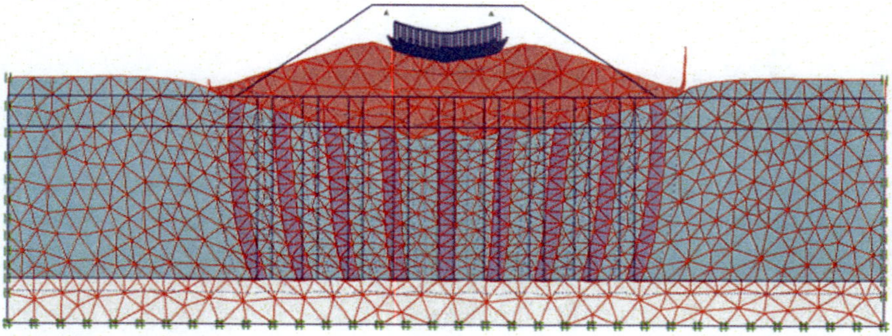

Fig. 8.4 Deformed mesh of model reinforced by sand column

reinforcement as sand columns transfer stresses to the hard strata below. Strain in both direction gets reduced as all stresses are safely transferred to the hard ground below the foundation soil (Fig. 8.4).

Stone Columns

Highly compressible marine clay fails to withstand the imposed load and undergo large deformation. Stone columns have been widely used in soft clays and are effective in reducing settlements to great extent.1 m diameter circular stone columns closely spaced in regular pattern are provided below the embankment. Stone being stiff, durable, and strong takes up the compressive load due to overlying embankment and safely transfers the load to hard strata lying below the foundation soil. Use of stone columns accelerates consolidation process and minimize settlements. This decreases stress and improves stability. Properties of stone columns are taken from [6].

There is enormous decrease in settlement found when stone column is used. Stone being strong and least cohesive has high stiffness and takes up large induced loads. Stone columns transfer the stress to deep strata, hence, stress developed in interface section between embankment and sub-grade soil is low. Due to decreased stress concentration in the sub-grade, stresses developed in both Cartesian directions decrease. However, stone column is most effective in reducing both total strains and Cartesian strains in xx and yy direction due to its nature and stress transfer behavior (Fig. 8.5).

The graphical comparison of performance of ground improvements like basal reinforcement layer, stone column, and sand column with respect to virgin model is presented from Figs. 8.6, 8.7, 8.8, 8.9, 8.10, 8.11, 8.12, 8.13, 8.14, 8.15, 8.16 and 8.17.

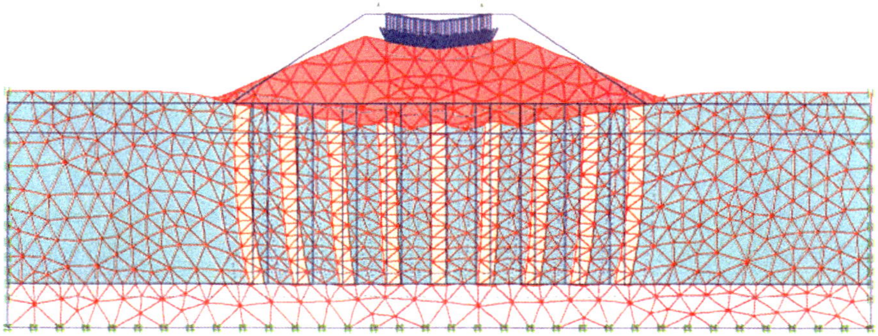

Fig. 8.5 Deformed mesh of model reinforced with stone columns

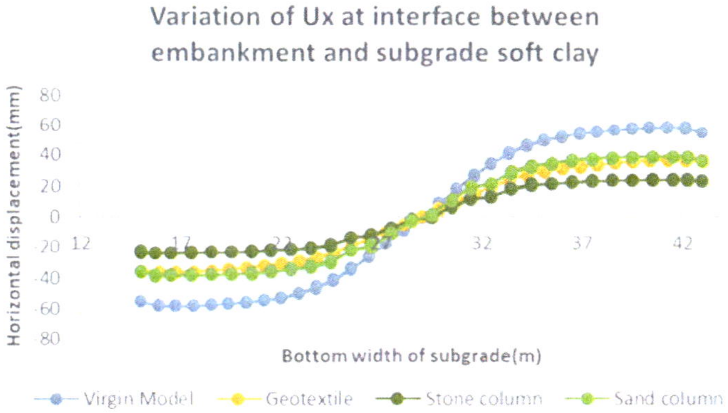

Fig. 8.6 Horizontal displacements

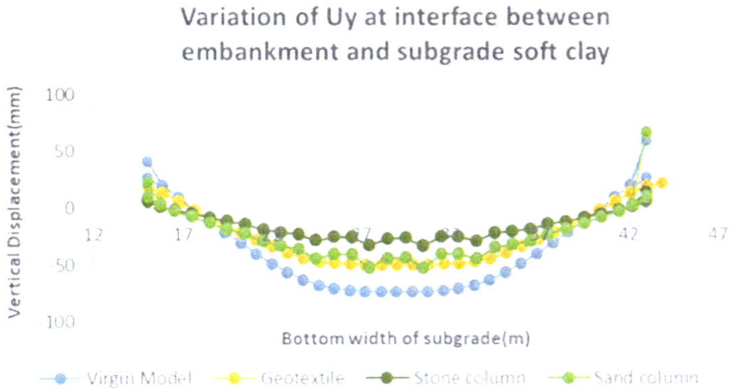

Fig. 8.7 Vertical displacements

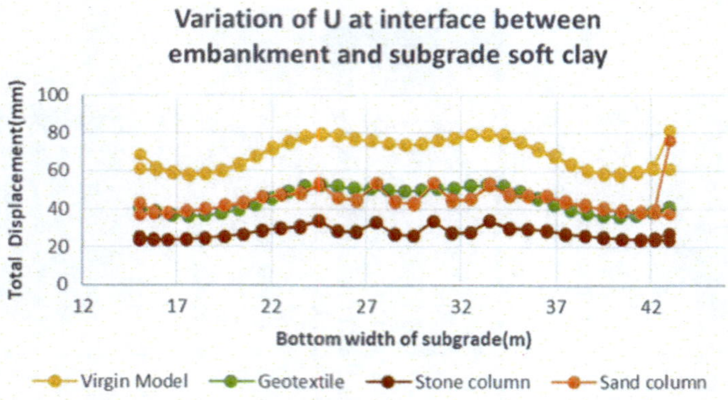

Fig. 8.8 Total displacements

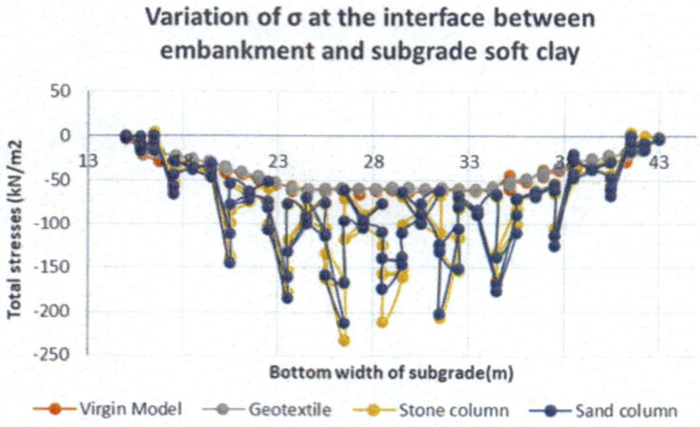

Fig. 8.9 Total stress

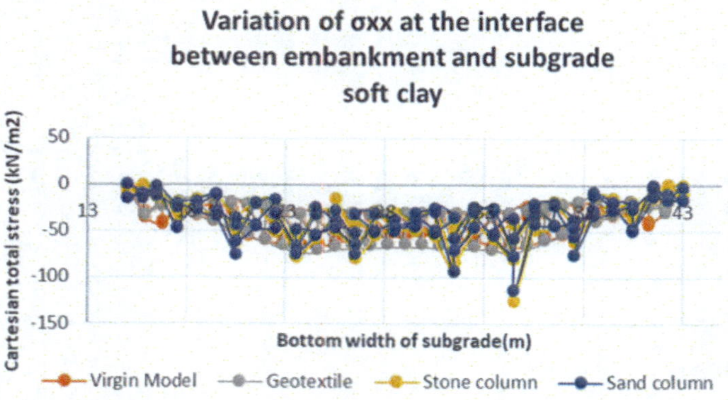

Fig. 8.10 Cartesian total stress σ_{xx}

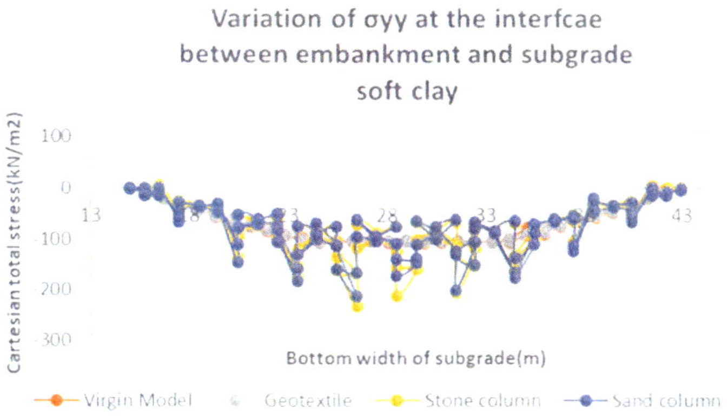

Fig. 8.11 Cartesian total stress σ_{yy}

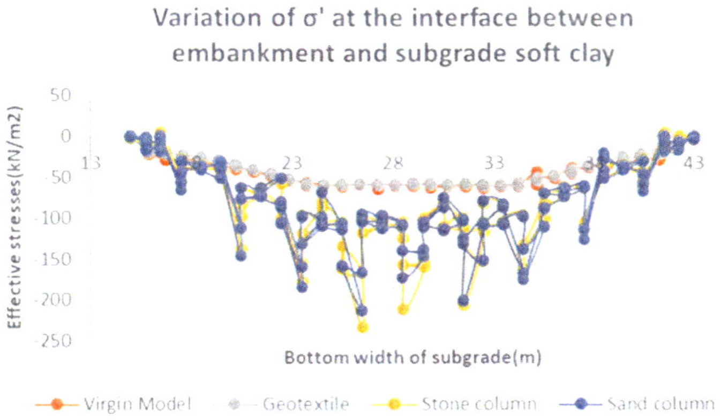

Fig. 8.12 Effective stress σ'

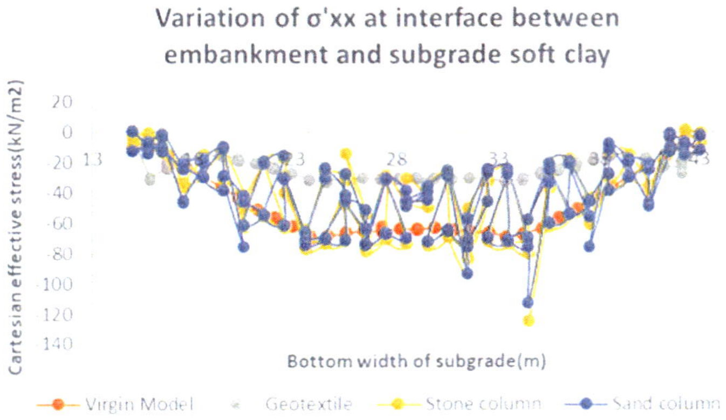

Fig. 8.13 Cartesian effective stress $\sigma_{xx'}$

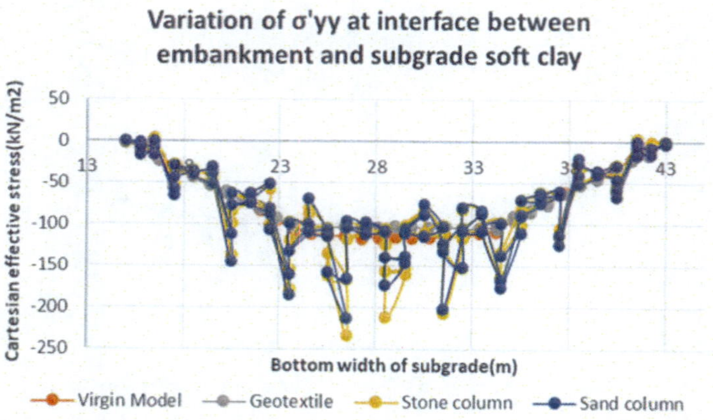

Fig. 8.14 Cartesian effective stress σ_{yy}'

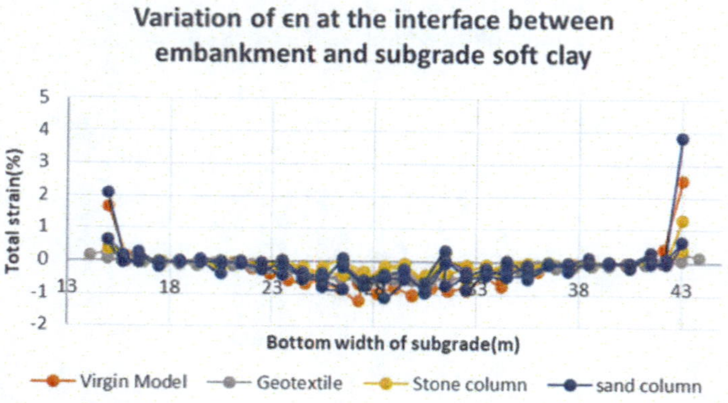

Fig. 8.15 Total strain

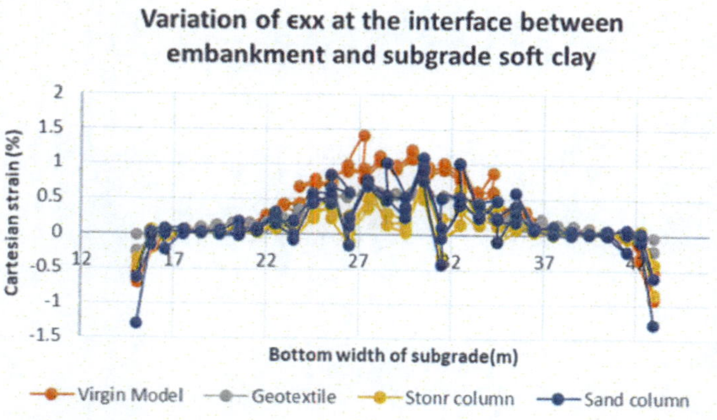

Fig. 8.16 Cartesian strain ϵxx

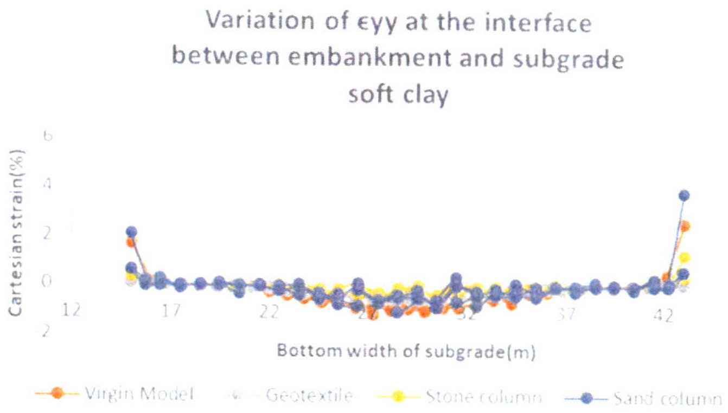

Fig. 8.17 Cartesian strain ϵyy

Conclusion

Basal reinforcement layer helps in equitable distribution of induced stress within the foundation soil. Stone and sand columns transfer the stresses deep into the bearing layer composed of sand which increases stability. It is observed that, sand columns reduce settlements by 32%, basal reinforcement reduces settlement by 35%, and stone columns decrease the settlement by 58%. All these methods are effective toward restricting settlements but selection of proper mitigation measure should meet safety and economical requirements.

References

1. Kaisim F, Marto A, Othman BA, Bakar I, Othman MF (2013) Simulation of safe height of embankment on soft ground using PLAXIS. Elsevier
2. IRC: SP:73–2007 (2007) Manual of Standards and Specifations for 2 laning of State Highways Indian Road Congress
3. PLAXIS (2013) User's manual
4. Abhishek SV, Madhav MR (2013) Embankments on Soft ground an overview. ICTDMU-1
5. Ariyarathne P, Liyanapathirana DS (2015) Review of existing design methods for geosynthetic reinforced pile-supported embankments. Jpn Geotech Soc. Elsevier
6. Miranda M, Ruiz JF, Castro J (2021) Critical length of encased columns. Elsevier, Geotextiles and Geo Membranes
7. Wu PC, Yin JH, Feng WQ, Chen WB (2019) Experimental study on geosynthetic reinforced sand fill over Marine clay with or without deep cement mixed soil columns under different loadings. Underground Space 4:340–347
8. Kamesh WE, Han J (2014) Displacement of column supported embankment over soft clay after widening considering soil consolidation and column layout numerical analysis. Soils Found 54(6):1054–1069, Elsevier
9. Purushothama Raj P Ground improvement techniques.; textbook, Laxmi Publications

Chapter 9
An Experimental Study on Development of the Bearing Capacity of Soft Clay Soil Using Stone Column with Bamboo Sheet Plate

Soumitra Biswas and Nirmali Borthakur

Introduction

Soft soils are generally not capable of sustaining high loads, but due to rapid urbanization and increase in population, nowadays, it becomes necessary to utilize this type of soil for construction purposes. Many researchers are trying their level best to give solution to address the issues related to soft soil, but still enough room is available for further study. Large areas of soft cohesive soils are present in India, especially in north-eastern region, which create problems in construction industry. Hence, ground improvement techniques are necessary to increase the load bearing capacity of soft clayey soil. Stone column is one of the recent techniques that can be used not only for soft clayey soil but also for silts and loose silty sand. The stone column concept was first applied in France in the year 1830 to enhance a native soil. Studies on enhancement of load carrying capacity of stone column induced clay soil are performed by different investigators in different times. Laboratory model experiment and numerical model analysis were conducted on reinforcement of single as well as group of stone columns in clay soil by researchers such as Ambily and Gandhy [1]; Shahu and Reddy [2], etc. Studies were also conducted on improvement of bearing capacity of clay soil using encasement of stone column with geogrid by Ajay and Kumar [3], vertically enclosed stone columns under a sand bed reinforced with geogrid by Debnath and Dey [4] and stone columns encased with jute geotextile by Roy et al. [5]. In another approach, a cement-mixed soil layer was placed at the top of the stone column groups to improve the load resistance property of soft clay soil [6].

In the present study, bamboo sheet plates are used in different positions of the stone columns as reinforcement because it can resist the lateral displacement of stone column in soft clayey soil. The aim is to decrease the bulging of the stone column

S. Biswas (✉) · N. Borthakur
National Institute of Technology Silchar, Silchar, Assam 788010, India
e-mail: soumitrab20@gmail.com

K. Muthukkumaran et al. (eds.), *Ground Improvement Techniques*, Lecture Notes in Civil Engineering 297, https://doi.org/10.1007/978-981-19-6727-6_9

and to increase the load bearing capacity of the soil. From the economical point of view, bamboo is cheaper than the geotextile, geogrid or geocell. The potential use of bamboo mat or sheet in different soil stabilization and soil bioengineering work such as reinforcement, erosion control, and slope protection are increasing due to its easy availability, low cost, efficiency, and easy construction [7, 8]. Seasoning of bamboo increases the resistance power against fungal and bacterial decay and reduces the drying and shrinkage property [8]. In the experimental work, to increase the longevity of bamboo sheet, it was treated with the kerosene oil.

Materials Used

Study Materials

In the present experimental work, diameter of the single stone column 50 mm and length 300 mm was constructed in a clay soil bed filled inside a steel tank in the laboratory. So, the materials necessary for the study includes clay soil, stone aggregates, and bamboo sheet plate.

Clay. Clayey soil was collected from a paddy field located in the neighborhood of the NIT Silchar campus. Different properties of soil were determined in the laboratory and given in Table 9.1. Hydrometer analysis was used to find out the particle size distribution of the soil. Particle size distribution graph is shown in Fig. 9.1 and percentages of silt and clay are found to be 39.05% and 55.85%, respectively. Soil was classified according to the Indian Standard Soil Classification System (ISSCS) and found to be CH. Proctor compaction test was also performed on the collected soil and the graph is shown in Fig. 9.2. Clay soil were then dried properly, converted to powdered form, then sieved by 425μ sieve to keep ready for the preparation of clay bed.

Stone aggregates. The stone aggregates were collected from a nearby quarry. The specific gravity of the stone aggregates was 2.64 and aggregate crushing value was 38.86%. The size of the aggregates was then reduced to 2 mm–6 mm size and shown in Fig. 9.3.

Bamboo sheet plate. In the study, bamboo sheet was woven in 50 mm diameter plates as shown in Fig. 9.4. The bamboo sheet plates were used to decrease the bulging of stone columns. The tensile strength of the bamboo sheet was determined with the help of universal testing machine and found out to be 405.6 Mpa. Bamboo sheets were then immersed in kerosene oil for 3 days and then dried to increase its durability and reduce fungus attack.

Table 9.1 Properties of clayey soil

Properties	Values
Liquid limit	63.05%
Plastic limit	30.12%
Plasticity index	32.93%
Specific gravity	2.61
MDD	17.10 kN/m^3
OMC	30.40%
Permeability	4.665×10^{-8} cm/sec
Compression index	0.443
Coefficient of consolidation	5.28×10^{-4} cm^2/sec
Gravel	0%
Fine sand	5.1%
Silt	39.05%
Clay	55.85%
Classification as per ISSCS	CH

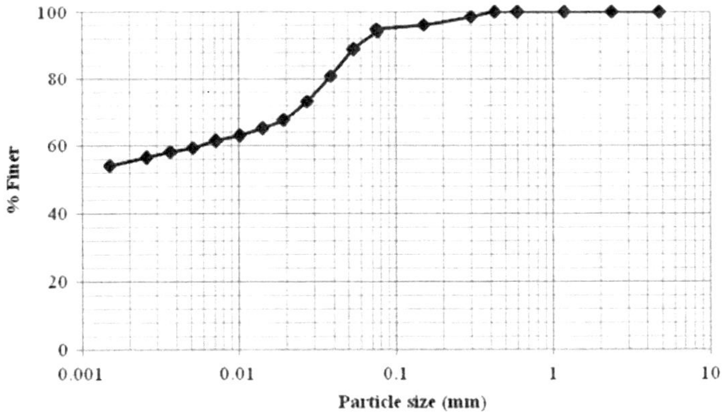

Fig. 9.1 Particle size distribution of the clay soil

Experimental Investigation

Preparation of Clayey Soil Bed and Construction of Stone Columns

First of all, unconfined compressive strength tests were conducted on clay soil samples at different water contents such as 20, 25, 30, 35, and 40%. A graph has been plotted for unconfined compressive strength (UCS) values against moisture content

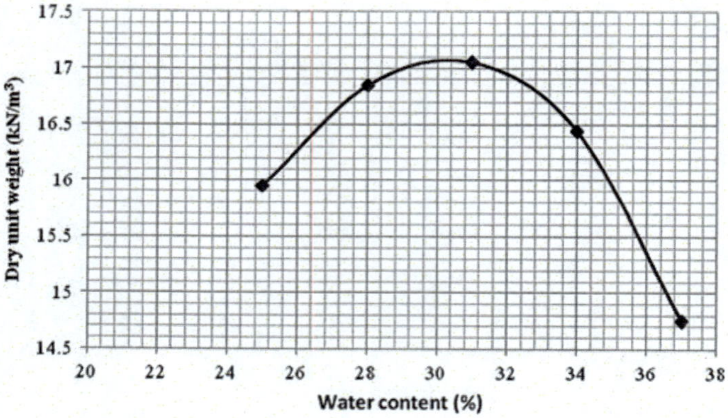

Fig. 9.2 Compaction curve of the clay soil

Fig. 9.3 Sample of stone aggregates

Fig. 9.4 Bamboo sheet plate

as shown in Fig. 9.5. It is observed from the graph that soil becomes soft attaining a UCS value of 50 kPa at a water content of 34%. The decrease of UCS values upon the increase in water content is less beyond this point. So, for the convenience is was decided to compact the clay soil bed at water content of 35% such that it remains soft. The clay soil bed was prepared inside a steel tank of dimensions 1 m × 1 m × 1 m in size. A polythene sheet was placed firmly in all the inside walls of the steel tank to reduce side friction. Then the fine dried clay soil was mixed thoroughly with 35% of moisture content by weight and compacted in the tank in thin layers of approximately 10 cm. Holes were made in the clay soil bed at specific locations by a 50 mm diameter hand auger. Stone columns were constructed in triangular pattern as shown in Fig. 9.6, as per the specification given in IS 15284 (Part I) [9]. A term scale factor is used to represent the actual field dimensions by the laboratory model dimensions. In this study, a 50 mm diameter model stone column was used to represent a 500 mm diameter field stone column having scale factor of 10. Generally, size of the stones used in the construction of stone columns in the field ranges 25–50 mm. So, it was reduced to size range of 2–6 mm to maintain the scale factor 10–12 [6].

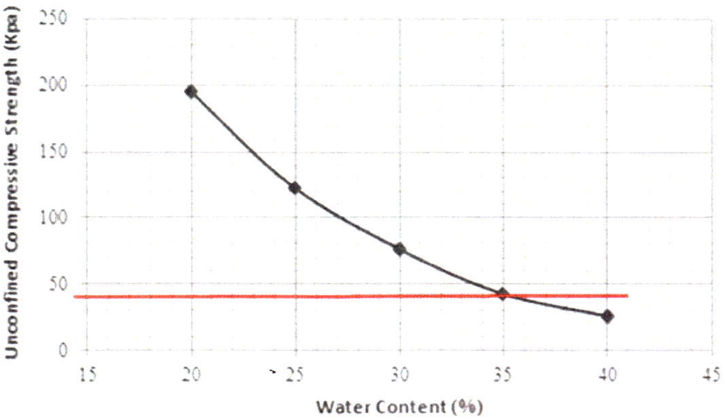

Fig. 9.5 Unconfined compression strength versus water content

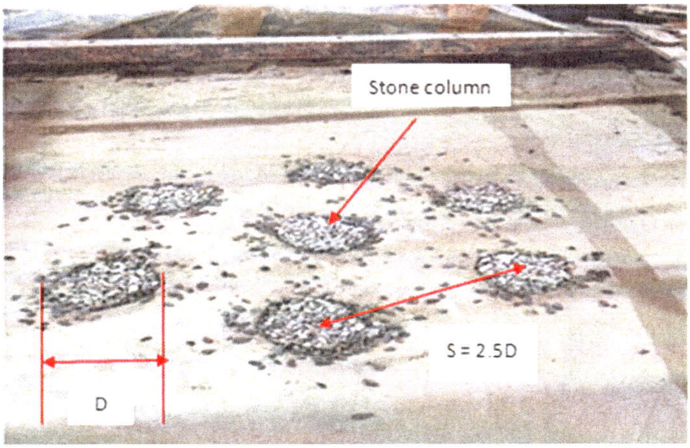

Fig. 9.6 Clay bed with stone columns

Experimental Modeling

All experiments were conducted in a steel of size 100 cm × 100 cm × 100 cm. Clay bed was prepared first and then a plate load test was performed with a circular plate of diameter 100 mm to determine the load bearing capacity of the soft clay soil. Then simple stone columns were installed in triangular pattern as shown in Fig. 9.6. A circular plate of 100 mm was placed above the middle stone column, and load test was conducted. In the third series of experiments, bamboo sheet plate was installed inside the stone columns at different locations of the stone column in a single layer such as $L/3$, $L/2$, and $2L/3$ from the top where L represents the length or depth of the stone column. Fourth series of experiment was conducted by placing two bamboo sheet plates together; one at the depth of $L/3$ and another at the depth of $2L/3$. The arrangement of bamboo sheets in the stone columns is shown in Fig. 9.7. In all the cases, the lengths of the stone columns were maintained at 300 mm. A 100 mm diameter and 20 mm thick circular metal plate was placed over the central stone column to act as a footing and distribute the load uniformly. Load increments were applied on the footing with the help of a pressure gauge and hydraulic jack assembly that was attached with the steel tank. A proving ring was connected with the hydraulic jerk to measure the applied load. The least count of the proving ring is that one small division of the proving ring is equal to 0.00732KN. Settlements was measured by two dial gauges with least count 0.01 mm, placed diagonally opposite positions by independent datum bars. Settlements were recorded at ½, 1, 2, 4, 8, 15, and 30 min and then at hourly intervals till the settlements becomes less than 0.01 mm/min or 2 h whichever occurs earlier, then settlement value was calculated by averaging of the two dial gauge readings. Then the next load increment was applied and settlement was recorded by same approach. This procedure was continued till failure that means settlement was continued to record under different load increments until large settlement were reached under a small load increase. A photographic view of the experimental setup is shown in Fig. 9.8.

Test Results and Discussions

Load Test Results

The load settlement plot for plate load test on plain clay soil is shown in Fig. 9.9. The ultimate load bearing capacity is calculated with respect to 25 mm settlement [10]. In the experiment, total dead weight of proving ring plate was 0.054KN. It is observed from the graph that at 25 mm settlement, load bearing capacity is 0.5KN. So, the total load bearing capacity of the plain clay soil is calculated to be 0.554 KN.

Figure 9.10 shows the load penetration graph for single stone column on soft clay with. At 25 mm settlement, the load bearing capacity of the clay soil with stone column is 0.866KN and the dead load is 0.054KN. So, the ultimate load bearing

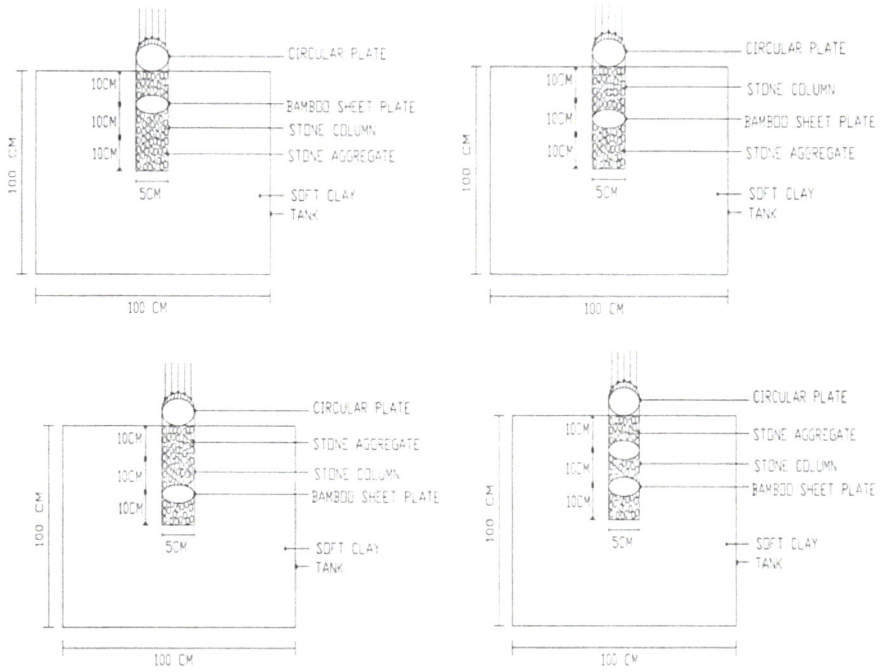

Fig. 9.7 Schematic diagram of bamboo sheet plate position within stone column

Fig. 9.8 Experimental setup

capacity of the soft clay with stone column is 0.92 KN. So, the capacity of the clay soil increases to 66.06% with single stone column.

The load penetration curve for the soft clay with stone column having bamboo sheet plate at L/3 from the top is shown in Fig. 9.11. Load bearing capacity corresponding to 25 mm settlement is 1.01KN. The dead load is 0.054KN, so the ultimate

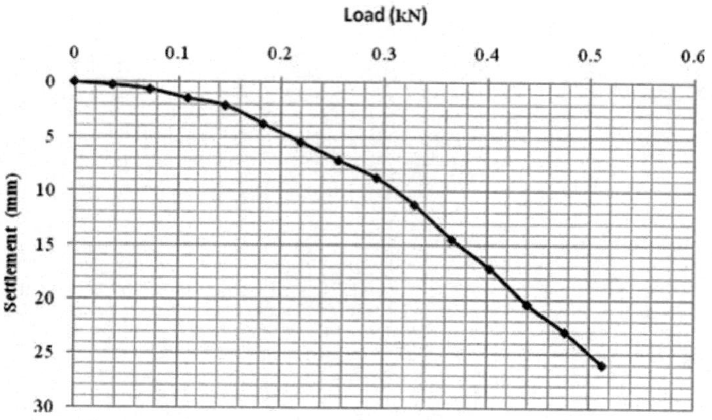

Fig. 9.9 Load settlement curve for plain clay soil

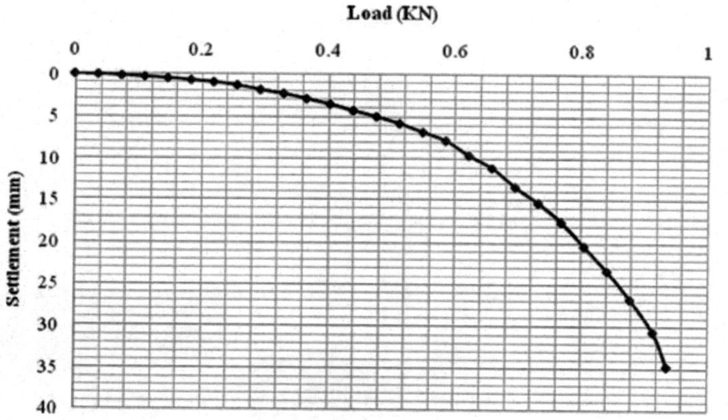

Fig. 9.10 Load penetration curve for clay soil with stone column

load bearing capacity is 1.064KN. The increment in capacity is 92.06% and 15.65% corresponding to the plain clay soil and single stone column, respectively.

Figure 9.12 shows the load versus penetration plot for the stone column with bamboo sheet plate placed at L/2 from the top. Here bamboo sheet plate has shifted to middle of the stone column and observed the load bearing capacity. From the graph, at the 25 mm settlement, the load bearing capacity is 1.02KN so the ultimate load bearing capacity is 1.074KN. The increment of the load bearing capacity with respect to plain clay soil is 96.86% and stone column is 16.74%.

Figure 9.13 shows the load penetration curve for soft clay with stone column with bamboo sheet plate placed at 2*L*/3 from the top. Load bearing capacity using bamboo sheet plate placed at the 2*L*/3 is 1.04KN and dead load is 0.054KN. So, the total load bearing capacity is 1.094KN. The increment of the load bearing capacity with respect to the plain clay soil is 97.47% and the single stone column is 18.91%.

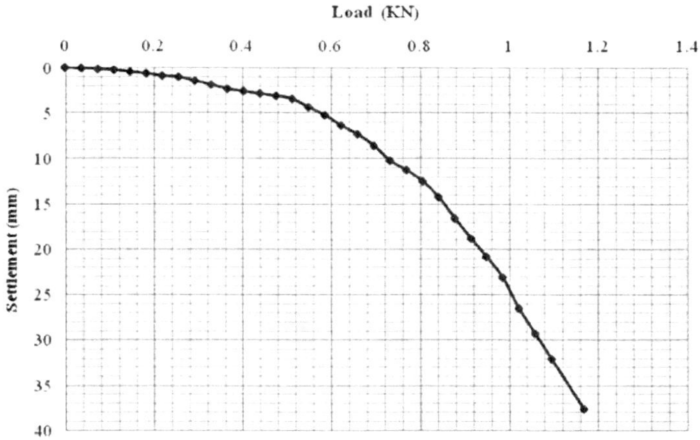

Fig. 9.11 Load penetration curve for clay soil with stone column with bamboo sheet plate placed at L/3 from the top

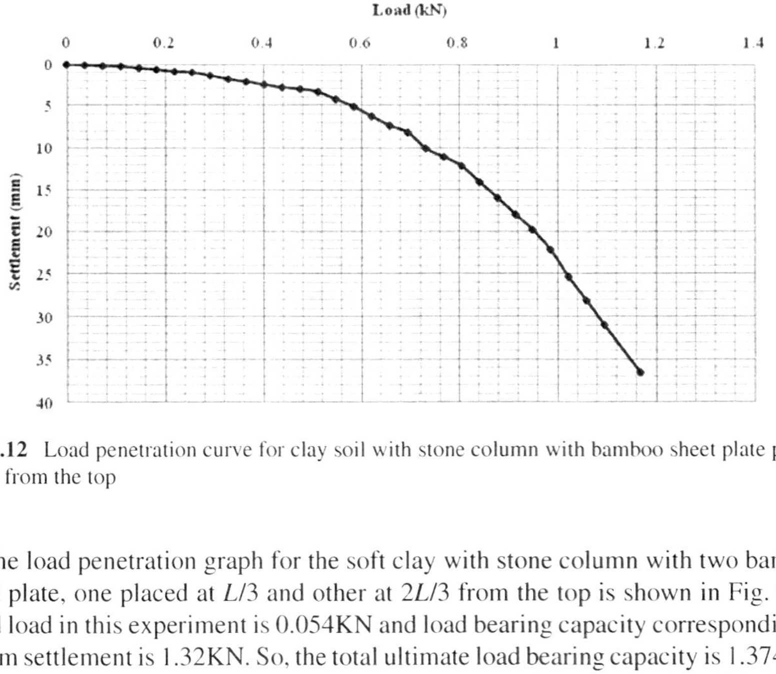

Fig. 9.12 Load penetration curve for clay soil with stone column with bamboo sheet plate placed at *L*/2 from the top

The load penetration graph for the soft clay with stone column with two bamboo sheet plate, one placed at *L*/3 and other at 2*L*/3 from the top is shown in Fig. 9.14. Dead load in this experiment is 0.054KN and load bearing capacity corresponding to 25 mm settlement is 1.32KN. So, the total ultimate load bearing capacity is 1.374KN. Respect to the plain clay soil, increment of the load bearing capacity is 148.01% and respect to the clay soil with single stone column is 49.35%.

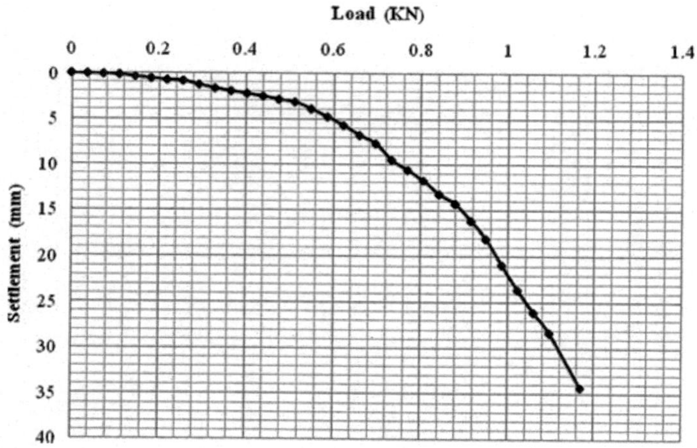

Fig. 9.13 Load penetration curve for clay soil with stone column with bamboo sheet plate placed at 2*L*/3 from the top

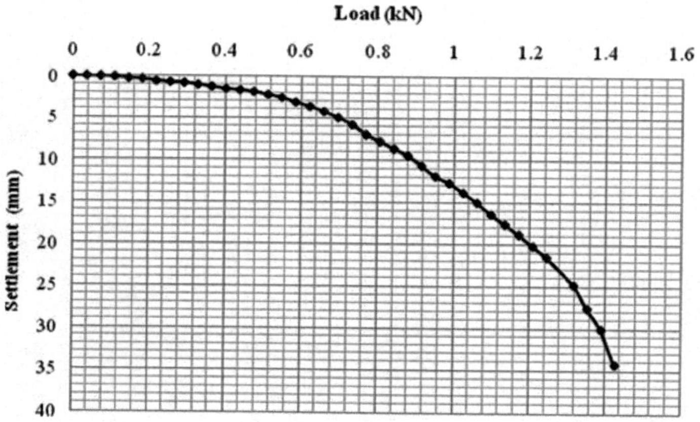

Fig. 9.14 Load penetration curve for clay soil with stone column with two bamboo sheet plate, one is at *L*/3 and another is place at 2*L*/3 from the top

Interpretation of Test Results

Figure 9.15 shows the load penetration curve drawn for all the load tests by putting a 10 cm diameter circular plate at the top of stone column constructed on soft clay with varying depth of bamboo sheet plate and with two bamboo sheet plate, one placed at *L*/3 and another at 2*L*/3 from the top. Ultimate load bearing capacity in each case is 0.554KN, 0.92KN, 1.064KN, 1.074KN, 1.094KN, and 1.374KN, respectively. With respect to plain clay soil, increment of the load bearing capacity is 66.06, 92.06, 96.86, 97.48, and 148.01. The increase in load bearing capacity is 1.66, 1.92, 1.94, 1.947,

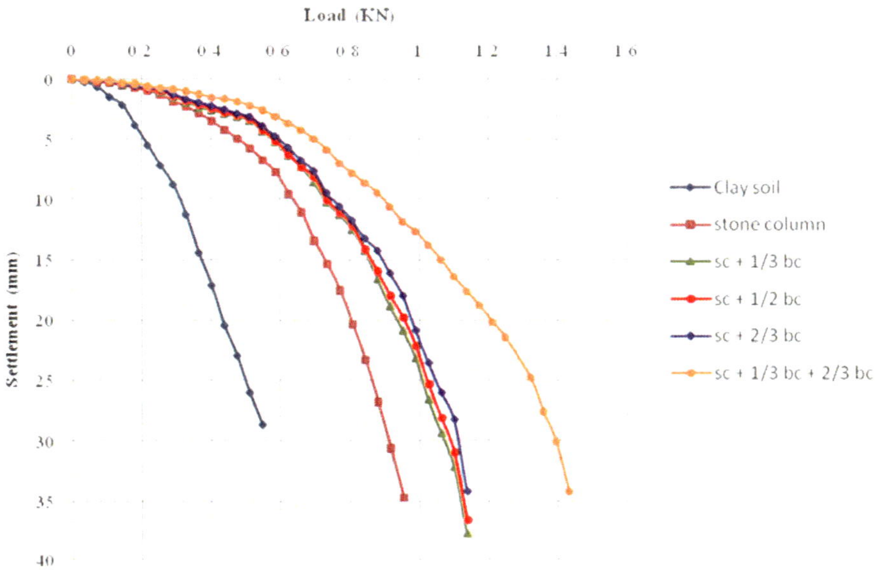

Fig. 9.15 Load penetration curve for clay soil with stone column with varying depth of bamboo sheet plate and with two bamboo sheet plate

and 2.48 times the load bearing capacity of the plain clay soil alone. It is observed from the study that maximum load bearing capacity occurs when two bamboo sheet plate are used with stone column.

Generally, for long stone columns, both end bearing and floating, bulging pattern of failure occurs [7]. Increase in load carrying capacity due to installation of discrete bamboo sheet disks at certain spacing within the stone column length indicates the reduction in size and pattern of bulging. However, initial load tests must be performed at the particular site to evaluate the effectiveness and load penetration behavior of proposed soil-stone column-bamboo sheet composite system.

A Comparative Study With Others Researchers

A comparative study has been performed between the presents work and Ajay and Kumar [3]. They used geogrid as reinforcement at varying depths of *L*/4, *L*/2, 3*L*/4, and *L* to improving the load carrying capacity of the clay soil. Load settlement curve obtained by Ajay and Kumar [3] is shown in Fig. 9.16. The load bearing capacity of the plain clay soil at 25 mm settlement with and without stone column are to be 0.471 KN and 0.696 KN and with stone column with geogrid at varying depths of *L*/4, *L*/2, 3*L*/4 and L are as 0.868KN, 0.99KN, 1.06KN, and 1.15KN, respectively. The increase in load bearing capacity is 1.48, 1.84, 2.1, 2.25, and 2.44 times the load bearing capacity of the plain clay soil alone. Table 9.2 gives the comparison between the present study and reference study.

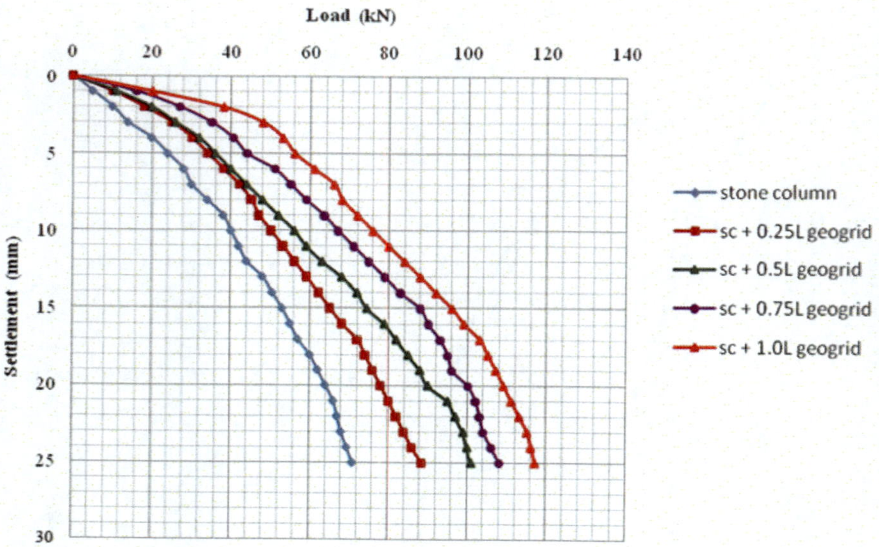

Fig. 9.16 Load penetration curve for clay soil with stone column with varying depth of geogrid reinforcement at L/4, L/2, 3L/4, and L [3]

Table 9.2 Comparison between present study and reference paper

Data of the present study		Data of the reference thesis paper	
Topics	Load (kN)	Topics	Load (kN)
Plain clay soil	0.554	Plain clay soil	0.471
Clay soil + stone column (sc)	0.92	Clay soil + stone column (sc)	0.696
Clay soil + sc + L/3 bc	1.064	Clay soil + sc + L/4 geogrid	0.868
Clay soil + sc + L/2 bc	1.074	Clay soil + sc + L/2 geogrid	0.99
Clay soil + sc + 2L/3 bc	1.094	Clay soil + sc + 2L/3 geogrid	1.06
Clay soil + sc + L/3 bc + 2L/3 bc	1.374	Clay soil + sc + L geogrid	1.15

Conclusion

Following conclusions have been drawn from the present study.

1. The load bearing capacity of the soft clay soil increases to 1.66 times by using single stone column without bamboo plate reinforcement. Bulging of stone columns occurs at failure which restricts the further increase of capacity.
2. Bamboo sheet plate that was used as reinforcement with the stone column reduces the bulging of the stone column due to which load carrying capacity increases. It is observed from the study that for single layer of bamboo sheet plate, placed at 2L/3 from the top has developed highest load bearing capacity, 1.947times more than the capacity of clay soil.

3. The development of maximum load bearing capacity, i.e., 2.48 times of the plain clay soil is observed when two layers of bamboo sheet plate were used together, one at L/3 and other 2L/3 from the top.
4. It is observed that in the reference study that increment of load bearing capacity is 143.75% of soft clay soil by using stone column with geogrid reinforcement and in the present study, increment of load bearing capacity is 148.01% of soft clay soil by applying stone column with bamboo sheet plate.

From the above discussions, it may be founded that the stone column with the bamboo sheet plate provides effective development of the bearing capacity of the soft clay soil, so it may be used as reinforcement with the stone columns.

References

1. Ambily AP, Gandhi SR (2007) Behaviour of stone columns based on experimental and FEM analysis. J Geotech Geoenviron Eng 133(4):405–415
2. Shahu JT, Reddy YR (2011) Clayey soil reinforced with stone column group: model tests and analyses. J Geotech Geoenviron Eng 137(12). © ASCE
3. Ajay CH, Kumar PMSS (2017) An experimental study on encasement of stone column with geo-grid in clayey soils. Int J Innovative Res Sci Eng Technol 6(2)
4. Debnath P, Dey AK (2017) Bearing capacity of geogrid reinforced sand over encased stone columns in soft clay. Geotext Geomembr 45:653–664
5. Roy, Chand, Soni (2019) In Clay soil ground improvement by using jute geotextile reinforcement in stone column. Int J Innovative Res Sci Eng Technol 8(01)
6. M. Das and A. K. Dey. (2020). "*Use of Soil–Cement Bed to Improve Bearing Capacity of Stone Columns.*" International Journal of Geomechanics. Vol. 20, issue 6. © ASCE.
7. Akhil KS, Sankar N, Chandrakaran S (2020) Use of bamboo mat as a potential soil reinforcement material–An experimental study. Materials Today: Proc 31(2020):S335–S339
8. Rauch HP, Florineth F, Lammeranner W, Wibmer S (2002) Soil bioengineering slope protection investigations in Nepal. Final report for the Austrian commission of development research
9. I.S. 15284 (Part I) (2003) Indian standard design and construction for ground improvement guidelines, Part I—Stone Columns
10. Bowles JE (2001) Foundation analysis and design. The Mc Graw-Hill Companies Inc., Electronic version

Chapter 10
Evaluation of Heave Behavior by Numerical Modeling of Granular Pile Anchor in Expansive Soil

Renuka Roy, Ch. Nageshwar Rao, and S. Sasanka Mouli

Introduction

Expansive soil shows drastic change in volume (swell-shrink) on water absorption and reduction, respectively [1, 2]. Volume change in these expansive soils causes distress to the foundation. Structures such as foundations, pavement, canal bed, and residential buildings constructed on expansive soil undergo severe damage. Foundations built on expansive soil face threat of uplift forces on rise in moisture content. As a result, loss of several billion dollars had been recorded all around the world. Due to the swelling nature of expansive soil, it is always suggested to have deep foundation to reduce the risk of failure. The swelling/shrinking soil requires creative and innovatory methods for improving the ground and to reduce the detrimental effect of swelling/shrinking on structures. One among such solutions is granular pile anchor (GPA) which is highly recommended method compared to the traditional methods such as sand cushion [3, 4], replacement with cohesive-non-swelling (CNS) layer [5], and under-reamed piles and belled pier [2]. In GPA system, anchorage is achieved through the anchor rod provided at the center of the pile, which connects the footing plate with anchor plate (made up of steel) provided at bottom. This system can be effectively utilized for heave reduction caused by expansive clay bed [14]. In the recent years, many experimental investigations have been undertaken to examine the effectiveness and promising performance of granular pile anchor (GPA) to control the heave behavior of footing in expansive soil [6–16]. Construction, erection procedures of GPA and mechanism, along with the forces imposed on it, were explained in detail with different studies [10, 11]. In the same context, to understand the heave behavior and efficacy of granular pile anchor in expansive soil, series of laboratory tests were

R. Roy (✉) · Ch. N. Rao · S. S. Mouli
Department of Civil Engineering, VNR Vignana Jyothi Institute of Engineering and Technology, Secunderabad, Telangana 500090, India
e-mail: renukashivroy428@gmail.com

K. Muthukkumaran et al. (eds.), *Ground Improvement Techniques*, Lecture Notes in Civil Engineering 297, https://doi.org/10.1007/978-981-19-6727-6_10

conducted for plain GPA and geogrid-encased GPAs for ground improvement [12–14]. Further FEM approach was adopted to analyze heave behavior of granular pile anchor foundation (GPAF) in expansive soil using numerical modeling [17–20]. In the present study, an attempt was made to analyze heave behavior plain GPA and geogrid-encased GPA embedded in expansive clay bed by numerical modeling using PLAXIS 2D software. Different pile dimensions and the soil properties were considered from Phanikumar et al. [13]. The obtained results of numerically analyzed model of plain GPA are compared with laboratory test results [13]. Further, to check the effect of roughness coefficient (R_{inter}), geogrid-encased GPA were also analyzed for different pile lengths.

Mechanism Involved

GPA technique can efficiently and effectively reduce the damaging effect on foundation by controlling the excesses volumetric change caused by expansive/shrinking soil. In a GPA, resistance to uplift is achieved mainly due to.

(a) The weight of GPA acting downward.
(b) The uplift resistance due to friction mobilized along the soil-pile interface.

Numerical Modeling

A numerical model was developed for the analysis of heave behavior of granular pile anchor in expansive soil for three different pile lengths with constant diameter as shown in Table 10.1. The overall heave behavior of GPA and geogrid-encased GPAs in expansive clay bed was analyzed using finite element program (PLAXIS-2D-edition V20.02). In order to model the circular pile resting on expansive soil layer, axi-symmetric condition type was used. The extent of expansive soil around the GPA was considered as 381 mm using borehole option available on the left-side (vertical) toolbar with suitable plan dimension. Total plan dimension for expansive soil was considered as 381 mm × 381 mm and depth as 381 mm.

Granular pile was modeled using poly-curve and extrude option available in structures mode. Plate elements such as footing plate and anchor plate were modeled as linear elastic model. In order to establish the connection between footing plate and anchor plate with anchor rod, node-to-node-anchor option was used. Complete

Table 10.1 Total plan size and pile dimensions

Total dimensions of sample (mm)	Diameter of the GPA (mm)	Diameter of footing plate (mm)	Pile lengths (mm)
381 × 381 × 381	76.2	127	124.4, 185.4 and 246.4

arrangement of structural elements along with GPA are shown in Fig. 10.1. Geogrid-encased GPA was modeled using create geogrid option. Material model type was selected as Mohr–Coulomb model for expansive soil, granular pile, and also for defining the interface material between granular pile and soil.

The values of shear parameters for soil-pile interface material used as 13 kPa and 30° and for soil-pile-geogrid encasement interface material considered as 13.3 kPa and 35°. Modeling of elements was done using 15-noded triangular elements. The material properties and parameters of soil, GPA, and interface material along with the details of structural elements are shown in Tables 10.2 and 10.3. Axial stiffness of geogrid is provided in Table 10.4.

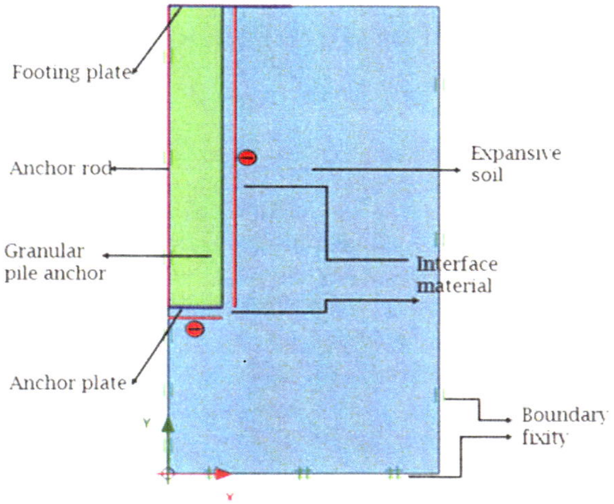

Fig. 10.1 Numerical model of GPA

Table 10.2 Material properties used for numerical modeling

Soil	Expansive soil	Granular pile	Interface material
Model	Mohr–coulomb	Mohr–coulomb	Mohr–coulomb
Type	Undrained A	Drained	Undrained A
γ_{usat} (kN/m^3)	14	17	14
γ_{sat} (kN/m^3)	16	20	16
E_{ref}, Modulus of elasticity (MPa)	5	50	5
Poisson's ratio	0.35	0.3	0.35
Cohesive strength (C in kPa)	115	0.1×10^{-3}	13, 13.3
Angle of shearing resistance φ(°)	5	40	30, 35
Permeability K_x(m/day)	0.0001	1	0.0001
Permeability K_y(m/day)	0.0001	1	0.0001

Table 10.3 Structural elements used for numerical modeling

Plate elements	Footing plate	Anchor plate	Anchor rod
Model type	Linear elastic	Linear elastic	Linear elastic
EI (kNm^2/m)	4×10^4	1×10^4	–
EA (kN/m)	5×10^6	5×10^6	3×10^6
Poisson's ratio (ν)	0.15	0.15	0.15

Table 10.4 Material property of geogrid

Type of material	Model type	Axial stiffness (EA) in kN/m
Geogrid	Linear elastic	200

Boundary Conditions

In order to simulate in-situ conditions (i.e., clay bed boundary), standard fixity conditions were assigned to all numerical models. For left and right side of model, horizontal displacement was considered as fixed and vertical displacement considered as free. In the bottom portion of model, both horizontal and vertical displacement was considered as fixed. These assumptions were considered in the context of surrounding soil having large horizontal extent in nature. Automated mesh was generated by considering element distribution as coarse as shown in Fig. 10.2 mesh generated of 537 elements, 4565 nodes with 6444 stress points.

Heave Calculations

In the present study, heave behavior for different pile lengths was analyzed in three phases. Out of three phases, first two phases were plastic, and the third phase was considered as consolidation phase given in Fig. 10.3. First phase indicated as initial phase in which only soil layer was activated for the model. Second phase was pile (GPA) activation phase, in which pile and all other components are activated such as footing plate, anchor plate, anchor rod, and interface. To ensure the time effect, third phase was considered as consolidation phase for the time period of 7 days, which depicts same saturation time period maintained in the laboratory test [13].

Hence, in order to evaluate heave behavior at different time period.

Results and Discussions

A positive volumetric strain is applied at a constant rate of 0.01–1.4% uniformly over the whole thickness of the clay bed to simulate the swelling behavior of expansive

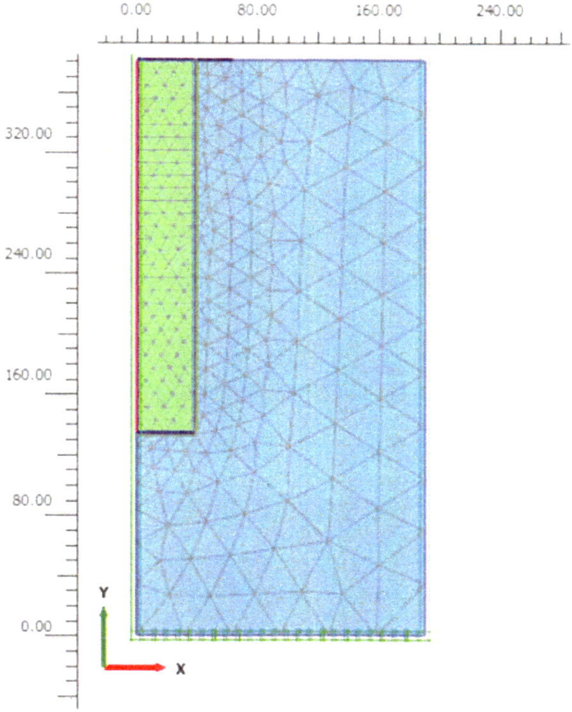

Fig. 10.2 Generated mesh of single GPA

soil, which depicts the actual free swelling nature of the expansive soil utilized for this study. In real field, the presence of moisture content and magnitude of overburden pressure is responsible for swelling of expansive soil. Therefore, in PLAXIS, 2D volumetric strain depicts the free swell behavior of the expansive soil. As the boundary condition to the model was arrested by fixing the horizontal displacement and only allowing the vertical displacement to take place, due to which volumetric strain act in vertical direction to produce the amount of heave.

Numerical Analysis of Granular Pile Anchor (GPA)

Volumetric strain was applied gradually to three different pile length model of GPA with and without geogrid encasement. The vertical displacement at the top of the pile (GPA) was found to be decreasing, i.e., amount of heave decreased as the pile length increased for plain GPA as well as for the case of geogrid-encased GPA as shown in Figs. 10.4 and 10.5. The heave values of reinforced expansive clay bed with plain GPA and geogrid-encased GPA are given in Table 10.5. The rate of heave (mm) was then plotted in form of graph with variation of the amount of heave on Y-

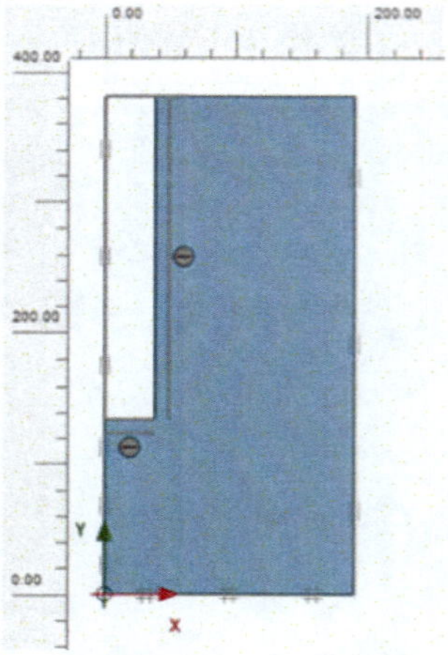

(a) Initial Phase

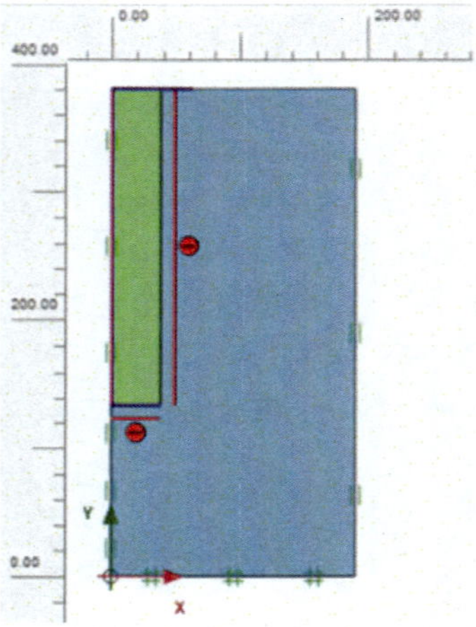

(b) Pile Activation

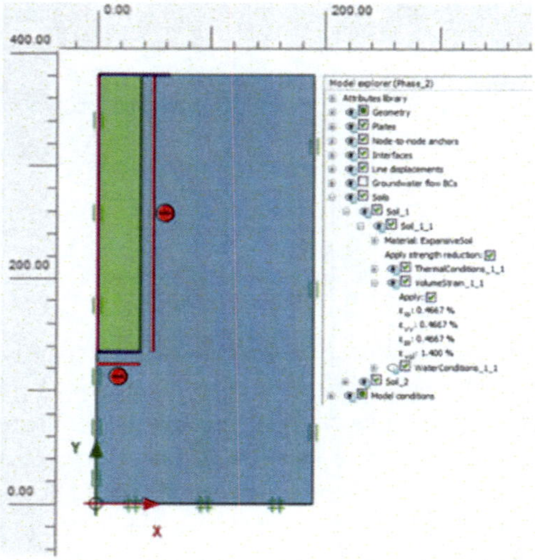

(c) Heave Measurements

Fig. 10.3 Heave calculations in PLAXIS analysis

axis in arthematic scale and time (mins) on X-axis in logarithmic scale. Figures 10.6 and 10.7 show graphical representation of heave (mm) measured at top of the pile for various pile lengths of plain GPA and geogrid-encased GPA.

The above graphs show effectiveness of plain and geogrid-encased GPAs in expansive soil as the amount of heave constantly decreasing with increase in pile length and provides proper anchorage against volume change.

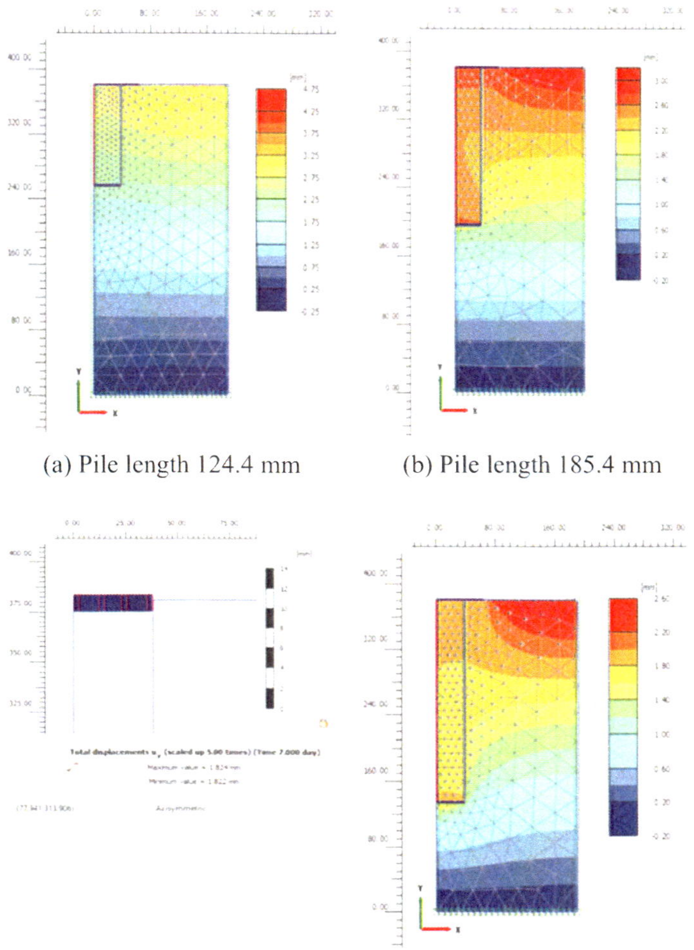

(a) Pile length 124.4 mm (b) Pile length 185.4 mm

(c) Pile length 246.4 mm

Fig. 10.4 Vertical displacement for plain GPA by PLAXIS analysis

Comparison of PLAXIS Model of Plain GPA with Literature [13]

The heave data of reinforced clay bed of plain GPA from literature [13] were compared with numerically analyzed data for two different pile lengths of 124.4 mm and 185.4 mm as shown in Fig. 10.8. It is observed that, the results from the numerical model are well in comparison with experimental data. The maximum percentage deviation was about 2%; this can be due the empirically assumed input data of material properties utilized for modeling.

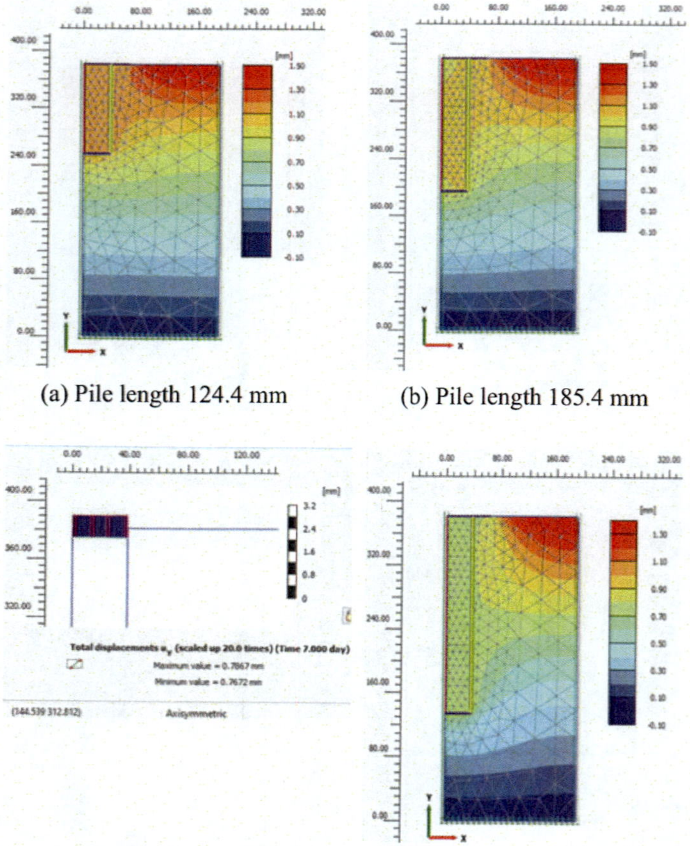

(a) Pile length 124.4 mm (b) Pile length 185.4 mm

(c) Pile length 246.4 mm

Fig. 10.5 Vertical displacement for geogrid-encased GPA by PLAXIS analysis

Table 10.5 Heave data of numerical analysis

Pile dimensions		Heave of plain GPA (PLAXIS) in mm	Heave of geogrid-encased GPA (PLAXIS) in mm
Diameter (mm)	Length (mm)		
76.2	124.4	3.33	1.2
	185.5	2.67	1
	246.4	1.82	0.78

Effect of Roughness Coefficient on Geogrid-Encased GPA

Roughness coefficient resembles the roughness value between the geogrid encasement around the pile and surrounding soil medium. In general, the interface and

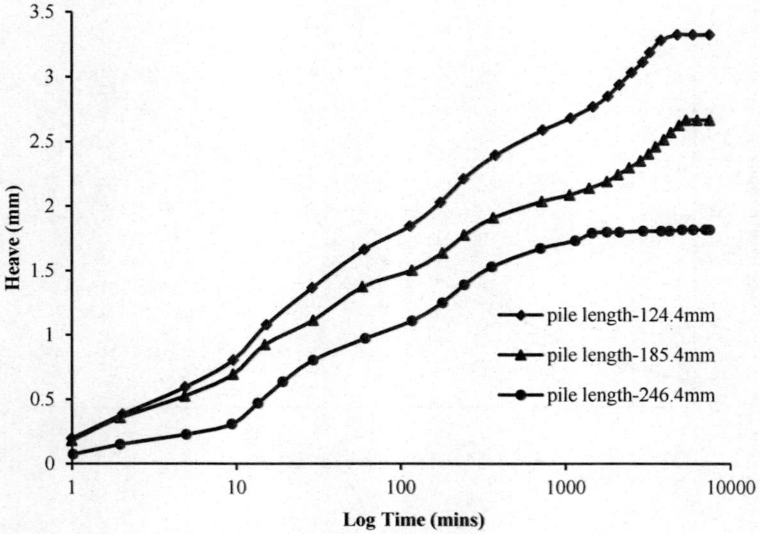

Fig. 10.6 Rate of heave for plain GPA by PLAXIS analysis

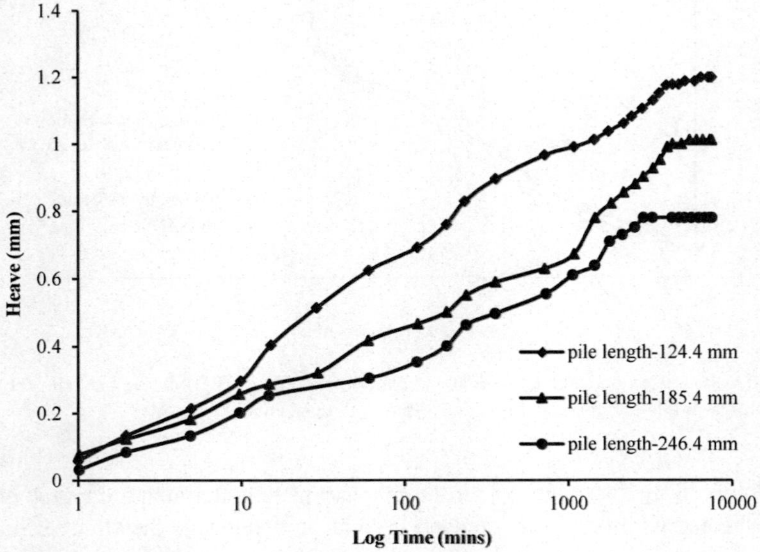

Fig. 10.7 Rate of heave for geogrid-encased GPA by PLAXIS analysis

interaction between the soil and the structure is weaker and more flexible than the surrounding soil; therefore, R_{inter} should be smaller than one. R_{inter} is directly proportional to the normal and shear stiffness of the interface element; therefore, mobilized shear resistance along the interface may be represented by varying the value of R_{inter}.

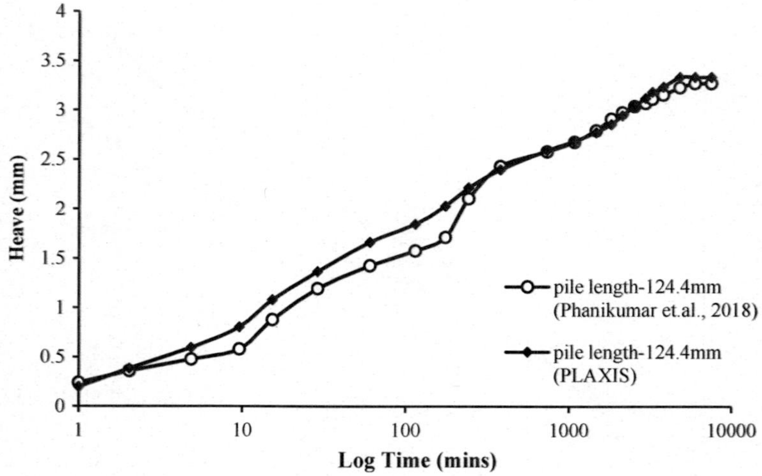

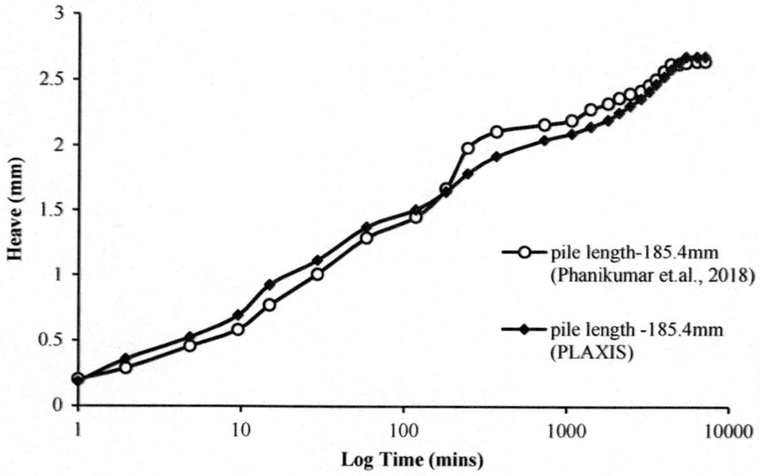

Fig. 10.8 Comparison between numerical and experimental data [13] of heave behavior for plain GPA of different pile lengths 124.4 mm and 185.4 mm, respectively

It is influenced by the roughness of the pile and plate material and the type of the contact created with the soil in contact. In order to understand the effect of roughness coefficient on heave behavior, three pile lengths of geogrid-encased GPA were analyzed by varying the R_{inter} from 0.2 to 0.8. From the result, it is observed that, the amount of heave is decreasing simultaneously with increase in the roughness coefficient which is given in Table 10.6. Figure 10.9 shows the variation of heave corresponding to roughness coefficient.

Table 10.6 Surface heave corresponding to different roughness coefficient (R_{inter})

Pile lengths (mm)	Surface heave (mm) with different R_{inter} value			
	$R_{inter} = 0.2$	$R_{inter} = 0.4$	$R_{inter} = 0.6$	$R_{inter} = 0.8$
124.4	1.61	1.47	1.36	1.2
185.4	1.3	1.22	1.1	1
246.4	1.2	1.1	0.95	0.78

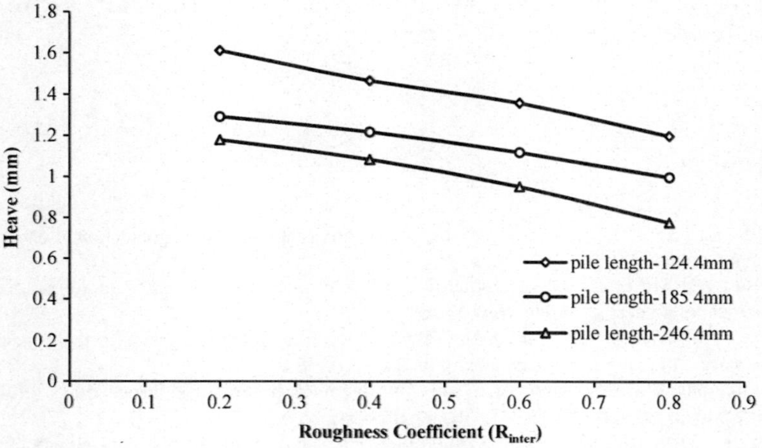

Fig. 10.9 Surface heave corresponding to roughness coefficient (R_{inter})

Conclusions

Expansive soils are subjected to swelling and shrinkage upon continues wetting and drying, which cause detrimental effect to the structure built on it. Hence, various techniques are advised as mitigation measures. Granular pile anchor is the latest innovative and effective mitigation technique recommended for expansive soil. In this present study, numerically analyzed data were obtained on heave behavior of plain GPA and geogrid-encased GPA in expansive soil. The heave (mm) was also obtained by varying the roughness coefficient (R_{inter}) on geogrid-encased GPA. The main conclusions drawn from the numerical study are as follows:

1. Granular pile anchor controls the heave phenomenon of expansive soil due to self-weight of pile and friction mobilized along the soil-pile interface.
2. In case of plain GPA frictional force mobilized along GPA-soil interface and for geogrid-encased GPA frictional resistance mobilized along the GPA-geogrid-soil interface in downward direction contributing to the greater reduction in heave.
3. Heave behavior of expansive clay bed reinforced with plain GPA was found to be decreasing by increasing pile length, i.e., 3.33 mm, 2.67 mm, and 1.82 mm for pile length of 124.4 mm, 185.4 mm, and 246.4 mm, respectively.

4. Further, heave decreases for expansive clay bed when reinforced with geogrid-encased GPA, i.e., 1.2 mm, 1 mm, and 0.78 mm for pile length of 124.4 mm, 185.4 mm, and 246.4 mm, respectively.
5. Overall, it is observed that the reduction in heave (mm) achieved up to 64% by providing geogrid encasement around the pile length.
6. Effect of roughness coefficient on heave behavior of geogrid-encased GPA was also analyzed, and it was found that heave decreases significantly with increases in roughness coefficient (R_{inter}).
7. Maximum percentage reduction in heave (mm) is found to be 35% for GPA length, when R_{inter} value varies from 0.2 to 0.8.

References

1. Subba Rao KS (1999) Swell–shrink behaviour of expansive soils—geotechnical challenges. Indian Geotech J 30(3):1–69
2. Nelson DJ, Miller JD (1992) Expansive soils—problems and practice in foundation and pavement engineering. Wiley, New York
3. Mowafy YM, Nagy AH (2001) The effect of reinforced sand cushion on the behavior of expansive soil using large scale apparatus. J Geotech Engg
4. Phanikumar B (2009) Expansive soils—problems and remedies. Proceedings of Indian Geotechnical Conference (GEOTIDE), pp 907–913
5. Katti R (1979) Search for solutions for problems in black cotton soils. Indian Geotech J 9(1):1–80
6. Phanikumar BR, Ramachandra Rao N (2000) Increasing pull-out capacity of granular pile anchors in reactive soils using base geosynthetics. Can Geotech J 37(4):870–881
7. Nusier OK, Alawneh AS (2004) Micropile technique to control upward movement of lightweight structures over expansive soils. J Geotech Geol Eng 22:89–104
8. Cooke RW, Price G (1973) Strains and displacements around friction piles. In: Proceedings of 8th ICSMFE, vol 2, Moscow, pp 53–60
9. Alawneh AS, Malkawi AIH, Al-Deeky H (1999) Tension tests on smooth and rough model piles in dry sand. Can Geotech J 36:746–753
10. Phanikumar B (1997) A study of swelling characteristics of and granular pile anchor foundation system in expansive soils. Ph.D. Thesis, Jawaharlal Nehru Technological University, Hyderabad, India
11. Phanikumar BR, Sharma RS (2004) Effect of fly ash on engineering properties of expansive soils. J Geotech Geoenviron Eng 130(7):764–767
12. Phanikumar BR (2016) Influence of geogrid reinforcement on pull out response of granular pile-anchors (GPAs) in expansive soils. Indian Geotech J
13. Phnanikaumar BR, Sai Raghuram AS, Sriramarao A (2018) Improving expansive clay beds with granular pile anchors (GPAs) and geogrid-encased GPAs. Proc Inst Civ Eng—Ground Improvement 173(4):237–248
14. Phanikumar BR, Rao AS, Suresh K (2008) Field behaviour of granular pile anchor in expansive soil. Ground Improvement 161(G14)
15. Muthukumar M, Shukla SK (2017) Comparative study on the behaviour of granular pile anchor and helical pile anchor in expansive soil subjected to swelling. Int J Geotech Eng
16. HariKrishna P (2006) A study on the use of granular anchor piles to control heave of footings resting on expansive soil. Ph.D. Thesis submitted to Kakatiya University, Warangal
17. Ismail MA, Shahin MA (2011) Finite element modelling of innovative shallow foundation system for reactive soils. Int J GEOMAT 1(1) (Sl. No. 1):78–82

18. Ismail MA, Shahin MA (2012) Numerical modelling of granular pile-anchor foundations (GPAF) in reactive soils. Int J Geotech Eng, 149–156
19. Ibrahim SF, Aljorany AN, Aladly AI (2014) Heave behavior of granular pile anchor-foundation (GPA-foundation) system in expansive soil. J Civ Eng Urbanism 4(3):213–222
20. Sangeetha S, Krishna PH (2019) Analysis of heave behaviour of expansive soil provided with granular pile anchors using Plaxis. Adv Comput Methods Geomech IACMAG Symp 1:391–404

Chapter 11
A Study of Load Distribution Between Soil and Stone Columns

Vamja Shreya and E. C. Nirmala Peter

Introduction

Large areas of land covered with soft clay soil having low bearing capacity and high compressibility made the ground unsuitable for construction activities. This has led to modification of such grounds by reinforcing them with stone columns/granular piles as one of the ground improvement techniques. Stone columns are vertical columns that are installed in soft soil by removing the soil in boreholes made with certain diameter and depth, and replacing the same with aggregates and sand compacted to a suitable density. Stone column being a stiffer material increases the overall stiffness of the ground in comparison with the untreated ground. Upon application of vertical load on the ground treated with stone columns, there is sharing of the load between stone column and the surrounding soil. The increase in the load carrying capacity and reduction in settlement of the treated ground increase with increase in the stiffness of the stone columns. These inclusions in the ground also act as a drainage medium in accelerating the process of consolidation of the surrounding soil.

Several experimental and analytical works were carried out to study the improvement of the foundation soil by using stone columns and also the factors on which the performance of stone column depends. Shahu and Reddy [1] have conducted model tests on group of stone columns and found that the major parameters on which the response of group of stone columns depends upon are area ratio, length to diameter ratio, overconsolidation ratio, relative stiffness, and stress ratio of the clayey soil. Marto et al. [2] from their analysis using finite element program (PLAXIS 2D) found out that as the diameter of stone column increases the load carrying capacity of stone column increases. A technical note given by Hanna et al. [3] has found from the parametric study that the load concentration on stone column depends upon the diameter of stone column and modulus ratio E_s/E_c. Rani and Kumar [4] have also

V. Shreya (✉) · E. C. N. Peter
Jawaharlal Nehru Technological University, Hyderabad, India
e-mail: shreyavamja@gmail.com

K. Muthukkumaran et al. (eds.), *Ground Improvement Techniques*, Lecture Notes in Civil Engineering 297, https://doi.org/10.1007/978-981-19-6727-6_11

found that the load carrying capacity increases with increase in diameter of stone column from their model study. Choobbasti et al. [5] conducted numerical evaluation on performance of stone columns in soft clay and found that it mainly depends upon column spacing.

Keeping all these factors in view the present paper presents a parametric study carried out to determine the load distribution between stone column and soil in terms of load distribution ratio.

Methodology

The load distribution ratio which is the ratio of load on stone column and load on soil is determined by considering vertical force equilibrium equation and equal strain equation in a unit cell as mentioned below.

The equilibrium of vertical forces in the unit cell gives

$$q_{\text{total}} = q_{\text{gp}}\, A_r + q_s(1 - A_r) \tag{11.1}$$

where

q_{gp} = stress on granular pile/stone column.

q_s = stress on surrounding soil.

A_r = area ratio = $(d/d_e)^2$

d = diameter of stone column/granular pile.

d_e = diameter of unit cell = 1.05*spacing for triangular pattern and 1.13*spacing for square pattern.

The vertical displacement of stone column/granular pile is given by

$$S_{\text{gp}} = \frac{q_{\text{gp}}}{E_{\text{gp}}} \times L \tag{11.2}$$

where E_{gp} = modulus of elasticity of stone column and L = length of the stone column (Fig. 11.1).

The vertical displacement of the soil is given by

$$S_{\text{soil}} = \frac{C_c}{1 + e0} \times H \times \log(\frac{q_s}{q0}) \tag{11.3}$$

where C_c = compression index of soil, e_0 = void ratio, H = thickness of clay layer, q_s = stress on soil, and q_0 = initial stress on soil.

As the vertical displacement of stone column/granular pile is equal to vertical displacement of surrounding soil, we get

$$S_{\text{gp}} = S_{\text{soil}} \tag{11.4}$$

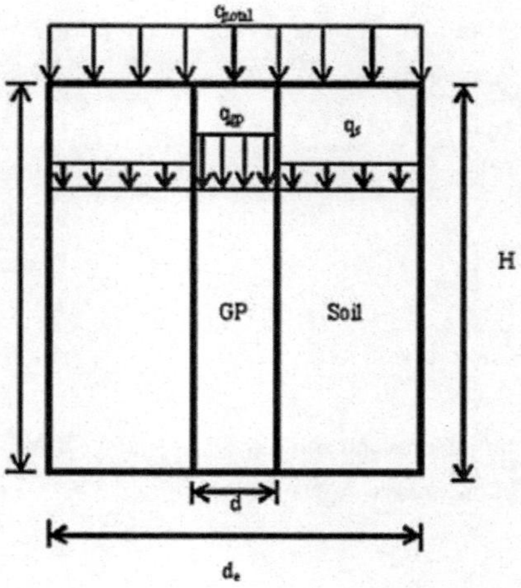

Fig. 11.1 Unit cell

$$\frac{q_{gp}}{E_{gp}} \times L = \frac{C_c}{1+e_0} \times H \times \log\left(\frac{q_s}{q_0}\right) \tag{11.5}$$

$$q_s = q_0 \times 10^{(X \times q_{gp})} \tag{11.6}$$

where

$$X = \frac{(1+e_0) \times L}{(C_c \times E_{gp} \times H)}$$

By solving the Eqs. 11.1 and 11.6, we obtain the stress on stone column/granular pile (q_{gp}) and the stress on soil (q_s), respectively. These stresses q_{gp} and q_s are multiplied with their respective areas, i.e., area of stone column and area of soil in order to get the load on stone column and load on soil to determine the load distribution ratio.

The values of the parameters that are to be defined in order to determine the load distribution ratio (Q_{gp}/Q_s) as observed from Eqs. 11.1 and 11.6 are specified in the Table 11.1.

Parametric Study

The variation of load distribution ratio (Q_{gp}/Q_s) is studied by varying parameters such as area ratio (A_r), relative stiffness (E_{gp}/E_s), and stress intensity (q_{total}). The following are the discussions below for each parameter studied (Figs. 11.2, 11.3, 11.4, 11.5, 11.6 and 11.7).

Table 11.1 Parameters considered for the study

Parameters	Values considered
Area ratio (A_r)	0.07 to 0.64
E_{gp} [6]	Dense condition—175 and 190 MPa
	Medium dense condition—120 and 150 MPa
	Loose condition—75 and 100 MPa
E_s	15 MPa
Compression index (C_c)	0.45
Void ratio (e_0)	1.7
Initial stress on soil (q_0)	1t/m^2
Total stress (q_{total})	5,10,15,25,40 and 50t/m^2

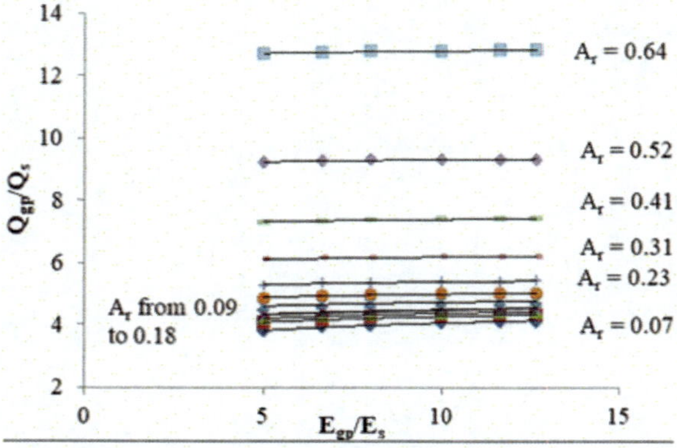

Fig. 11.2 Q_{gp}/Q_s versus E_{gp}/E_s for different area ratios for $q_{total} = 5$t/m^2

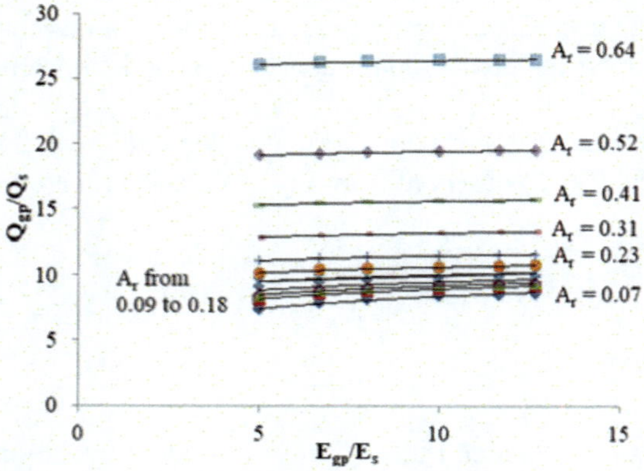

Fig. 11.3 Q_{gp}/Q_s versus E_{gp}/E_s for different area ratios for $q_{total} = 10$t/m^2

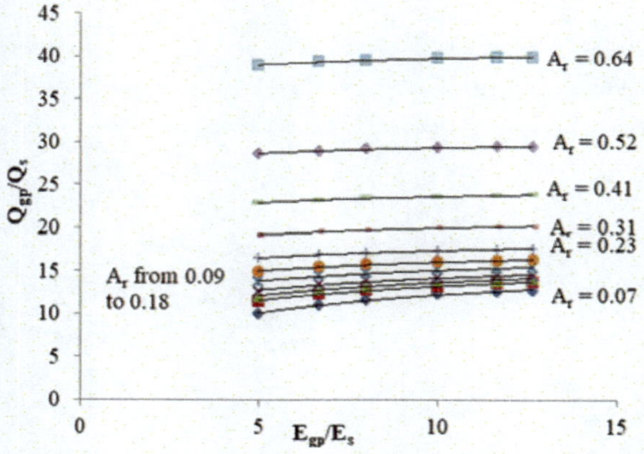

Fig. 11.4 Q_{gp}/Q_s versus E_{gp}/E_s for different area ratios for $q_{total} = 15t/m^2$

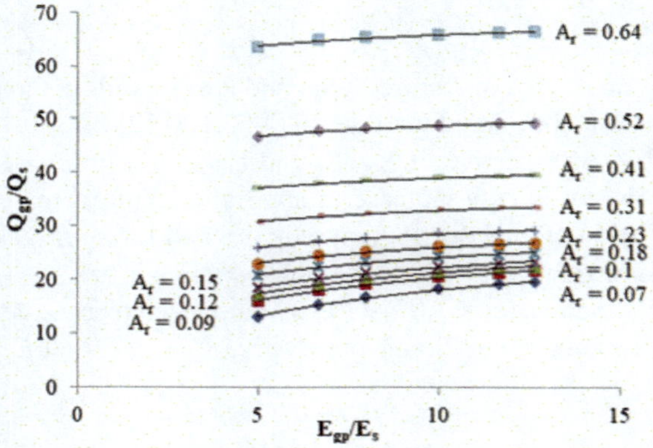

Fig. 11.5 Q_{gp}/Q_s versus E_{gp}/E_s for different area ratios for $q_{total} = 25t/m^2$

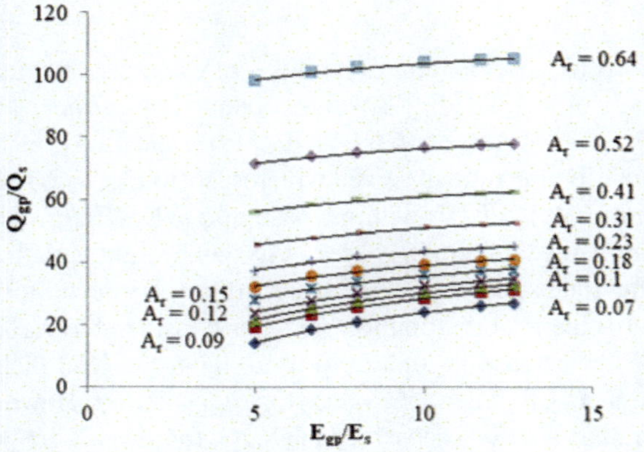

Fig. 11.6 Q_{gp}/Q_s versus E_{gp}/E_s for different area ratios for $q_{total} = 40t/m^2$

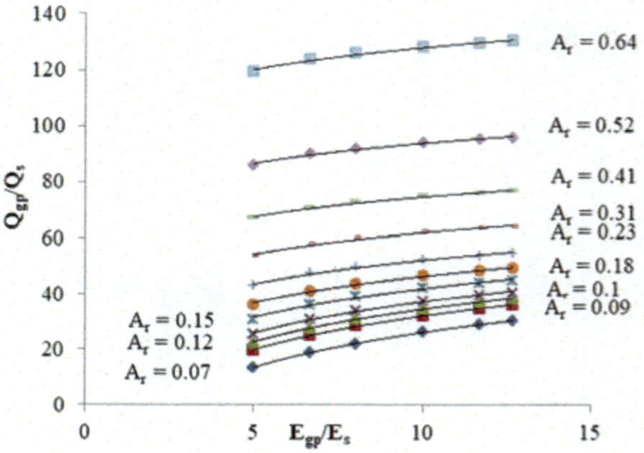

Fig. 11.7 Q_{gp}/Q_s versus E_{gp}/E_s for different area ratios for $q_{total} = 50t/m^2$

Effect of Area Ratio (A_r)

Area ratio is the ratio of area of stone column and area of soil. In the present study, the area ratios that are considered are 0.07, 0.09, 0.1, 0.12, 0.15, 0.18, 0.23, 0.31, 0.41, 0.52, and 0.64. Irrespective of E_{gp}/E_s and q_{total}, it is observed that the load distribution ratio increases with increase in area ratio. With increase in area ratio $A_r \leq 0.18$, the variation of load distribution ratio is small for all E_{gp}/E_s values up to $q_{total} = 15t/m^2$. For higher stresses, considerable variation is observed for all E_{gp}/E_s values with more variation for $E_{gp}/E_s < 10$. By further increase in area ratio from 0.23 to 0.64, a considerable variation is observed in Q_{gp}/Q_s for all E_{gp}/E_s and q_{total} values.

Effect of Relative Stiffness (E_{gp}/E_s)

The relative stiffness (E_{gp}/E_s) is ratio of modulus of elasticity of stone column and modulus of elasticity of soil. The E_{gp}/E_s values obtained by considering E_{gp} from 75 to 190 MPa and $E_s = 15$ MPa are 5, 6.67, 8, 10, 11.67, and 12.67, respectively. The variation of load distribution ratio observed with respect to E_{gp}/E_s is logarithmic. For a particular area ratio $(A_r) = 0.07$, with increase in E_{gp}/E_s ($E_{gp}/E_s \geq 10$), the load distribution ratio remains same for $q_{total} \leq 15$ t/m^2 but a slight variation is observed in Q_{gp}/Q_s for higher total stresses $q_{total} = 40$ and $50t/m^2$. However, for lower E_{gp}/E_s values ($E_{gp}/E_s < 10$) the load distribution ratio has considerable variation even for $q_{total} \geq 15t/m^2$. With increase in area ratio up to 0.64, the load distribution ratio remains more or less same for all E_{gp}/E_s values for $q_{total} \leq 25t/m^2$. However, for higher stresses a considerable variation is observed in Q_{gp}/Q_s for lower E_{gp}/E_s ($E_{gp}/E_s < 10$).

Stress Intensity (q_{total})

The total stresses considered for the present study are 5, 10, 15, 25, 40, and 50t/m^2. From the above graphs, it is observed that with increase in q_{total} the load distribution ratio also increases. However, it is observed that there is slight increase in Q_{gp}/Q_s with respect to q_{total} for smaller area ratios. But with increase in area ratio $A_r = 0.64$, it is observed there is considerable increase in load distribution ratio with increase in q_{total} for all E_{gp}/E_s.

Conclusions

From this study on load distribution between stone column and soil, the following conclusions are drawn.

1. Irrespective of area ratio (A_r) and q_{total}, the load distribution ratio (Q_{gp}/Q_s) varies logarithmically with $\text{E}_{\text{gp}}/\text{E}_{\text{S}}$.
2. For $q_{\text{total}} \leq 15\text{t/m}^2$,

   (a) The curves Q_{gp}/Q_s v/s E_{gp}/E_s are very flat indicating very small variation of Q_{gp}/Q_s with E_{gp}/E_s for total stress up to $\leq 15\text{t/m}^2$ for all the area ratios.
   (b) For $A_r < 0.31$, there is slight increase in Q_{gp}/Q_s with increase in the corresponding area ratios.
   (c) For $A_r \geq 0.31$, there is considerable increase in the Q_{gp}/Q_s; however, these values corresponding to the A_r values remained more or less constant irrespective to the stiffness of the material.

3. For $q_{\text{total}} \geq 25\text{t/m}^2$ and $A_r < 0.15$, the curve Q_{gp}/Q_s v/s $\text{E}_{\text{gp}}/\text{E}_{\text{s}}$ is steep indicating considerable variations in Q_{gp}/Q_s with $\text{E}_{\text{gp}}/\text{E}_{\text{s}}$ with values being almost same for corresponding A_r values. For higher A_r values, the steepness decreased shows less variation, especially curves corresponding to $q_{\text{total}} > 25\text{t/m}^2$ became flatter.
4. This analysis indicated that the most dominating factor is the area ratio (A_r).

References

1. Shahu, Reddy (2011) Clayey soil reinforced with stone column group: model tests and analyses. J Geotech Geoenviron Eng 137(12):1265–1274. ASCE
2. Marto et al (2013) Performance analysis of reinforced stone columns using finite element method. Electron J Geotech Eng 18:315–323
3. Hanna et al (2013))Mode of failure of a group of stone columns in soft soil. Int J Geomech 13(1):87–96 ASCE
4. Rani SS, Kumar PSP (2016) A study of behaviour on stone column in homogenous soil. Int J Sci Res (IJSR) 5:1215–1219

5. Choobbasti AJ et al (2011) Performance of stone columns in soft clay: numerical evaluation. Geotech Geol Eng 29:675–684
6. Bowles JE (1997) Foundation analysis and design, 5th edn. McGraw-Hill, Singapore

Chapter 12
An Experimental Study to Determine the Best Aggregate Mix for Stone Columns

Dipika Choudhury, RaiBahadur Reang, and Sanjay Paul

Introduction

Stone columns are an efficient, economic, and eco-friendly safe style for enhancing soft cohesive and cohesionless soils. Soil around the stone columns are less permeable compared to stone column. The high permeability of the column causes radial drainage and quick removal of excess pore water pressure. It replaces 10–30% weak soil with comparatively stiff and high strength granular materials [6]. High column stiffness reduces vertical stress on the soil body. Granular column also acts as a pile transferring the foundation load to the greater depth.

This type of method widely used to improve ground, this method enhance bearing capacity and reduce settlement of construction sites. This technique is mostly preferred in constructions of road embankments, liquid storage tank and raft foundation, factories that can undergo some extent of settlement. The load settlement behavior of individual floating stone column in a group is different and the stiffness of unit cell rise with the rise of the area ratio(ratio of cross-sectional area of columns to cross-sectional area of foundation soil)but beyond a certain point the length of the column does not give any advantage in results [15]. Failure stress of the foundation rises and downward vertical displacement reduces with the rise of the area ratio [14]. A long column which have length more than the critical length (length to diameter ratio greater than 4) fails by bulging [7]. Mostly single column fails by bulging whereas group of columns fail by shear or bulging. General, local and punching shear failure of groups would happen depending upon the soil characteristics and geometry of the group [6]. The radius of smear zone is a factor of four to five times of the radius of central drain and the stiffness of the enhance ground increases due to quick removal of excess pore water pressure [3–5].

D. Choudhury (✉) · R. Reang · S. Paul
National Institute of Technology Agartala, Agartala, Tripura, India
e-mail: deepikachoudhury11@gmail.com

K. Muthukkumaran et al. (eds.), *Ground Improvement Techniques*, Lecture Notes in Civil Engineering 297, https://doi.org/10.1007/978-981-19-6727-6_12

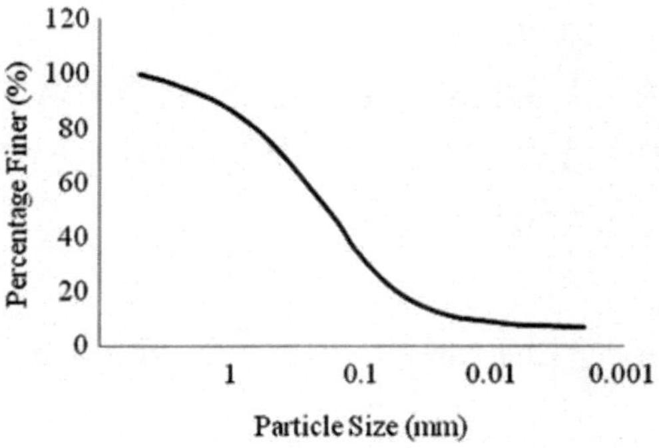

Fig. 12.1 Particle size distribution curve of the soil

The previous researchers did not work on the stone column using different sizes of granular materials to observe their influence on the behavior of the unit cell. Therefore, in this work an effort has been made to obtain best aggregate for stone column out of several aggregate mixtures and load displacement behavior of unit cell studied. Here, along with experimental works 3D finite element analyses were performed to validate the work.

The load settlement behavior of a unit cell under the whole area loading condition is linear. The stiffness of improved and unimproved soil can be obtained from the load settlement curve. The object of this study is to analyze the nature of stone columns in clay soil using different types of granular mixtures and also find out factor influencing their behavior. The load carrying capacity of different soils is different, and this is due to different lateral confinement. The consolidation stress on stone columns rises with time as stress on soil reduces with time (Fig. 12.1).

Material Used

Clay

Soil was extracted from the site of NIT Agartala, Tripura, India. The sieve of size 4.75 mm used to remove the coarser particles. The samples were dried in the oven at the temperature of 105–1100 °C. Some remolded characteristics of clay shown in Table 12.1. The classification of soil is CI-MI and liquid limit is 42%, plastic limit is 21%, maximum dry density is 17.16 kN/m^3, optimum moisture content is 27.5% and other details are shown in Table 12.1.

Table 12.1 Material properties of sand and clay soil

Material	G	$OMC(\%)$	MDD (kN/m^3)	C_u (kPa)	Υ_{bulk}(kN/m^3)	Υ_d(kN/m^3)	Φ(deg.)
Clay	2.69	27.5	17.16	29	18.6	15.23	--
Sand	2.67	–	–	–	16.7	14.87	35

Table 12.2 Constituent of stone aggregates used

Stone mixture	Constituent
Mixture A	10% of 3.0 mm + 45% of 5.6 mm + 45% of 8.0 mm
Mixture B	30% of 3.0 mm + 35% of 5.6 mm + 35% of 8.0 mm
Mixture C	40% of 3.0 mm + 30% of 5.6 mm + 30% of 8.0 mm

Sand

Clean river sand of size less than 4.75 mm were collected for this study from the nearby site of the institute. All these aggregates were washed and dried in the oven at the temperature of 1100 °C. The maximum and minimum densities of sand are 17.1 and 15.5 kN/m^3, respectively. The direct shear test used to determine the value of angle of internal friction of the sand, the internal friction angle of the sand found was 35° (Table 12.2).

Stone Aggregate

All aggregates were washed and oven dried at the temperature of 1100 °C. Crushed stones of 3.0 mm, 5.6 mm, and 8.0 mm size were used. Poisson's ratio was calculated from a drained triaxial test. Modulus of elasticity (E) of the clay soil and is the inverse of coefficient of compression (a_v) which is determined from consolidation test at the pressure 100–200 kPa. Standard Proctor test was carried out by preparing samples at different water content. Crushed stone of size 2–8 mm used. Direct shear box (300 mm × 300 mm × 100 mm) used to determine dilation angle and angle of internal friction of aggregates. Crushed stone of sizes 2–8 mm is considerable to study a small stone column [1], and in this study the stone aggregate sizes of 3 to 8 mm were selected to study the influence of several aggregate mixtures on the contribution of a stone column.

Test Setup

A test setup for single stone column is shown in Fig. 12.2. The height and diameter of the tank are 400 mm and 350 mm, respectively. Length to diameter ratio (l/d) of 8 was used for making stone columns. A mandrel of 38 mm inside diameter and 2 mm of thickness were used. A uniform vertical load was applied over the entire top surface using direct load and dial gauge was used to measure the corresponding settlement. The load was applied at 7 kPa each with 8–10 equal loading increments.

Preparation of Clay Bed

The pulverized clay was dried in oven for 24 h and the adequate quantity of water was mixed to the sample. The optimum moisture content (OMC) received from the standard Proctor test was used to make soft clay paste and the mixture was kept in an airtight polythene bag for 24 h to obtain uniform consistency. A thin layer of grease and polythene sheet were used along the inside wall of the tank to reduce friction between the tank wall and clay. The clay soil was filled in five equal layers in the unit cell tank. Quantity of clay was measured by weight and each layer was compacted properly with a temper of weight 1.75 kg and height of 50 cm. Tank was filled upto the height of 320 mm.

Fig. 12.2 Preparation of soil bed and stone column

Construction of Stone Column

Replacement technique was used to create the stone columns for the current study. A seamless open ended steel pipe of 40 mm outside diameter and 2 mm thickness pushed into the center of the clay bed to construct the stone column. Grease was applied to the both inside and outside of the steel pipe for smooth penetration into the clay. The clay inside the pipe was scoped out and stones were poured into the steel pipe in layers and compacted with a 1.5 kg circular steel rod. Pipe was uplifted in stages to secure a minimum penetration depth of 10 mm below the top layer of stone aggregate. Light compaction effort was used. This method was repeated to complete the column. Sand blanket of 30 mm thickness was given on the surface of the clay bed.

Test Procedure

We can divide the entire test procedure in four phases. The first phase includes preparation of clay and clay bed in the unit cell tank. Second phase includes installation of stone columns step by step with required compaction by tempering. The third phase includes application of load on the entire area gradually. In the fourth phase, dial gauge used to calculate the load and corresponding settlement. The load settlement characteristics of the stone column and whole unit cell was considered by employing vertical load. The load on the unit cell applied thorough thin plate of thickness 7 mm and diameter 5 mm less than the inside diameter of the unit cell tank. A 30 mm thick layer of sand was provided above the clay layer before loading which acted as a sand blanket. The main objective of the sand blanket is to serve the load evenly and to act as a drainage path. An incremental load of 7 kPa applied and settlement measured with respect to time and loading. This same test procedure is continued for other granular mixtures. Load was applied up to the failure or upto 8 mm settlement in the unit cell.

Method to Obtain Deformed Shape of Stone Column

The failure condition of the stone column was observed by giving plaster of Paris slurry into the stone column because plaster of Paris slurry cannot penetrate through clay. When the slurry becomes hard and strong enough to withstand its self-weight the nearby soil was cleansed and the shape or figure of the stone column was obtained. This process was followed repeatedly for a number of times for individual models. Figure 12.3 shows a typical stone column after loading, where B is showing the stone column of Mixtures B and A is for other mixtures. The figure shows that Mixture B exhibits good results then others as bulging is less and hence failure load is more.

Fig. 12.3 Stone column after loading

Finite Element Analysis

In this study a single end bearing rigid-based stone columns in soft soil under vertical displacement were considered. The geometry of the material was considered and axisymmetric analysis was carried out using the Modified Drucker-Prager model. The tank used in the analysis is 400 mm high and 350 mm in diameter. A stone column with diameter of 40 mm and height of 320 mm was used. The finite element mesh was made using 16 node triangular element, coupled pore water-displacement, and elastoplastic behavior of clay. Hinged support was assumed along the circumference of the tank as settlement of the soil surrounding the column can occur only in vertical direction in which horizontal deformation is restricted. The bottom of the tank was made as fixed supported condition since the bottom cannot move in both vertical and radial direction (Fig. 12.4).

Modeling and Analysis of Stone Column

The analysis was implemented for vertical load when the whole area loaded for l/d ratio equal to 8 and cohesion of 29 kPa. Figure 12.5 shows the correlation between time and settlement for different sizes of granular mixtures. Figure 12.6 shows the correlation between load and settlement obtained from model test. Young's modulus (E), angle of internal friction (Φ) and cohesion (c) of all the materials determined from the laboratory test and value of relevant input parameter shown in Table 12.3. The value of dilation angle are determined from the relation $\Psi = \Phi - 30°$ and Poisson's ratio used as per Bowels [2]. Along the circumference of the tank only vertical displacement considered. Both radial and vertical displacement at the base of the tank considered, no interface element considered between stone column and clay.

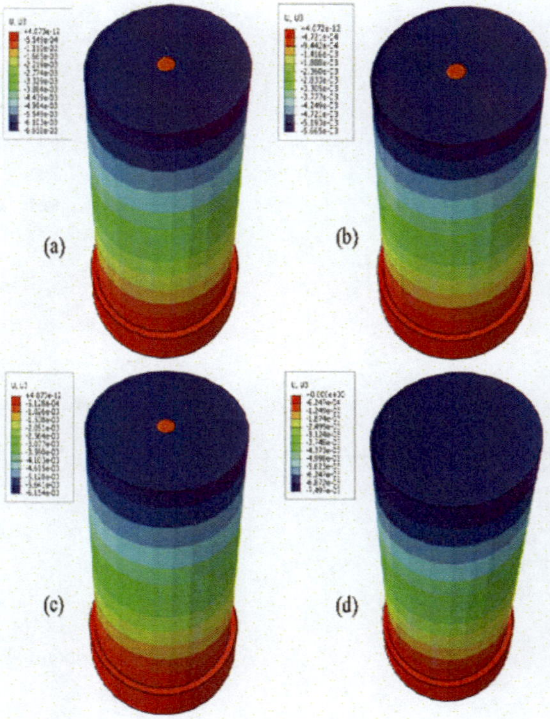

Fig. 12.4 Finite element analysis using ABAQUS for **a** Mix A, **b** Mix B, **c** Mix C, and **d** only soil (without stone column)

Here mixture B has minimum vertical displacement compared to others. The value of angle of internal friction and relative density for granular mixture B are 37° and 58%, respectively. When porosity of the material decreases the angle of internal friction increases. The internal friction angle is the angle between the tangent to Mohr envelope at failure and normal stress axis, angle of internal friction is the function of porosity (Perkins and Weingarten 1998). When the Poisson's ratio of the material rises the compressibility of the material also reduces.

Results and Discussion

This work was carried out to determine the outcome of different aggregate mixtures on stone columns embedded in clay soil. Surcharge load intensity varies from zero to failure of the granular columns. When the stiffness of the column increases the settlement decreases and with the change of porosity the internal friction angle of material also changed [13].

Figure 12.5 is showing that the maximum settlement without a stone column is 7.13 mm and when we used stone columns its settlement reduced to 5.2 mm. From experimental studies the settlement of unit cell using mix A, B, C and without

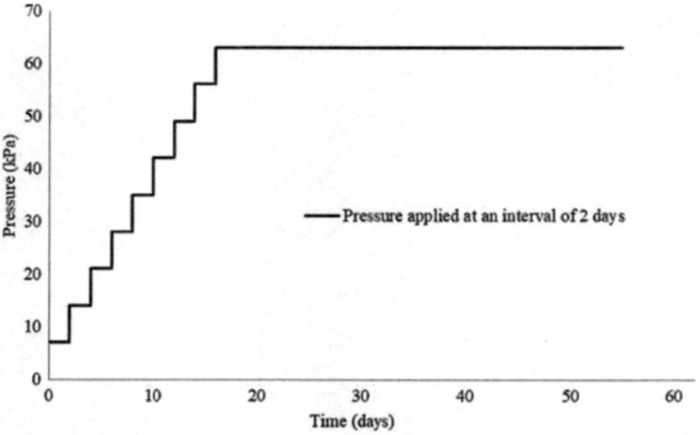

Fig. 12.5 Correlation between time and settlement for different sizes of mixtures

Table 12.3 Physical properties of stone aggregate used

Parameter	Types of mixture		
	Mix A	Mix B	Mix C
Specific gravity	2.65	2.66	2.675
Angle of internal friction	38	42	39
γ_{max}(kN/m^3)	16.9	17.5	17.6
γ_{min} (kN/m^3)	15.1	15.3	15.5

stone column found to be 6.2 mm, 5.2 mm, 5.67 mm, and 7.6 mm, respectively. The differences of settlement results found between finite element and experimental test result were approximately 0.122 mm, 0.11 mm, and 0.131 mm for mix A, B and C, respectively. Figure 12.6 also reveals the load settlement characteristics of stone column with different mixtures of granular material. From the graph it is observed that stone column containing mixture B shows less settlement. The mixture B has maximum relative density compared to other mixtures.

Figure 12.5 also shows the results obtained from FEM analysis (ABAQUS 2014 package). The result found from the laboratory test was validated with FEM analysis results, but the result found from the experimental work is slightly less than that of the finite element result. The reason may be so as a finite element considers a constant permeability value. The load settlement nature of the unit cell under the whole area loading makes it possible to find the improved ground stiffness which depends on internal friction angle of stones. The lateral confining pressure exerted by the adjoining soil is assumed to be at rest lateral earth pressure, which is equal to radial stress acting on column by soil. Load carrying efficiency of the stone columns arising from the resistance given by adjoining soil in opposition to its lateral deformation (bulging). End bearing stone column increase the bearing capacity of unit

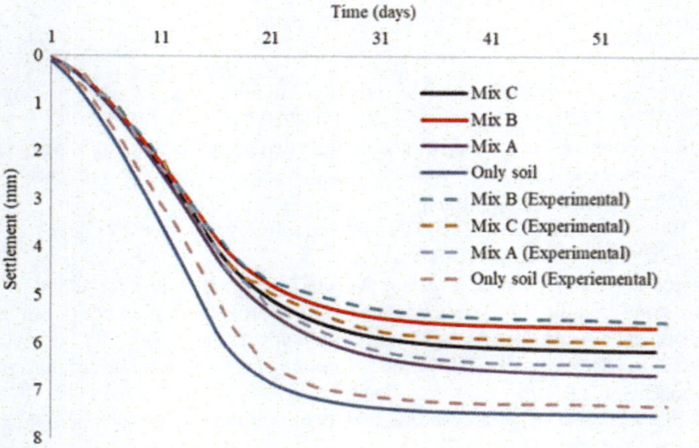

Fig. 12.6 Comparison of settlement results between laboratory and FEM models

cell compare to floating stone column but end bearing stone columns are expensive compare to floating stone column.

Conclusions

The present work describes experimental study and finite element analysis of settlement characteristics of stone columns for diverse mixtures of stone aggregates. Hence, the following conclusions can be made:

- The behavior of stone columns depends on the properties of granular material used. When the internal friction angle of granular material increases the settlement of the stone column decreases.
- Stone column used soil can carry extra load compared to soil without treatment.
- Load settlement behavior for entire area loading is almost linear which can be used to determine stiffness of improved ground.
- Enhanced soil stiffness can be obtained from the linear load displacement relationship and improved soil stiffness depends on the properties of granular materials
- It is found that mixture B is the best among all as it gives lower settlement and can carry more load than others

References

1. Ambily AP, Gandhi SR (2007) Behavior of stone columns based on experimental and FEM analysis. J Geotech Geoenviron Eng ASCE 133(4):405–415
2. Bowles JE (1998) Foundation analysis and design, 4th edn. McGraw-Hill, Singapore
3. Castro J, Sagaseta C (2007) Consolidation around stone columns: influence of column deformation. Int J Numer Anal Meth Geomech 33(7):851–877
4. Guetif Z, Bouassida M, Debats JM (2007) Improved soft claycharacteristics due to stone column installation. Comput Geotech 34(2):104–111
5. Gab M, Schweiger HF, Thurner R, Adam D (2007) Field trial toinvestigate the performance of floating stone columns. In: Proceedings of the 14th European conference on soil mechanics and geotechnical engineering. Amsterdam, Netherlands, 1311–1316
6. Hanna A, Etezad M, Ayadat T (2013) Mode of failure of a group of stone columns in soft soil. Int J Geomech 13(1):87–96
7. IS 15284-Part 1 (2003) Design and construction for ground improvement-Stone column, Bureau of Indian Standard, New Delhi
8. Indraratna B, Redana IW (1998) Development of the smear zone around vertical band drains. Int J Ground Improvement 2(4):180–185
9. IS: 2720-Part 3 (1980) Methods of test for soils—determination of specific gravity of fine-grained soils, Bureau of Indian Standards, New Delhi
10. IS: 2720-Part 2 (1980) Methods of test for soils—determination of moisture content using rapid moisture meter, Bureau of Indian Standards, New Delhi
11. IS: 2720-Part 4 (1985) Methods of test for soils—grain size analysis, Bureau of Indian Standards, New Delhi
12. IS: 2720-Part 5 (1985) Methods of test for soils—Determination of Atterberg's limit, Bureau of Indian Standards, New Delhi
13. IS: 2720-Part 7 (1980) Methods of test for soils—light/standard proctor compaction test of soil, Bureau of Indian Standards, New Delhi
14. Shahu JT, Reddy YR (2011) Clayey soil reinforced with stone column group: model tests and analyses. J Geotech Geoenviron Eng 137(12):1265–1274
15. Wood DM, Hu W, Nash DFT (2000) Group effects in stone column foundations: model tests. Geotechnique 50(6):689–698

Chapter 13
Static and Dynamic Study on the Performance of Modified Stone Column in Ahmedabad Soil

Milind Amin and Manendra Singh

Introduction

Certain projects may be required to be built in places with thick levels of sand deposits. Even when subjected to mild or moderate surface loads, these deposits have low shear strength and high voids ratio, resulting in excessive settlements. As a result, loose sands are regarded as difficult soils for foundations. To strengthen such soils, a variety of improvement strategies are used. Stone columns enhance load carrying capacity of the loose sand bed [1]. Stone columns are not only to operate as reinforcing components but also increases the overall strength and stiffness of deformable sand deposits. Study of behavior of soft soil is very important for ground improvement. Rashid et al. [2], Ye et al. [3], Sadaoui and Bahar [4], Jamal et al. [5] found that geosynthetic encased columns provide the best performance in terms of load settlement behavior of clayey soil. Mokhtari et al. [6] has studied the stone columns in soft cohesive soil and found that stone column could sustain isolated footing as well as wide raft footing. The majority of the studies in the literature have been carried out for better understanding and estimating the bearing capacity and compressibility of soft soil reinforced with stone columns. The effectiveness of the stone column for improving/densifying the loose cohesionless soil deposits has received very little attention. The effectiveness of stone columns in different types of soil is given in Table 13.1.

There are indeed a variety of approaches for analyzing stone column issues, ranging from closed-form analytical approaches to numerical approaches based on

M. Amin
Department of Civil Engineering, Indian Institutes of Technology Palakkad, Palakkad, India

M. Singh (✉)
Department of Civil Engineering, National Institute of Technology, Hamirpur, Himachal Pradesh 177005, India
e-mail: manendra@nith.ac.in

K. Muthukkumaran et al. (eds.), *Ground Improvement Techniques*, Lecture Notes in Civil Engineering 297, https://doi.org/10.1007/978-981-19-6727-6_13

Table 13.1 Effectiveness of stone column in different type of soil [7]

Ground type	Relative densification	Relative reinforcement
Sand	Excellent	Very good
Silty sand	Very good	Very good
Non-plastic silts	Good	Excellent
Clay	Marginal	Excellent
Mine soils	Excellent depending on gradation	Good
Dumped fill	Good	Good
Garbage	Not applicable	Good

Finite Elements Method (FEM). Each approach has its own set of benefits and drawbacks. Only if all of the below circumstances are met, the solution will be correct [8].

a. External forces must be entirely equivalent to internal forces in order to achieve force equilibrium.
b. The relationship among deformations at various points in a continuum.
c. Material's stress–strain equations.
d. Boundary conditions.

The most extensively utilized approach for analyzing the stone column for improved soil behavior is FEM analysis. 3D modeling is the ideal technique to simulate real-world soft soil behavior; nevertheless, for practical reasons, either the axisymmetric or plane strain model is commonly utilized [9].

The stone column is effective for soil improvement under static and dynamic load. However, post-earthquake behavior of stone column load is needed to be understood and this area is little studied. In this paper, the behavior of modified stone column has been studied. The study also involves to find the effect of dynamic load, by applying pseudo-static acceleration of 0.1, 0.15, 0.2, and 0.25 g. In addition to this, the workability of the same stone column which has under gone the dynamic load before has been studied again. The complete study was done for the soil of Ahmedabad, Gujarat, India.

Material and Methodology

Site Selection

For the current study, the soil utilized is from Sarangpur area of Ahmedabad city. The surrounded water bodies near to the site are Kankaria Lake and Sabarmati River. Kankaria Lake is 1 km away in the south direction while Sabarmati is 2.5 km away in the west direction as shown in Fig. 13.1. The area falls under the seismic zone III.

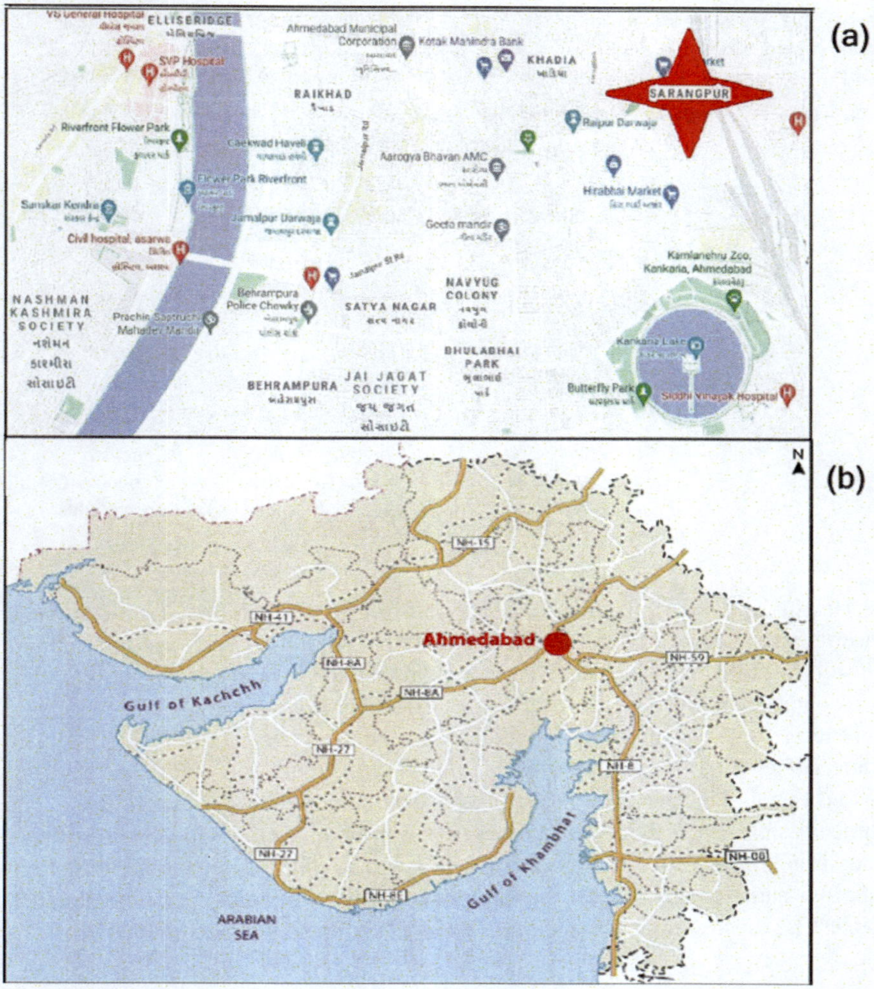

Fig. 13.1 Location of site on map (Photo: Google Map)

At the site, soil consisting of dark brown silty sand with clay and brick bats was found up to the depth of 2 m. These soil properties have been considered in this paper. And despite of nearby the water bodies, ground water is not even close to 15 m depth in the area.

Material Properties and Constitutive Model

PLAXIS 2D, a finite element software, has been used for mathematical modeling. In this study, a fictitious case study has been created. A 2.5 m × 20 m (Width

Table 13.2 Parameters for stone column and model used [10]

Parameters	Symbol (unit)	Soil	Stone column
Model	–	MC model	MC model
Soil condition		Drained	Drained
Unit weight	Υ_d (kN/m^3)	16.28	14.9
	Υ_{sat} (kN/m^3)	19.91	16.6
Shear strength parameters	$C\prime$ (kN/m^2)	0	0
	$\varphi\prime$ (°)	28	46
	Ψ (°)	0	16
Soil stiffness parameter	E (kPa)	14,710	40,000
	ν	0.35	0.30

Note MC = Mohr–Coulomb model; Υ_d = dry density; Υ_{sat} = saturated density; $c\prime$ = effective cohesion; $\varphi\prime$ = effective friction angle; Ψ = dilatancy angle; E = elastic Young's modulus; ν = Poisson's ratio

× Height) size of axisymmetric soil layout has been taken and stone column is inserted at the center. For soil, Mohr–Coulomb's model has been used to examine elastoplastic behavior. Soil depth is taken as 15 m with no water table in between. The soil condition is considered as drained and effective strength parameters are used. In absence of water table, result of undrained and drained behavior of soil will remain same. Effective stress and total stress parameters will remain same if drained behavior is considered. For stone column also, Mohr–Coulomb's model has been adopted. The stone column was designed as a homogenous drained material with certain stiffness and strength properties. The properties of stone column have been adopted from Ghazavi and Afshar [10]. On the basis of the Mohr–coulomb constitutive model, Table 13.2 illustrates the essential geotechnical parameters utilized in this work.

Geometry and Boundary Condition

The objective of this research was to introduce new modified stone column and to study its behavior. Figure 13.2 shows the geometry of conventional stone column (1) and modified reshaped stone columns (2a, 2b, 2c, and 2d). The behavior of modified reshaped stone columns was compared with the conventional stone column. PLAXIS 2D software was used to create plane strain models. 15-noded triangular elements were used to discretize the soil domain. The gravity effect was used to accommodate for self-soil weight, with the usual gravity acceleration, g, set to 9.810 m/s^2. On the horizontal ends, the axisymmetric area boundary has been restrained by shear and radial development, and at the bottom both horizontal and vertical mobility were constrained.

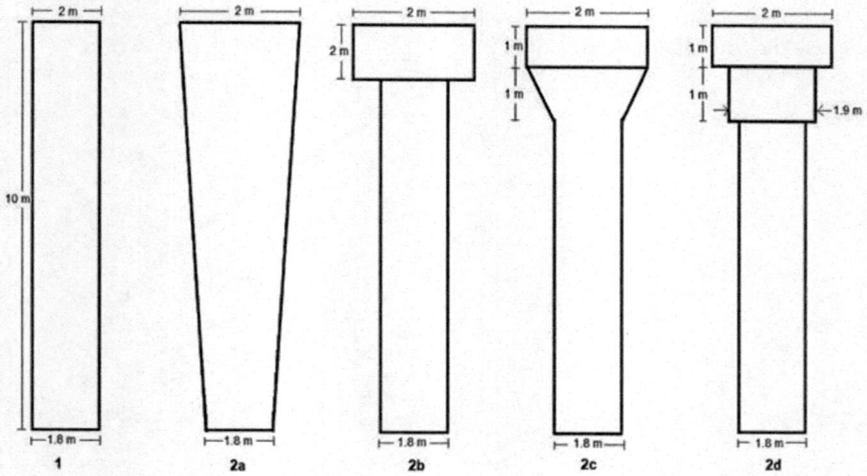

Fig. 13.2 Conventional (1) column and modified reshaped stone columns (2a, 2b, 2c, and 2d)

Result and Discussion

Effect of Modified Stone Columns

Figure 13.3 shows the relationship between load and settlement for different columns in static condition. It is surprising to observe that column 1 and column 2a results are not too different. However, column 2a experiences a 2.25% reduction in load carrying capacity. For column 2b, the slope is slightly lower but there is improvement of 9% in load carrying capacity. The behavior of column 2b, 2c, and 2d is more or less similar. The column 2b provides the best result among the all. Also, lesser material is utilized in this case. If the parametric study is done in future, then best fitted dimension ratios can be suggested for optimum results.

Effect of Pseudo-static Load

In this section, dynamic behavior of the stone column is discussed. The parametric study for pseudo-static load has been carried out by considering acceleration of 0.1, 0.15, 0.2, and 0.25 g. It is not unexpected to find that the load bearing capacity of the stone column decreases with the increase of pseudo-static load. Figure 13.4 shows the curve between load and settlement for five stone columns under the pseudo-static acceleration of 0.1, 0.15, 0.2, and 0.25 g. The effect of 0.1 g acceleration is negligible and that of 0.15 g is very little in all cases.

The effect of pseudo-static load has been compared on the basis of load bearing capacity. In case of conventional stone column (1), 3.5% decrement in load bearing

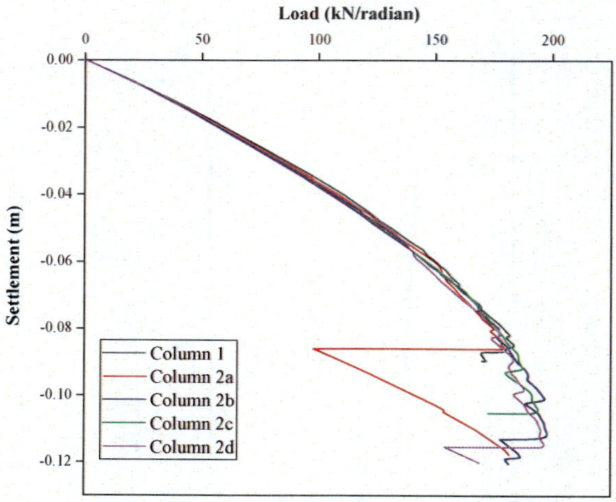

Fig. 13.3 Load versus settlement curve for different stone columns

capacity is observed for 0.2 g pseudo-static acceleration and 7.2% for 0.25 g pseudo-static acceleration, as shown in Fig. 13.4a. In case of stone column 2a, 4.8 and 7.4% decrement in load bearing capacity were observed for pseudo-static acceleration of 0.2 g and 0.25 g, respectively, as depicts in Fig. 13.4b. From Fig. 13.4c, it can be noticed that 2.9 and 5.3% decrement in load bearing capacity of stone column 2b has been found out for pseudo-static acceleration of 0.2 g and 0.25 g, respectively. In case of stone column 2c, 1.7% decrement in load bearing capacity has been observed for 0.2 g pseudo-static acceleration and 3.5% for 0.25 g pseudo-static acceleration, as shown in Fig. 13.4d. For stone column 2d, 4.0 and 6.5% decrement in load bearing capacity was found out pseudo-static acceleration of 0.2 g and 0.25 g, respectively, as obvious in Fig. 13.4e. Overall it was observed that with the increase in pseudo-static acceleration the Bearing capacity of stone column decreases.

Horizontal Displacement

All stone columns were distorted by bulging during the pseudo-static analyses. However, in case of modified stone columns higher bulging was found than the conventional stone column. Figure 13.5 shows the deformed meshing indicating punching failure of column 2c under the pseudo-static acceleration of 0.2 g. At the ground surface and the upper portion of the stone column, the bulges appeared. Because the stone tries to attain the uniformity and spread out and ultimately leads to bulge due to horizontal acceleration.

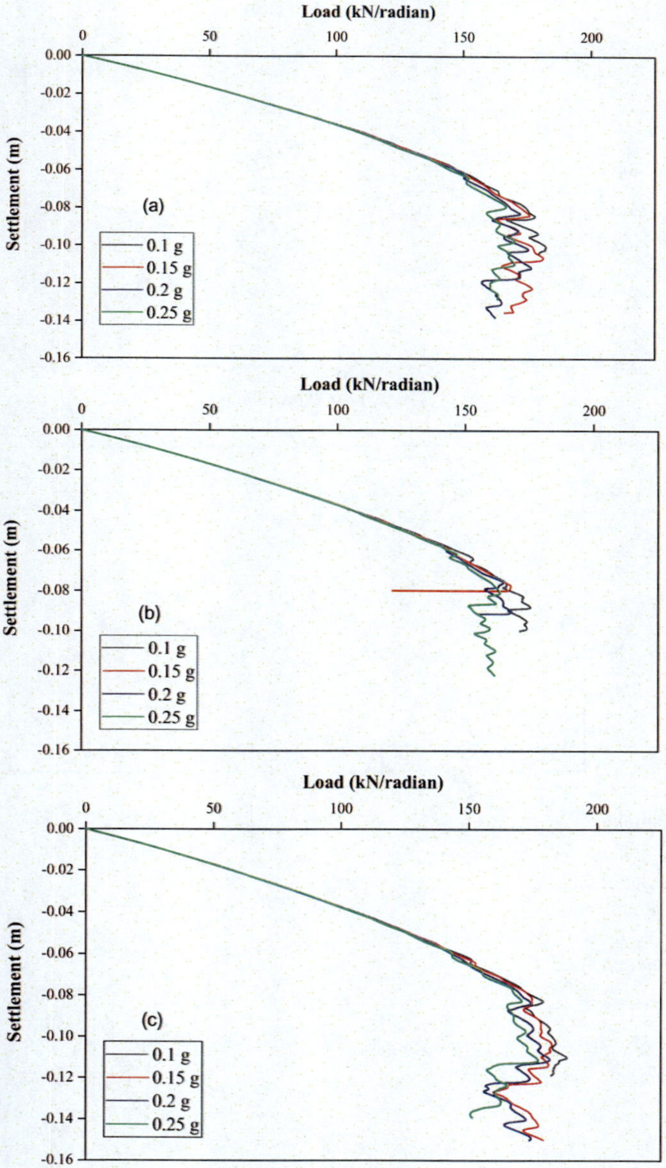

Fig. 13.4 Load bearing capacity under pseudo-static acceleration of 0.1, 0.15, 0.2, and 0.25 g for; **a** stone column 1; **b** stone column 2a; **c** stone column 2b; **d** stone column 2c; and **e** stone column 2d

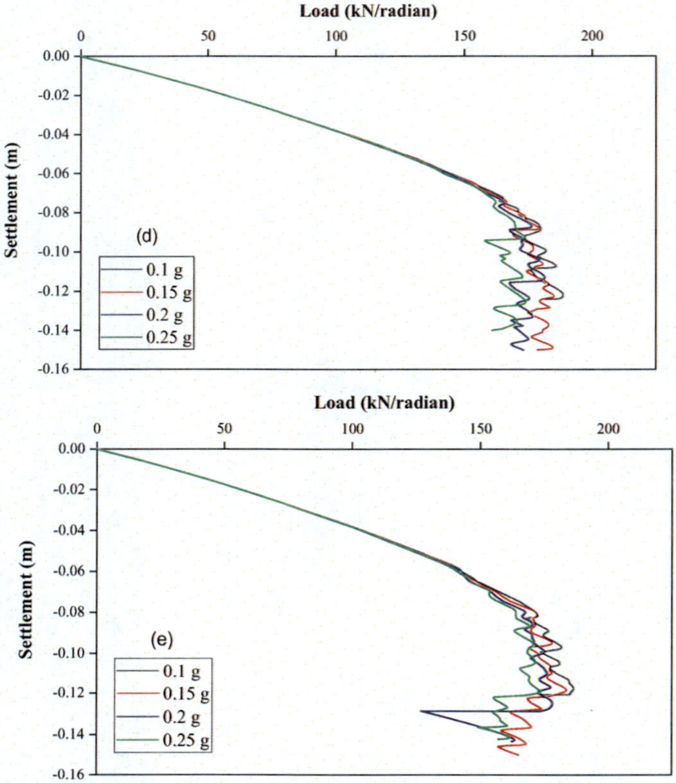

Fig. 13.4 (continued)

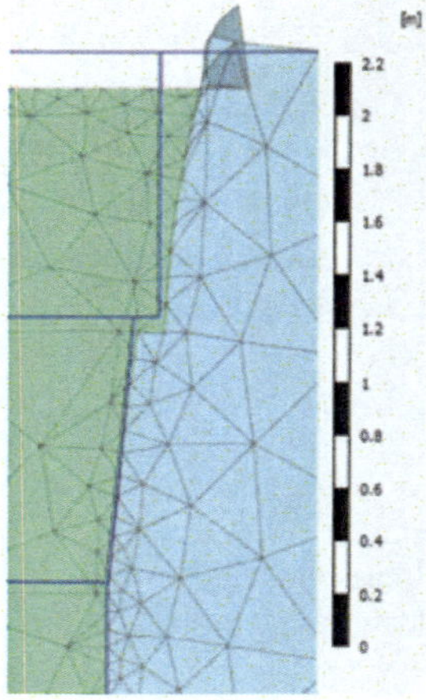

Fig. 13.5 Output of column 2c for 0.25 g pseudo-static acceleration, i.e., deformed mesh indicating punching failure

Table 13.3 Settlement of stone column after earthquake

Stone column type	Pseudo-static acceleration (g)	Current line load (kN/m)	Settlement (mm)
1	0.25	100	32.96
2a	0.25	100	36.30
2b	0.25	100	36.68
2c	0.25	100	40.20
2d	0.25	100	36.63

Post-earthquake Behavior

The effect of post-earthquake on the behavior of the stone column has been discussed in this section. In order to consider the extreme case, post-earthquake effect of 0.25 g pseudo-static acceleration is utilized. Table 13.3 presents the settlement of stone columns due to application of 100 kN/m of uniformly distributed load. It was observed that conventional stone column undergo least settlement, whereas column type 2c experience maximum settlement. The column 2b and column 2d shows more or less same behavior with moderate settlement; and column 2a shows slightly lesser deformation competitively. As per IS 15284: Part 1: 2003 [11] for a single column 10 to 12 mm settlement at design load is allowed.

Conclusion

The following are the conclusions reached after studying the influence of the different modified stone columns under static and pseudo-static load.

1. With little modification in the shape of stone column improved the performance of stone columns. Also, the modified shapes consume the lesser material than that of conventional columns and hence it economical as well.
2. Some bulging was observed in the non-conventional stone columns at the upper part of the column and the ground surface. However, it may be avoided by using appropriate dimensions or using some cementing agents at the upper part as remedies.
3. Not only in the static case but the load carrying capacity of modified stone columns has also enhanced under the pseudo-static acceleration.
4. The post pseudo-static behavior in terms of settlement for all the stone columns is almost similar. But modified reshaped stone column consume lesser material and hence it is more economical than the conventional column.

The results of the present study are dependent mainly on finite element simulations; more experimental research is needed to validate the findings.

References

1. Samanta M, Sawant VA, Ramasamy G (2010) Ground improvement using displacement type sand piles. In: Proceedings of IGC, December 16–18, IIT Bombay, pp 629–632
2. Rashid A, Safuan A, Black JA, Kueh AH, Mohamad H, Noor N (2017) Bearing capacity charts of soft soil reinforced by deep mixing. Proc Inst Civ Eng-Ground Improv 170(1):12–25
3. Ye G, Cai Y, Zhang Z (2016) Numerical study on load transfer effect of stiffened deep mixed column-supported embankment over soft soil. KSCE J Civ Eng 21(3):703–714
4. Sadaoui O, Bahar R (2017) Field measurements and back calculations of settlements of structures founded on improved soft soils by stone columns. Eur J Environ Civ Eng 23(1):85–111
5. Jamal MM, Tandel Y (2015) Group performance of geo encased sand column. Indian J Appl Res 5(1)
6. Mokhtari M, Kalantari B (2012) Soft soil stabilization using stone columns—a review. EJGE 17
7. Nanda P (2014) Load carrying capacity of encased stone column in compacted pond ash bed. M.Tech Thesis, National Institute of Technology Rourkela
8. Rajagopal K (2016) A primer on numerical and physical modelling in geotechnical engineering. Technical Indian Geotechnical Society, New Delhi
9. Barksdale RD, Bachus RC (1983) Design and construction of stone columns federal highway administration, RD-83/026
10. Ghazavi M, Afshar JN (2013) Bearing capacity of geosynthetic encased stone columns. Geotext Geomembr 26–36
11. IS 15284-Part 1 (2003) Design and construction for ground improvement—guidelines—stone columns. Bureau of Indian Standards, New Delhi

Chapter 14
Subgrade Strength Prediction Modeling On Fiber-Reinforced Expansive Soil Treated With Alkali Activated Binder

Mazhar Syed, Anasua GuhaRay, and Divyam Goel

Introduction

Expansive soft soil undergoes a high degree of settlement upon exposure to moisture, leading to weakening the strength bearing ratio causing early loss of pavement structures [1]. If these types of soils are not adequately addressed before the pavement layers are laid, major disasters may lead to millions of dollars in repairs and rehabilitation [2]. It is also seen that, after completion of pavement construction on expansive soils, several transportation authorities have addressed the appearance of linear shrinkage and desiccating cracks, which typically affect the asphalt or concrete layers' resistance to rutting and fatigue failures [3]. Usage of well-known lime and cement binder has been adopted to enhance strength bearing ratio and efficiently reduce the settlement as Portland cement's base exchange and cementing activity with clay are relevant to lime [4–6]. However, many studies have proved that the manufacture of these traditional binders has a major environmental issue and is energy-intensive. Many experts adopt chemical additives (full or partial replacement) to achieve the target strength in cement or lime soil. Fly ash, slag, silica fume, cement kiln dust, agro-waste, volcanic ash, industrial waste gypsum, and different fibers are among the additives [7–9].

Alkaline active binder (AAB) has grown in popularity as an alternative cement binder for weak sub-grade soil stabilization. AAB forms a lengthy polymeric sodium aluminosilicate molecule when dry pozzolanic precursors are activated. AAB lowers the cost of depositing fly ash-slag into landfills [10]. It also has superior mechanical strength, workability, and durability, as well as a low environmental impact (80% less CO_2 and 70% less potential for global warming than PC-based binder) [9]. Although the peak compression and the shearing tendency of alkaline sub-grade soil rose substantially, it still has a low flexural and tensile strength [11, 12]. Reinforcing

M. Syed · A. GuhaRay (✉) · D. Goel
BITS-Pilani Hyderabad Campus, Secunderabad, Telangana 500078, India
e-mail: p20170007@hyderabad.bits-pilani.ac.in

K. Muthukkumaran et al. (eds.), *Ground Improvement Techniques*, Lecture Notes in Civil Engineering 297, https://doi.org/10.1007/978-981-19-6727-6_14

the discrete fibers in the sub-grade layer can effectively deal with shrinkage and tension cracking. Although random discrete fiber inclusion does not prevent soil crack progression, it effectively reduces the frequency and depth of shrinkage cracks [13]. Moghal et al. [14] investigated the sub-grade strength behavior of lime mixed with fiber cast and fiber mesh soil mixtures. They found that 6% lime with 0.6% fiber cast samples had the maximum CBR penetration. With damage analysis prediction in the KENPAVE program, Lekha et al. [15] studied the UCS, CBR, durability, and fatigue behavior on low-volume pavement stabilized with 1% Arecanut coir fiber reinforced with 3% cement mixed lateritic soil.

This research focused on improving the durability of jute fibers by chemically treating them with 10 M KaOH. A combined mixture of fiber and slag in the alkaline mixed sub-grade soil can efficiently overcome the brittleness behavior. The study's primary goal is to determine the sub-grade strength characteristics (flexural and penetration) of fiber-reinforced soil that has been treated with lime and an alkaline mixture. In addition, the study provides regression equation models for estimating the penetration resistance of JF-alkaline-soil at various fiber and slag doses.

Material and Method

Soil

Expansive soil was obtained from the Warangal district of Telangana, India, for this investigation. The disturbed soil (excavated 10 cm below the ground level) was dark gray with 21.2% (approx.) in situ moisture content. The unified soil classification system classified the soil as highly clayey (CH) with 75% fine content less than 75 μ. JSW Cement Ltd. and NTPC, Ramagundam supplied blast furnace slag and class F fly ash (low calcium). As per ASTM C618, fly ash-slag and hydrated lime were acquired from Heico Ltd. Table 14.1 shows the properties of various raw materials.

Table 14.1 Engineering properties of soil and other raw materials

Material	Sg	pH	Is (%)	Ll (%)	γ (g/cc)	Ip (%)	Ls (%)	MC (%)	CBR (%)	UCS (kPa)
Soil	2.2	7.6	96	74		39	13.2	22.3	1.86	168
Slag	2.6	11.3	–	–		NP	–	–		–
FA	1.9	9.6	2.5	19		NP	–	–		–

Note CBR: California bearing ratio, FA: fly ash, Ip: plasticity index, Is: swelling index, Ll: liquid limit, Ls: linear shrinkage, MC: moisture content, pH: potential of hydrogen, P_S: swelling pressure, Sg: specific gravity, UCS: unconfined compressive strength

Table 14.2 Engineering properties of glass fiber

Properties	L (cm)	D (μm)	γ (g/cc)	C (%)	Hc (%)	Lg (%)	H_2O (%)	S_t (mPa)	E (mPa)
Jute fiber	2–3	35	0.94	65	10	12	10	490	2600

Note C: cellulose, D: diameter, E: Young's modulus, Hc: hemicellulose, coefficient of friction, H_2O: moisture content, L: length, Lg: lignin, St: tensile strength, γ: density of fiber

Jute Fiber (JF)

Jute fiber (Corchorus olitorius) of length 2–3 cm size having evergreen fiber industries, Kerala, supplied diameter 35 μm. Before inclusion in the soil as a reinforcement material, jute fibers were chemically treated with a potassium hydroxide solution for 3 days to increase the durability effectively. Fibers were soaked for at least three days to ensure a consistent reaction and eliminate waxes, lignin, and other alkali-soluble compounds around the fiber surfaces. AAB dosage was fixed at 7% (total soil mass) with varying slag content (0–20%) in the alkaline binder. Jute fiber was mixed at 0.2–1% (weight of soil). Table 14.2 showed the physico-mechanical characteristics of fiber provided by the manufacturer.

Alkaline Active Binder (AAB)

The sodium silicate (Na_2SiO_3), sodium hydroxide (NaOH), and dry precursors of fly ash-slag were mixed with water to make AAB. Hybrid Laboratories in Hyderabad provided sodium hydroxide and sodium silicate materials in the form of pellet and solution. Moreover, the mass ratio of Na_2SiO_3:NaOH: slag was maintained throughout the study with an optimum 0.4 water to solid ratio (w/s) in the AAB mixture [16]. To obtain the optimum binder mixture, the proportion of fly ash-slag was varied.

Sample Preparation

Oven-dried expansive soil was randomly mixed with different fiber dosages (0–1%) with 0.2% variation. Before fiber inclusion, untreated soil was mixed thoroughly with 7% hydrated lime and 7% AAB (total soil mass) with 0.4 w/s in the alkaline binder.

Different slag doses in alkaline mixed soils were physically compacted into three layers in different containers at a density of 90%. The compacted soil was covered with moist jute bags for 28 days to achieve a complete chemical reaction in the soil.

The fiber-reinforced soil specimens in AAB and lime were designated as samples $S.A_7(G_x)J_y$, $S.A_7(G_5)J_{0.2}$, and $S.L_7(G_0)J_{0.8}$ S = expansive sub-grade soil, A_7 = 7% alkaline binder in the soil, L_7 = 7% lime mixed in the soil, G = slag, x = slag percentage in the alkaline binder (0–20%), J = jute fiber, y = percentages of glass fiber (0–1%).

Results and Discussion

Fourier Transform Infrared Spectroscopy (FTIR)

The transmittance molecular spectra were examined by using K.BR. Pellet-based JASCO-4200 sets up in the 800–4000 cm^{-1} wavenumber. Figure 14.1 showed the FTIR spectrum curve of fiber mixed soil treated with hydrated lime and alkaline binder. The sharp IR spectrum peak at 1050 cm^{-1} attributed to asymmetric bending of the Si-Al–O group in untreated and lime-alkaline soil. The C–O–H carbon symmetric bonding vibration at around 1440 cm^{-1} was detected, primarily corresponding to the carbonation reaction [17]. Also, the C = O alkene and symmetric –CH2 lignin peaks intensity were reduced in lime-treated hemp fiber. The carbonyl broadband (C = O) at 1720 cm^{-1} was raised in fiber-alkaline soil relative to untreated soil. Furthermore, in alkaline soil, the H bonded free O–H stretching and C–H methyl groups were found at 3640 cm^{-1}, 3450 cm^{-1}, and 2930 cm^{-1}, respectively [8, 9]. Due to the quick breakdown of pozzolanic precursors in the silica-rich soil, these spectrums were only loosely described. As a result, the associated bonds were visible in the spectra peak with a molecular movement of roughly 20 cm^{-1} in both treated and untreated fiber soil.

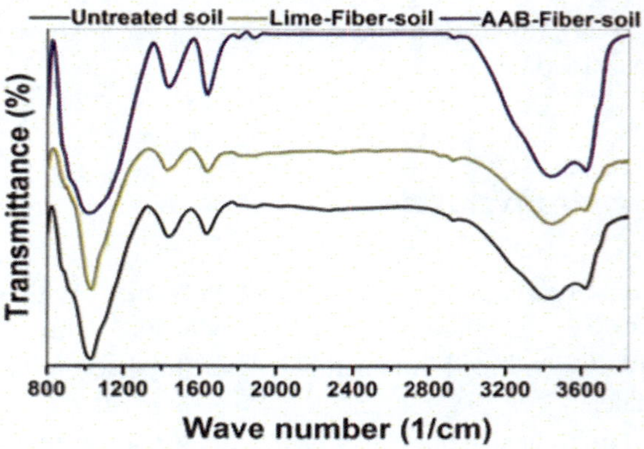

Fig. 14.1 FTIR profile of untreated, JF-reinforced lime, and alkaline mixed soil

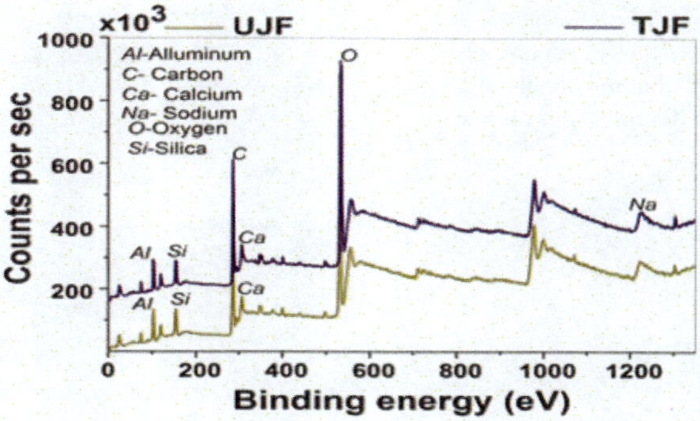

Fig. 14.2 XPS elemental profile of untreated and chemically treated jute fiber

X-Ray Photoelectron Spectroscopy (XPS)

The surface elemental composition in XPS peaks was performed using thermo scientific XPS setup with K–α rays ranging from 1400 to 100 eV. The 1800 hemispheric double focused analyzer with a 100 μm X-ray spot size is used throughout the study. Figure 14.2 displayed the elements of jute fiber before and after chemical treatment. As predicted, the primary constituents in UJF are Carbon (C), Oxygen (O), and Silica (Si). In the treated fiber, the peak level of Calcium (Ca), Oxygen (O), and Carbon (C) was increased [9]. TJF also revealed an additional peak at 1100 eV that corresponded to sodium (Na), which could be related to alkaline chemical breakdown in the fibers.

Modulus of Rupture (Flexural Strength: S_f)

The flexure strength of lime and alkaline fiber-reinforced soil was measured using an ASTM D-1635 three-point bending flexure machine. The soil beam was 28 cm long, 7 cm broad, and 7 cm thick. Both lime mixed and fiber-alkaline soil mixtures were compacted into five layers using a 3 kg steel rammer with a 31 cm free fall. Two bottom anchors supported the beam specimens with a gap of 10 cm with a strain rate of 0.1 mm/min throughout the study. The flexure strength (modulus of rupture) in MPa was calculated using the following equation.

$$S_f = \frac{3p_{\text{ult}} * l}{2\text{bh}^2} \tag{14.1}$$

Figure 14.3 depicted the modulus of rupture (flexural strength) variance as a fiber and slag dose function. The curves revealed that increasing the slag volume in the AAB soil combination enhanced modulus and flexural crack resistance [18]. The flexural strength growth percentages increased with increasing fiber dosages in lime and fly ash-slag mixture, regardless of fiber-alkaline proportions. Comparable observations on flexural strength enhancement by strengthening fibers in soils

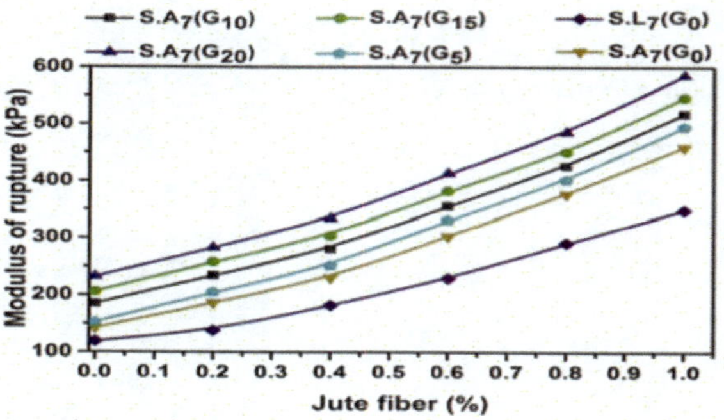

Fig. 14.3 Variations of modulus of rupture of JF-lime-alkaline soil at different slag proportions

treated with lime-cement and agro-alkaline binder were reported by Anggraini et al. [11] and Pourakbar et al. [19]. Furthermore, with a larger slag percentage in the TJF-AAB reinforcement combination, the compressive shear and flexural strength enhancement were more connected with pozzolanic-based geopolymerized soils. Interestingly, the slag dependent (G_{20}) was 26% higher than the pure fly ash-based (G_0) in the JF-alkaline-soil samples. Substitution fly ash with slag increased the rate of geopolymerization and actively generated amorphous silicate hydrates from the existing alkaline crystalline ions in the silica-rich alkaline soil. The overall percentage gain in flexural strength in fibers (beyond 0.6%)-reinforced soil was 62% in AAB and 34% in mixed lime soil compared to the unreinforced soil mixture. Combining slag with the optimum fiber dosage increased the soil-fiber interaction, resulting in a high interconnecting density. Thus, increasing fiber doses in alkaline soil mixtures with slag concentration improved the stiffness, shear, and flexural strength by restricting the proportional mobility of soil-fiber.

California Bearing Ratio (CBR)

Soaked CBR testing on expansive raw soil and fiber-reinforced soil was carried out according to the ASTM D-1883 standard. Different fiber dosages in lime and alkaline binder mixed soil samples were deformed in a conventional CBR mold with a mean diameter 15 cm and a height of 17.5 cm at corresponding MDD-OMC values. Compacted soil-fiber samples were soaked in freshwater for 96 h with a 5 kg overcharge before being tested utilizing a loading frame and a 50-mm diameter plunger at a predetermined strain rate of 1.25 mm/min. Figure 14.4 showed soaked CBR penetration tests for fiber-reinforced lime and alkaline soil at varying slag doses. The penetration resistance was shown to rise as the slag and fiber doses were increased. The inclusion of lime and a 100% fly ash-based (zero slags) alkaline combination (G_0) to JF-reinforced soil enhanced the CBR values from 1.96 to 4.76% and 7.08%, respectively, as shown in the graph. The pozzolanic reactions that catalyzed the flocculation and geopolymerization processes during the soaking phase might be

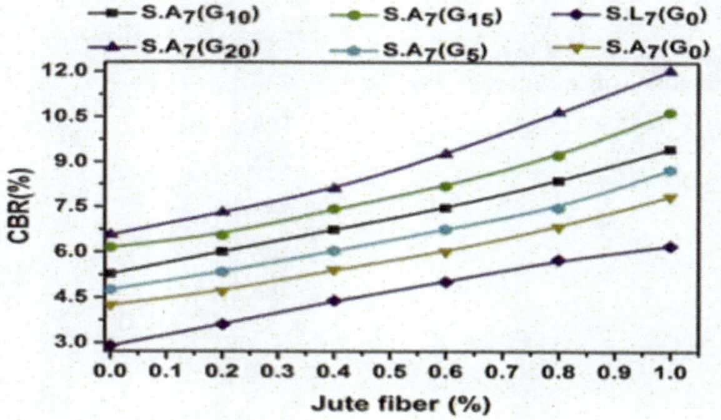

Fig. 14.4 Variations of CBR of JF-reinforced lime and alkaline soil at different slag proportions

attributed to the significant enhancement in the sub-grade strength value. The addition of JF and slag to AAB-treated soil was beneficial as it had effectively increased confinement bonding and stretching resistance during load penetration [9, 20]. As a result, clay particles around JF found it more difficult to move position from one step to the next, increasing penetration resistance. JF could also efficiently disperse in the soil specimen and withstand the pulling tension due to a greater linking effect [21]. It was also discovered that between 0.2 and 0.6% fiber doses, the influence of CBR on lime and AAB soils rose significantly, regardless of slag proportions. Hence, the rapid changes in soil-fiber intercoupling intensity and clay interface structure could improve CBR.

Tensile Strength (Indirect) (ITS)

The tensile test was performed in a Marshal stability machine with a 12.5 mm loading strip attached to the load frame under a 50.5 mm/min strain rate. Both lime and alkaline soils were compacted in a 10 cm diameter and 8 cm height cylindrical mold. Figure 14.5 showed the variation of ITS of both lime and AAB soil-reinforced different JF doses. Only a couple of ITS data (G0: zero slags, 100% fly ash, and G20: 20% slag and 80% fly ash in the AAB) were displayed in Fig. 14.5 for comparison to prevent data clustering. The combination addition of slag-fly ash-fibers has been shown to influence the tensile strength of sub-grade soil substantially. ITS increase was proportional to flexural and penetration resistance values (Figs. 14.3 and 14.4). The JF-lime-soil mixture had the lowest ITS value compared to a sodium aluminosilicate compound (greater slag dosage) based on the JF-alkaline soil mixture [12, 19]. Thus, the combined blending of slag and fiber mixture in the alkaline soil successfully controlled the interfacial bonding mechanism (soil particle grasping), which absorbed considerable deforming energy.

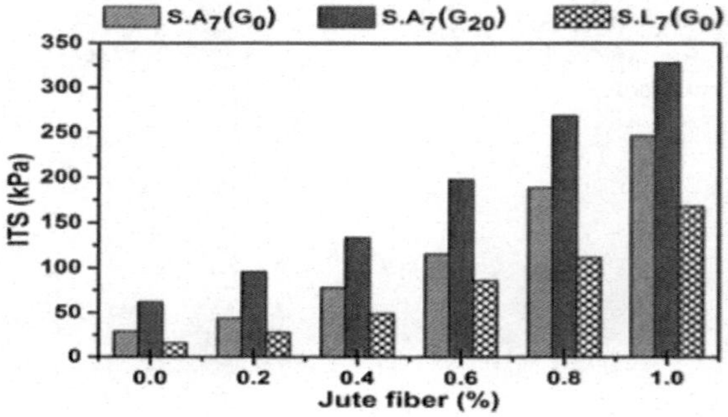

Fig. 14.5 Variations of ITS of JF-reinforced lime and alkaline soil at different slag proportions

Regression Analysis

A regression model is one of the most often used traditional approaches for relating variables and parameters in statistical conditions [9, 14]. The current investigation described the effect of fiber and slag doses in the alkaline soil samples on penetration and modulus of rupture (flexure) using non-linear equations. The input parameters for predicting CBR and S_f were chosen to be slag and jute fiber doses, respectively. For regression analysis, the statistical program included in the commercially accessible software Python was employed. The regression equation for a response function (y) and variables (D_{Fiber}, D_{Slag}) impacting is expressed as follows:

$$y = a \times (D_{Fiber}) + b \times \left(D_{Slag}\right) + K_0 \tag{14.2}$$

where a and b = regression coefficient; K_0 = constant; D_{Fiber} = jute fiber dosage; D_{Slag} = dosage of slag in the alkaline binder.

$$\text{CBR} = 4.2080 \times (D_{\text{fiber}}) + 0.1470 \times \left(D_{\text{ggbs}}\right) + 3.8572 \quad \text{with } R^2 = 0.906 \tag{14.3}$$

$$S_f = 3.3759 \times (D_{\text{fiber}}) + 0.0499 \times \left(D_{\text{ggbs}}\right) + 1.1847 \quad \text{with } R^2 = 0.927 \tag{14.4}$$

Figure 14.6 showed the graphical representations among tested and predicted flexural and CBR data at different fiber and slag percentages in AAB soil samples. Figure 14.6a indicated the flexural strength tests for both tested and model-predicted results of JF-alkaline soil in the shape of a scatter chart and covering mesh. Figure 14.6b displayed the JF-AAB-reinforced soil for experimental and forecasted CBR values at varying amounts of slag in AAB in the form of straight bar charts and surface mesh.

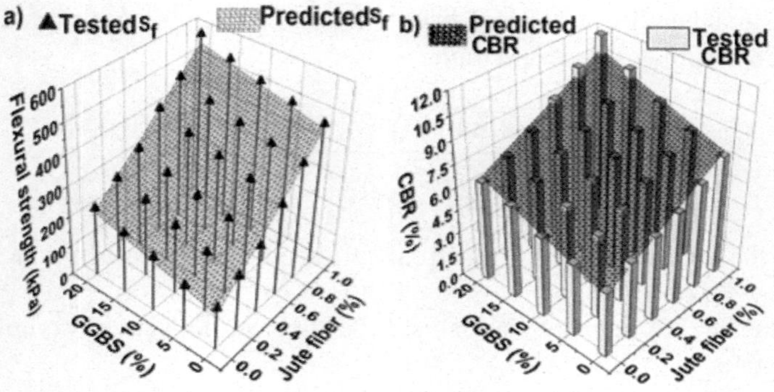

Fig. 14.6 Comparison of tested and predicted results **a** flexural strength, **b** CBR of JF-reinforced alkaline soil at different slag content

Case Study

A typical section of flexible pavement structure using fiber-alkaline treated soil was provided in Fig. 14.7.

An example of flexible pavement design (as per IRC-37-2017 guidelines) was demonstrated using the CBR regression equations as follows.

$$\text{CBR} = 4.2080 \times (D_{\text{fiber}}) + 0.1470 \times \left(D_{\text{ggbs}}\right) + 3.8572 \tag{14.5}$$

Here, $D_{\text{fiber}} = 0.4$ jute fiber, $D_{\text{Slag}} = 5$.

Substituting these values in the above equation, we got CBR = 6.24%.

Aim: To design the thickness of flexible pavement for a dual-lane single carriageway for 10 years.

Fig. 14.7 Typical section of flexible pavement using fiber-alkaline treated soil

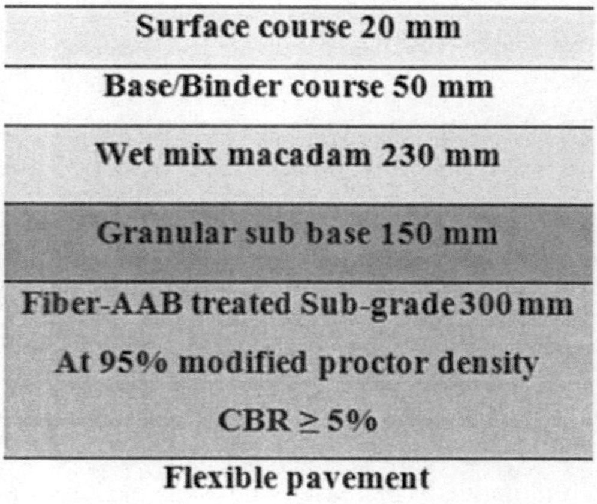

Sol. The total number of standard axles to be carried over the road's design period should be approximate the design traffic.

$$N_{\text{Des}} = \frac{365 \times \left[(1+r)^n - 1\right]}{r} \times A \times D \times F \tag{14.6}$$

N_{Des} = the total number of standard axles to be accommodated throughout the design period of "n" years.
A = initial traffic (commercial vehicles per day) in the year of completion of construction.

Assuming the expected traffic of 450 commercial vehicles per day (CVPD).

D = lateral distribution factor (LDF).
LDF for a two-lane single carriageway = 0.75 (75% of the number of commercial vehicles in each face).
F = vehicle damage factor (VDF) = 3.2 (Based on initial traffic volume 150–1500 CVPD, for plain terrain, IRC-37 has given 3.9) (3.2 < 3.9 it is acceptable limit).
n = design period, in years (n = 10 years).
r = decimal equivalent of the yearly growth rate of commercial vehicles.
Annual growth rate of traffic 5% (r = 0.05).

$$N_{\text{Des}} = \frac{365 \times \left[(1+0.05)^{10} - 1\right]}{0.06} \times 450 \times 0.75 \times 3.2$$

N_{Des} = 4.958205 msa, say 5 msa.

The total thickness required for sub-grade CBR = 6%, and traffic ≈ 5 msa is 450 mm.

Conclusions

The effects of chemically treated jute fiber in an alkaline active binder and lime mixed sub-grade soil were explored in this study. This investigation leads to the following conclusions:

- Microchemical analysis results confirmed the activation of new molecular bonding in the lime and alkaline JF-soil. Moreover, the surface elements of chemically treated fibers showed new elemental compounds compared to untreated fibers in XPS analysis.
- The rate of flexural and penetrating strength gain was strongly affected by the composition of pozzolanic precursors in active alkaline binder and JF; these values dramatically increased with slag and fiber content

- Fiber-AAB soil efficiently regulated the tensile and flexural cracking by transferring the plastic failure toward the elastic behavior through active geopolymerization and interfacial frictional bonding formation around the fiber-clay matrix.
- An empirical model was validated for flexural and CBR prediction for JF-reinforced alkaline soil, which could be used to estimate the sub-grade performance indicator by considering the effects of slag and fiber doses.

References

1. Ackroyd LW, Husain R (1986) Residual and lacustrine black cotton soils of north-east Nigeria. Géotechnique 36:113–118. https://doi.org/10.1680/geot.1986.36.1.113
2. Saride S, Puppala AJ, Chikyala SR (2013) Swell-shrink and strength behaviors of lime and cement stabilized expansive organic clays. Appl Clay Sci 85:39–45. https://doi.org/10.1016/j.clay.2013.09.008
3. Puppala A, Hoyos L, Viyanant C, Musenda C (2008) Fiber and fly ash stabilization methods to treat soft expansive soils, pp 136–145. https://doi.org/10.1061/40552(301)11
4. Bell FG (1996) Lime stabilization of clay minerals and soils. Eng Geol 42:223–237. https://doi.org/10.1016/0013-7952(96)00028-2
5. Al-Rawas AA, Hago AW, Al-Sarmi H (2005) Effect of lime, cement and Sarooj (artificial pozzolan) on the swelling potential of an expansive soil from Oman. Build Environ 40:681–687. https://doi.org/10.1016/j.buildenv.2004.08.028
6. Petry TM, Little DN (2002) Review of stabilization of clays and expansive soils in pavements and lightly loaded structures—history, practice, and future. J Mater Civ Eng 14:447–460. https://doi.org/10.1061/(asce)0899-1561(2002)14:6(447)
7. Jalal FE, Xu Y, Jamhiri B et al (2020) On the recent trends in expansive soil stabilization using calcium-based stabilizer materials (CSMs): a comprehensive review. Adv Mater Sci Eng 2020. https://doi.org/10.1155/2020/1510969
8. Miao S, Wei C, Huang X et al (2017) Stabilization of highly expansive black cotton soils by means of geopolymerization. J Mater Civ Eng 29:04017170. https://doi.org/10.1061/(asce)mt.1943-5533.0002023
9. Syed M, GuhaRay A, Goel D (2021) Strength characterisation of fiber reinforced expansive subgrade soil stabilized with alkali activated binder. Road Mater Pavement Des 20:1–25. https://doi.org/10.1080/14680629.2020.1869062
10. Davidovits J (1994) Properties of geopolymer cements. Alkaline Cem Concr Kiev Ukr 1–19. https://doi.org/10.1073/pnas.0811322106
11. Anggraini V, Asadi A, Farzadnia N et al (2016) Reinforcement benefits of nanomodified coir fiber in lime-treated marine clay. J Mater Civ Eng 28:1–8. https://doi.org/10.1061/(ASCE)MT.1943-5533.0001516
12. Tang C, Shi B, Gao W et al (2007) Strength and mechanical behavior of short polypropylene fiber reinforced and cement stabilized clayey soil. Geotext Geomembr 25:194–202. https://doi.org/10.1016/j.geotexmem.2006.11.002
13. Syed M, Guharay A (2020) Effect of natural fiber reinforcement on strength response of alkali activated binder treated expansive soil: experimental investigation and reliability analysis. Constr Build Mater 121743. https://doi.org/10.1016/j.conbuildmat.2020.121743
14. Moghal AAB, Chittoori BCS, Basha BM (2018) Effect of fibre reinforcement on CBR behaviour of lime-blended expansive soils: reliability approach. Road Mater Pavement Des 19:690–709. https://doi.org/10.1080/14680629.2016.1272479

15. Lekha BM, Goutham S, Shankar AUR (2015) Evaluation of lateritic soil stabilized with Arecanut coir for low volume pavements. Transp Geotech 2:20–29. https://doi.org/10.1016/j.trgeo.2014.09.001
16. Syed M, Guharay A, Agarwal Sagar KA (2019) Stabilization of expansive clays by combined effects of geopolymerization and fiber reinforcement. J Inst Eng Ser A 10:1–20. https://doi.org/10.1007/s40030-019-00418-3
17. Syed M, GuhaRay A, Goel D et al (2020) Effect of freeze–thaw cycles on black cotton soil reinforced with coir and hemp fibres in alkali-activated binder. Int J Geosynth Gr Eng 6. https://doi.org/10.1007/s40891-020-00200-7
18. Cristelo N, Cunha VMCF, Topa Gomes A et al (2017) Influence of fibre reinforcement on the post-cracking behaviour of a cement-stabilised sandy-clay subjected to indirect tensile stress. Constr Build Mater 138:163–173. https://doi.org/10.1016/j.conbuildmat.2017.02.010
19. Pourakbar S, Asadi A, Huat BBK, Cristelo N, Fasihnikoutalab MH (2016) Application of alkali-activated agro-waste reinforced with wollastonite fibers in soil stabilization. J Mater Civ Eng 29:04016206. https://doi.org/10.1061/(asce)mt.1943-5533.0001735
20. Sudhakaran SP, Sharma AK, Kolathayar S (2018) Soil stabilization using bottom ash and areca fiber: experimental investigations and reliability analysis. J Mater Civ Eng 30:1–10. https://doi.org/10.1061/(ASCE)MT.1943-5533.0002326
21. Mazhoud B, Collet F, Pretot S, Lanos C (2017) Mechanical properties of hemp-clay and hemp stabilized clay composites. Constr Build Mater 155:1126–1137. https://doi.org/10.1016/j.conbuildmat.2017.08.121

Chapter 15
Behavior of Jute Fiber-Reinforced Sand Using Direct Shear Test for Ground Engineering Application

Shashank Singh and Shiv Shankar Kumar

Introduction

Due to the unavailability of land, for many upcoming civil engineering projects, engineers are utilizing the land which might not having sufficient bearing capacity to sustain the superstructure load. Therefore, engineers utilizing the reinforcing techniques which are commonly utilized to improve the bearing capacity of weak or soft soil by improving their engineering properties. Several geosynthetics reinforcing materials such as geotextile, geogrid, geonet, geocell, polypropylene fiber, and natural fiber like coir fiber and jute fiber are being utilized to improve the engineering properties of soil in significant manner. The significance of the use of jute fiber, as well as its behavior, as a soil-reinforcement material over the use of geotextile and other soil-reinforcement materials has been explored by several researchers utilizing different testing techniques [1–7]. Though several literatures are available, the study on the behavior of jute fibers and sand mixes is scanty. Therefore, further study is needed for the proper explanation and justification while using jute fiber with soil. This study presents the utilization of natural jute fibers as a reinforcing material (since jute has a property to resist the tensile stresses which arises within the soil reinforcement during loading) enhances the engineering properties of the soil because of its economical, bio-degradable, and environment friendly conditions. The main aim of this paper is to investigate the behavior of jute fiber-blended sand under different testing conditions such as normal stress, relative density, water content, fiber content, and length of fiber.

S. Singh (✉) · S. S. Kumar
Department of Civil Engineering, National Institute of Technology, Patna 800005, India
e-mail: shashanks.pg19.ce@nitp.ac.in

S. S. Kumar
e-mail: k.shiv.ce@nitp.ac.in

K. Muthukkumaran et al. (eds.), *Ground Improvement Techniques*, Lecture Notes in Civil Engineering 297, https://doi.org/10.1007/978-981-19-6727-6_15

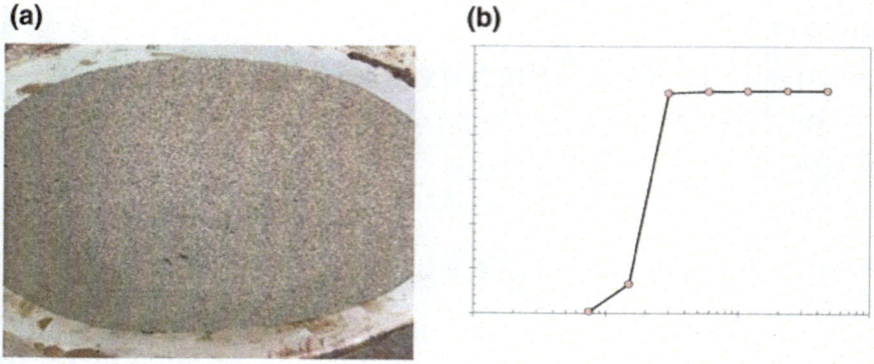

Fig. 15.1 **a** Photograph of Ganga sand and **b** grain size distribution

Materials, Methodology, and Investigating Parameters

Sand

The locally available sand was collected from the bed of Ganga River, near NIT Patna. Since sand is mostly utilizing as a construction material, the characterization based on their properties either physical or engineering is utmost important. Figure 15.1a presents the photo of sand reflecting small particle size and gray in color. The graphical representation of sieve analysis called as the particle size distribution presented in Fig. 15.1b, shows the percentage finer against the particle size on a logarithmic scale. Based on the grain size analyses [8], it can be seen that the soil exhibits fine particles significantly, as shown in Fig. 15.1b. Based on the USGS soil classification system [9], the soil can be characterized as a fine sand and further classified as poorly graded sand (SP). The value of coefficient of uniformity 'C_u' = 1.23 and coefficient of curvature 'C_c' = 0.8 was evaluated based on the measurement of $D_{10} = 0.13$, $D_{30} = 0.17$, and $D_{60} = 0.23$. The specific gravity of sand was found to be 2.62 [10]. The minimum and maximum dry unit weight of Ganga Sand was found to be 13.85 kN/m^3 and 16.84 kN/m^3, respectively [11, 12].

Jute Fiber

The jute fiber used in the present study was in the length of 10, 20, and 40 mm, and the color of the fiber was observed to be brown as shown in Fig. 15.2. Jute was also used in different percentage by weight of the dry sand such as 0.25, 0.5, 1, and 1.5%. The tensile strength jute fiber was found to be in the range of 98–102 MPa. Arunavathi et al. [13] have reported that the tensile strength jute fibers may vary in the range of 71.7 ± 9.5 MPa to 104.9 ± 8.8 MPa. However, Eldesouky et al. [14]

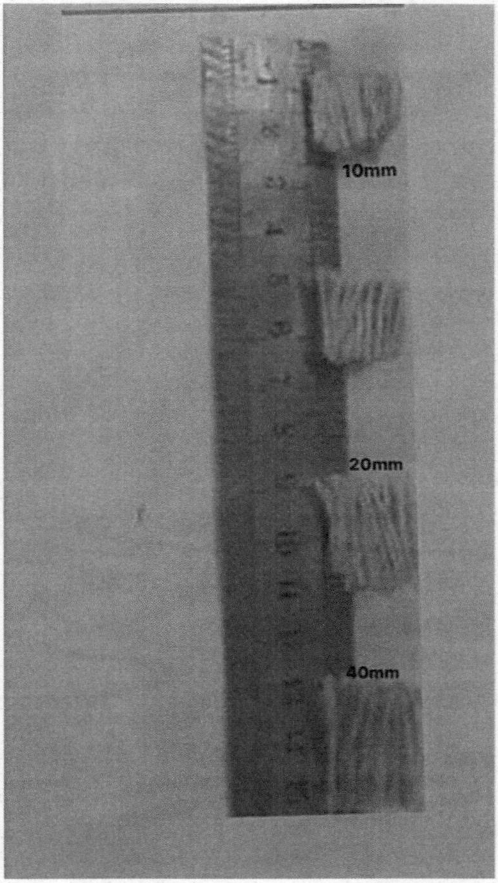

Fig. 15.2 Photograph of jute fibers

have reported that the tensile strength of fiber 350 MPa at 165 °C. The density of jute may vary in the range of 1.42–1.47 g/cc [15].

Jute Fiber-Sand Mixture

Commonly, reinforcing material mixes with the soils at the construction site to improve the bearing capacity of soil. For the present study, different percentage contents of jute fiber were mixed randomly to the sand soil, as shown in Fig. 15.3a, on which further compaction tests were done [16], as shown in Fig. 15.3b. It was observed that MCC was found to be 1.67 g/cc corresponding to OMC = 11.4% at 0% fiber content. Further, it was noticed that MDD decreased with the addition of fiber contents from 0.25 to 1%; however, OMC marginally increased. This may be

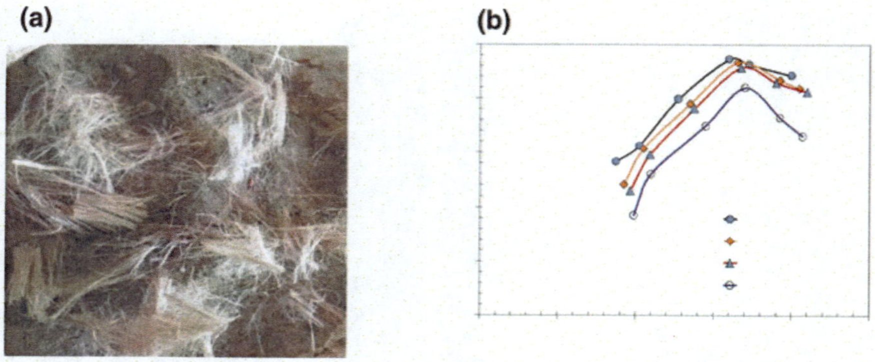

Fig. 15.3 **a** Randomly mixed jute-sand, **b** compaction curve of jute-sand mix

due to a fact that some portion of soil is replaced by fiber because of its less specific gravity, and hence, the MDD of jute-sand mixes decreased.

Testing Methodology

Direct shear testing methodology [17] is one of the geotechnical engineering laboratory experimentation to determine the shear strength parameters such as cohesion © and angle of internal friction (ϕ) of cohesionless soils or jute-sand-reinforced soil, which can be further utilized in the design and construction. During shearing, the soil as well as soil-reinforced specimens subjected to a normal load and sheared across the pre-determined failure plane. A displacement rate of 1.25 mm/min was applied to shear the specimen, and the displacement and the corresponding load were recorded which was further used to find out the shear stress and horizontal strain at different investing parameters such as relative density, fiber content, length of fiber, water content, and normal stress.

Investigating Parameters

The sand soil specimens for the direct shear tests were prepared at different relative densities 30 $\pm$ 2%, 50 $\pm$ 2% and 80 $\pm$ 2% reflecting loose, medium, and dense conditions, as shown in Table 15.1. However, the specimens for jute-sand mixes were prepared at different OMC and MDD conditions. The stress–strain response of sand as well as jute-sand mixes were observed at three different normal stresses 50, 100, and 150 kPa. In this study, the jute fiber length of 10, 20, and 40 mm whereas, the range of fiber contents 0–1.5% by weight were considered. Further, the sand as well as jute-reinforced sand specimens, prepared at OMC and MDD, were

Table 15.1 Investing parameters for the present study

Parameters	Values
Water content (by weight) (%)	0, 20, 60, 100
Fiber contents (f_c) (%)	0, 0.25, 0.50, 1.0, 1.5
Fiber length (l_f) (mm)	10, 20 and 40
Normal stress (σ_n) (kPa)	50, 100 and 150
Relative density (D_r, %)	30 (±2), 50 (±2), 80 (±2)
Density (ρ) (g/cc)	1.4, 1.5 and 1.6

sheared at different water content ranging from 0 to 100%. The displacement rate of 1.25 mm/min was used to perform the test at the different investigating parameters. Table 15.1 lists the investigating parameters for the present study.

Results and Discussion

The results obtained from the direct shear tests at different investigating parameters, mentioned in Table 15.1, are presented. Further, the results have been discussed in terms of stress–strain response and shear strength parameters of sand as well as the jute-sand mixes.

Effect of Normal Stress on Dry Sand

The stress–strain response and the failure envelope based on peak shear stress of dry sand at different σ_n (50, 100, and 150 kPa) and D_r (30, 50, and 80%) were analyzed. σ_n applied on the soil specimens reflects the stresses in the field at approximate depth ranging from 3 to 10 m. Figure 15.4a and b presents the stress–strain response and failure envelope based on peak shear stress of dry sand for $D_r = 50\%$. Figure 15.4a reflects that the resistance to achieve the failure of a soil specimen increases with increasing σ_n. This is attributed to the fact that the confinement within the soil mass increases with increase in normal stress. Figure 15.4b shows that $C = 0.39$ kg/cm^2 whereas ϕ was observed to be 32.84° at $D_r = 50\%$. Similar responses were observed at $D_r = 30$ and 80%, when shared at $\sigma_n = 50$, 100, and 150 kPa. C and ϕ were observed to be 0.382 kg/cm^2 and 31.7°, respectively, at $D_r = 30\%$. Further, Table 15.2 presents the shear strength parameters C and ϕ for $D_r = 30$, 50, and 80%. Figure 15.5a presents the failure envelope obtained at different D_r considering σ_n = 50 kPa, 100 kPa, and 150 kPa whereas, Fig. 15.5b depicts the variations in ϕ at different D_r for $\sigma_n = 100$ kPa. It can be seen from Fig. 15.5b that ϕ increased by nearly 9% with the increase of D_r from 30 to 80%, this may be due to the increase in compactness of soil with increasing D_r.

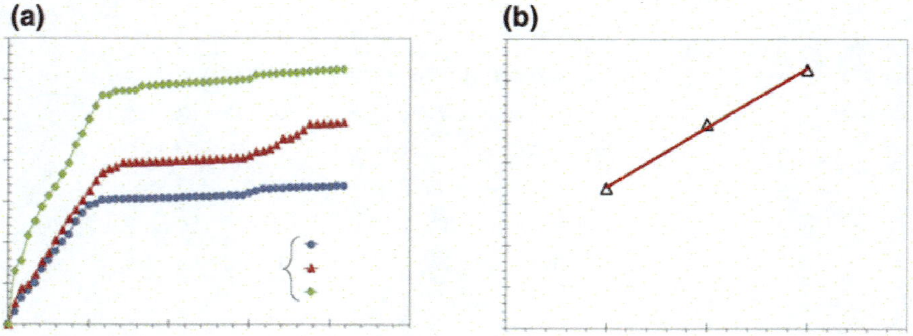

Fig. 15.4 **a** Stress–strain response, **b** failure envelope based on peak shear stress of dry sand at different σ_n and $D_r = 50\%$

Table 15.2 Shear strength parameters C and ϕ for different D_r

D_r (%)	C (kg/cm^2)	ϕ (°)
30	0.382	31.7
50	0.39	32.84
80	0.357	34.44

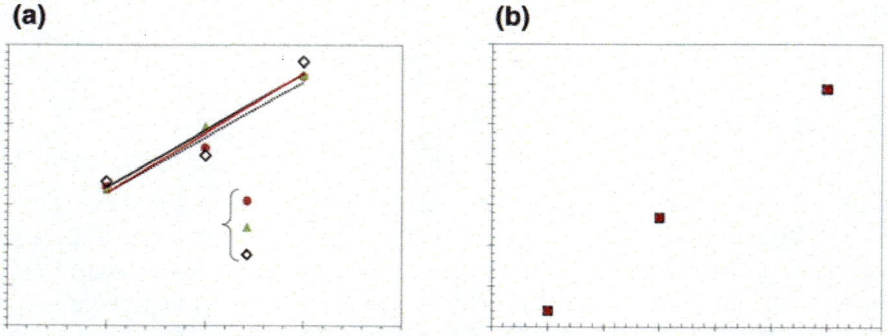

Fig. 15.5 Variation of angle of internal friction at varying **a** different D_r and σ_n **b** at different D_r for $\sigma_n = 100$ kPa

Effect of Relative Density on Dry Sand

This section deals with the response of dry sand in terms of stress–strain response and failure envelope at different D_r. D_r reflects the degree of compactness of soil. Figure 15.6a and b presents the stress–strain response and failure envelope at different D_r and $\sigma_n = 50$, 100, and 150 kPa. Figure 15.6a reflects that the effect of D_r on the stress–strain responses is minimal at $\sigma_n = 100$ kPa. Figure 15.6b presenting the variations in peak shear stress corresponding to different D_r at $\sigma_n = 100$ kPa, which shows that though increasing D_r depicts increasing degree of compactness, the effect

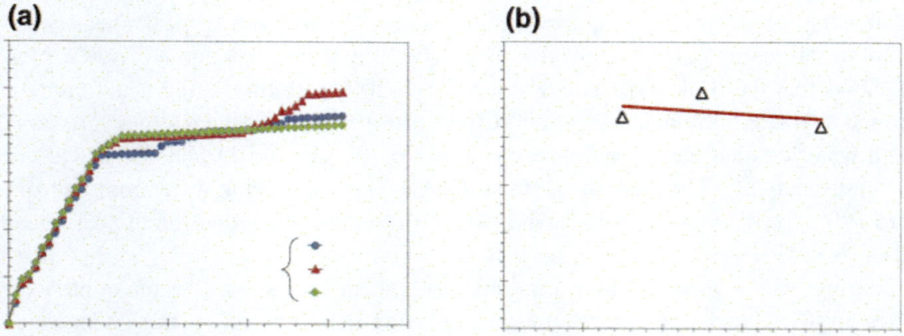

Fig. 15.6 **a** Stress–strain response, **b** failure envelope based on peak shear stress of dry sand at different D_r and $\sigma_n = 100$ kPa

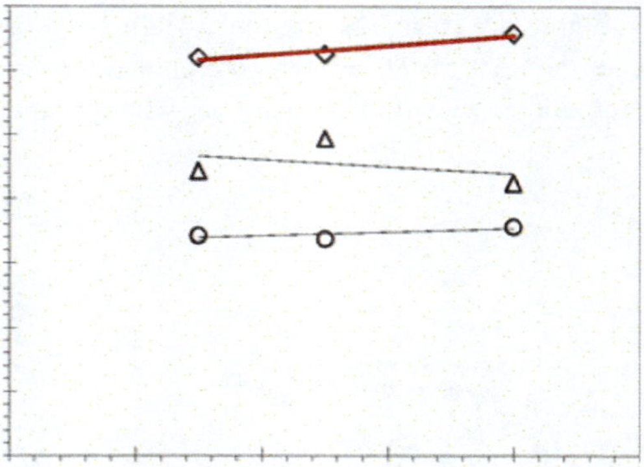

Fig. 15.7 Failure envelope based on peak shear stress of dry sand at different D_r and σ_n

of compactness is very minimal (~2–3%) from $D_r = 30\%$ to 80%. Similar responses were observed at different D_r for $\sigma_n = 50$ kPa and 150 kPa. From Fig. 15.7, it is also noticed that the effect of D_r at higher $\sigma_n = 150$ kPa is relatively higher (i.e., ~6% from $D_r = 30\%$–80%) in comparison to the effect of D_r at $\sigma_n = 50$ kPa (which is ~3% from $D_r = 30$–80%,) or at $\sigma_n = 100$ kPa (which is ~2% from $D_r = 30$–80%).

Effect of Water Content on Shearing Response of Sand

The amount of water presents in the soil mass may lead to the reduction of strength as well as the bearing capacity of soil. Therefore, the behavior of soil with the presence of different level of water content is to be estimated prior to the design and construction of any soil-supported structure. Hence, the behavior of sand with varying range of water content, reflecting different level of degree of saturation, has been estimated at different D_r and σ_n. Figure 15.8a presents the stress–strain response at $D_r = 50\%$ and $\sigma_n = 100$ kPa with varying range of water content, i.e., from 0 to 100% by weight. It can be observed that the stress–strain response significantly reduced with increasing water contents. Figure 15.8b presents the variations of peak shear stress with variations in water content from 0 to 100% at $D_r = 50\%$ and $\sigma_n = 100$ kPa. It shows that the peak shear stress reduced by ~20% with increasing water content from 0 to 100%. Similar stress–strain response as well as peak shear stress variations with water content have been observed at $\sigma_n = 100$ kPa, for $D_r = 30\%$ and $D_r = 80\%$, respectively. The peak shear stress decreased, with varying water content from 0 to 100%, by 40% at $D_r = 30\%$ whereas, at $D_r = 80\%$, the reduction in peak shear stress is ~5%.

Figure 15.9a presents the variations peak shear stress with water content at different D_r whereas, Fig. 15.9b presents the variations peak shear stress with water content at different σ_n for $D_r = 50\%$. It can be observed from Fig. 15.9a that the peak shear stress is significantly affected by D_r. At, $D_r = 80\%$, the peak shear stress is higher in comparison to the stress obtained at $D_r = 30\%$ and 50%; however, the peak shear stress reduces with increasing water content at all D_r.

Figure 15.10 represents the variations in peak shear stress with σ_n at different water content for $D_r = 50\%$. It can be observed from Fig. 15.10 that the peak shear stress increases with increasing σ_n whereas, the same reflect reduction in peak stress

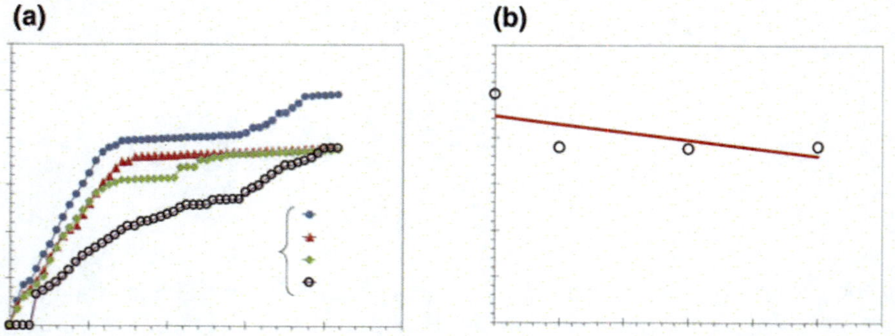

Fig. 15.8 **a** Stress–strain response, **b** peak shear stress with different water content at $D_r = 50\%$ and $\sigma_n = 100$ kPa

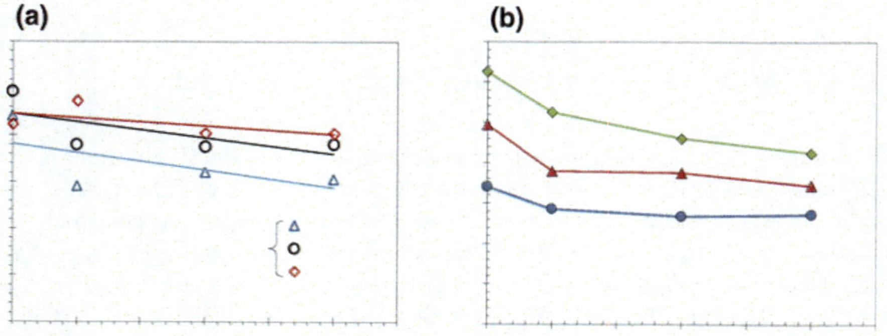

Fig. 15.9 Variations in peak stress with different water content at **a** different D_r, **b** different σ_n

with increasing water content for any one particular value of σ_n. It can also be seen that ϕ reduces with increasing water content. Similar responses have been observed for D_r = 30 and 80%. Figure 15.11a and b presents the stress–strain response at different σ_n and D_r = 50% for water content (w_c) 20% and 100%, respectively. It can be observed from Fig. 15.11a that, at w_c = 20%, with the increase in σ_n, the stress–strain response increases. The stress–strain response at σ_n = 150 kPa is higher than that of 50 kPa and 100 kPa. Similar response can be observed in Fig. 15.11b at w_c = 100%. However, from both the figs., it can be noted that the magnitude of stress–strain response at all σ_n is reduced with increasing w_c from 20 to 100%. From Fig. 15.11b, it can be seen that the initial portion of stress–strain response is flatter may be due to the fact that the soil is not taking load up to the horizontal strain of 2.5%.

Figure 15.12a and b presents the stress–strain response at different D_r and σ_n = 100 kPa for water content (w_c) 20% and 100%, respectively. It can be observed from Fig. 15.12a that, at w_c = 20% for σ_n = 100 kPa, with the increase in D_r, the stress–strain response increases. The stress–strain response at D_r = 80% is higher

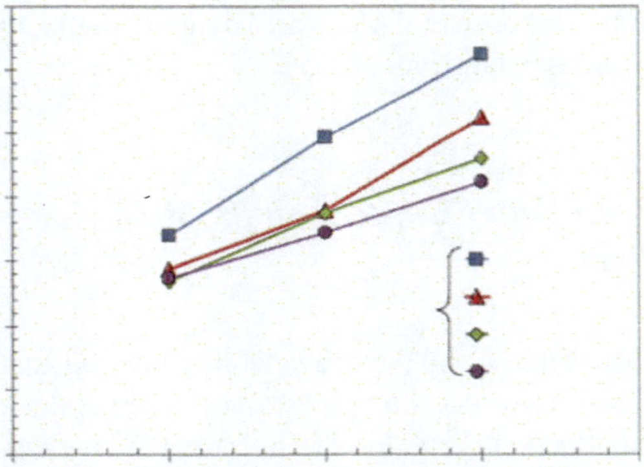

Fig. 15.10 Variations in peak shear stress with σ_n at different water content

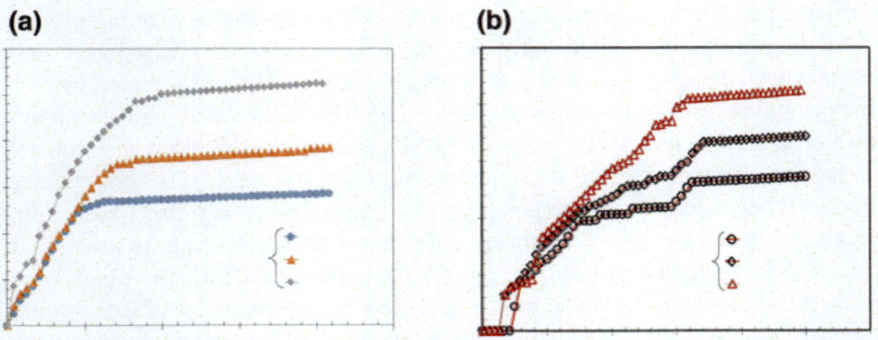

Fig. 15.11 Stress–strain response at different σ_n for **a** w_c = 20%, **b** w_c = 100%

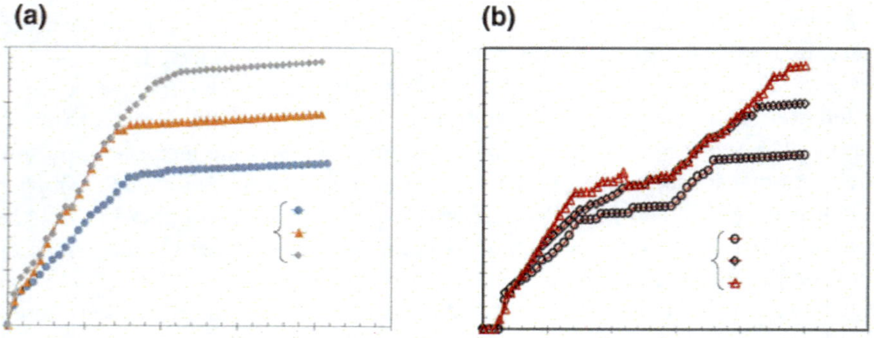

Fig. 15.12 Stress–strain response at different D_r for **a** $w_c = 20\%$, **b** $w_c = 100\%$

than that of 30% and 50%. Similar responses have been observed in Fig. 15.12b at $w_c = 100\%$ for $\sigma_n = 100$ kPa. However, from both the figs., it can be noted that the peak magnitude of stress–strain response at all D_r is almost nearly same while increasing w_c from 20 to 100%. From Fig. 15.12b, it can be seen that the initial portion of stress–strain response is flatter may be due to the fact that the soil is not taking load up to the horizontal strain of 1.5%.

Effect of Fiber with Water Content on the Behavior of Jute-Sand Mixes

Due to increase in urbanization and the unavailability of good bearing capacity land for many civil engineering projects, engineers are utilizing the weak land, where the application of jute fibers inclusion in the soil mass has been reported by several researches to stabilize or to improving the bearing capacity the weak or soft soil [1]. Several soil-reinforcing techniques such as geotextile, geo-membrane, geogrid, and polypropylene fibers have been utilized by engineers worldwide to improve the bearing capacity of soil. Though these materials improve the engineering properties of reinforced soils significantly, the jute fiber can also be utilized as a soil-reinforcement material since it has also a property to resist the tensile stresses which arises within the soil reinforcement during loading. This section deals with the importance of the use of jute fiber, as well as its behavior, as a soil-reinforcement material based on results obtained from the direct shear test. Figure 15.13 presents the effect of fiber length on the response of jute-sand mixes using direct shear test. Figure 15.13a and b presents the stress–strain response and variations of peak shear stress at different fiber lengths for $D_r = 50\%$, fiber content = 0.50% (by weight) and $\sigma_n = 100$ kPa. It can be observed from Fig. 15.16a that the amplitude of stress–strain response increases with the increase in fiber length up to 20 mm, and thereafter, the amplitude decreased as the fiber length increased from 20 to 40 mm. This might be attributed to the fact that the interaction between the soil particles sand jute fiber is appropriate which might further lead to the increase in interface angle between the soil particles sand jute fiber. However, with increasing fiber length might reflect the slippage of the soil particles over the jute fiber, and hence, amplitude of stress–strain reduced. Figure 15.13b shows the variations of peak shear stress with fiber length and depicts that peak shear stress increased by more than 50% with increasing fiber length from 0 to 20 mm whereas, the same is decreased by nearly 20% when the fiber length increased from 20 to 40 mm. Figure 15.14 presents the effect of water content on the jute-sand mixes for $D_r = 50\%$, fiber content = 0.25% (by weight), and $\sigma_n = 100$ kPa. From Fig. 15.14a, it can be observed that by increasing water content from 0 to 20%, the amplitude of stress–strain response increased significantly whereas, further increase in water content from 20 to 100% the stress–strain response decreased drastically. Figure 15.14b shows the variations of peak shear stress with water content, which reflects that the peak shear stress increased by nearly 25% with increasing water content from 0 to 20% whereas, the same is decreased significantly by 54% (approximately) when the fiber length increased from 20 to 40%.

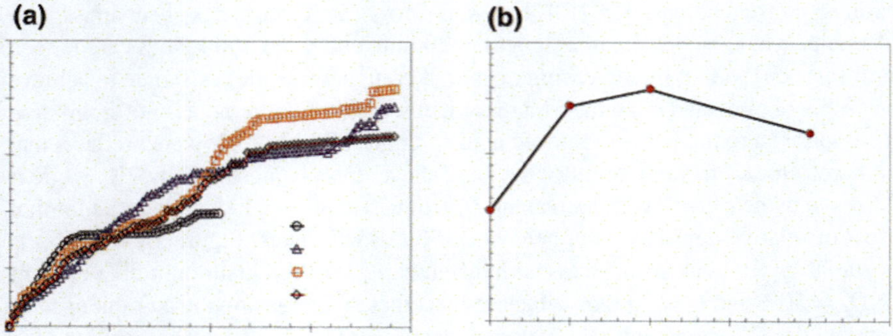

Fig. 15.13 **a** Stress–strain response, **b** variations of peak shear stress with fiber length at $D_r = 50\%$ and $\sigma_n = 100$ kPa

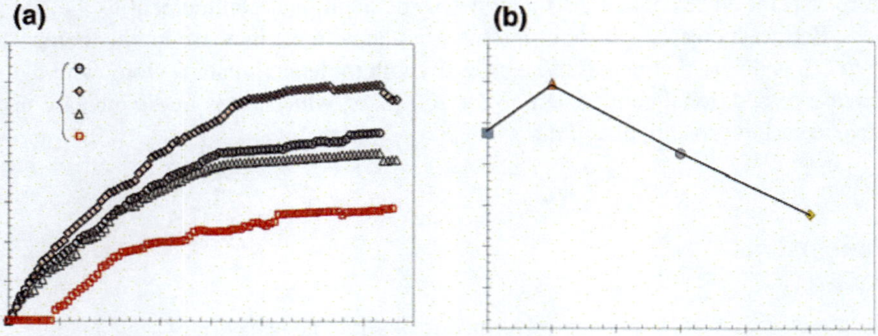

Fig. 15.14 **a** Stress–strain response, **b** variations of peak shear stress with water content at $D_r =$ 50% and $\sigma_n = 100$ kPa

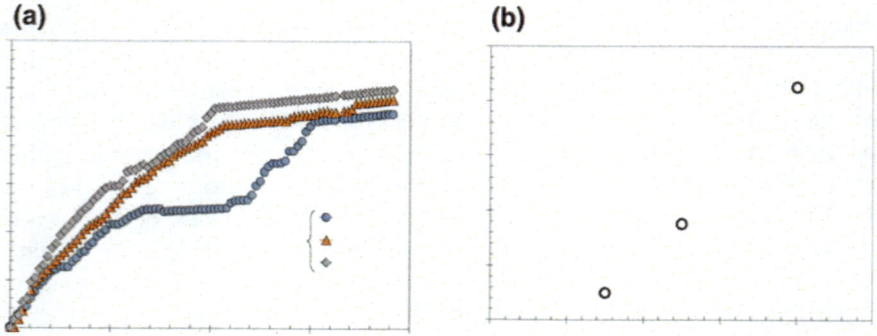

Fig. 15.15 **a** Stress–strain response, **b** variations of peak shear stress with D_r at $f_c = 0.25\%$, fiber length = 20 mm and $\sigma_n = 100$ kPa

Figure 15.15 presents the response of jute-sand mixes at $D_r = 50\%$ with $f_c = 0.25\%$, fiber length = 20 mm and $\sigma_n = 100$ kPa. Figure 15.15a presents the stress–strain response whereas, Fig. 15.15b presents the variations of peak shear stress with different D_r at $f_c = 0.25\%$, fiber length = 20 mm, and $\sigma_n = 100$ kPa. It can be seen form Fig. 15.15a that by increasing degree of compactness the resistance to achieve the failure of jute-sand mixes also increased; however, this increment is not significant in comparison to the failure achieved at different σ_n. Figure 15.15b also reflects that by increasing D_r from 30 to 50%, the peak shear stress is increased by 2%, while it increased by nearly 4% with increasing D_r from 50 to 80%, which reflects a minimal effect of relative density on the behavior of jute-sand mixes. Figure 15.16 presents the variations of peak shear stress with different σ_n at different amount fiber content for $D_r = 50\%$ and $l_f = 20$ mm, which reflects that by increasing the amount of fiber contents the shear strength parameters cohesion and angle of internal friction also increases. The same has been listed out in Table 15.3. The increase in cohesion with the increase of fiber content up to 1% is attributed to the increase of interlocking of soil grains which further restrict the movement of soil particles along the fiber length. Moreover, the fiber content also provides some additional confinement to the soil mass. Further increase in angle of internal friction with the increase in fiber content up to 1% is due to the increase in tensile strength to the soil matrix, along with the increase of additional confinement to the soil mass, which leads to the increase in the load caring capacity of soil mass.

Conclusions

The direct shear tests have been performed to observe the behavior of sand and jute-sand at different investigating parameters such as relative density, normal stress, fiber contents, fiber length, and degree of saturation. The following conclusions have been drawn from this study:

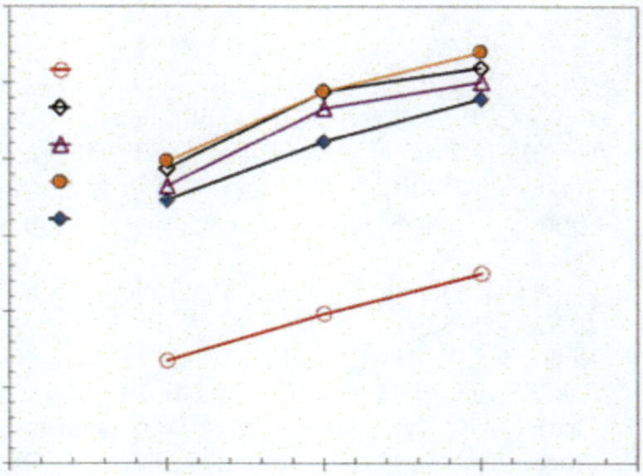

Fig. 15.16 Variations of peak stress with σ_n at different fiber content at D_r = 50%, l_f = 20 mm

Table 15.3 Shear strength parameters C and ϕ for different amount of fiber content

f_c (%) by weight	C (kg/cm^2)	ϕ (°)
0	0.39	32.84
0.25	1.61	37.38
0.50	1.5	38.92
1.00	1.65	40.8
1.50	1.4	38.40

1. With increasing normal stresses, the stress–strain responses of dry sand (without jute fiber) increased significantly whereas, the effect of relative density on the stress–strain responses is minimal for a particular σ_n.
2. At $D_r = 50\%$ and $\sigma_n = 100$ kPa, the peak shear stress reduced by ~20% with increasing water content from 0 to 100%. The peak shear stress decreased by 40% at $D_r = 30\%$ whereas, at $D_r = 80\%$, the reduction in peak shear stress is ~5%. Moreover, the peak shear stress increases with increasing σ_n for a particular D_r.
3. The amplitude of stress–strain response increases with the increase in fiber length up to 20 mm, and thereafter, the amplitude decreased as the fiber length increased from 20 to 40 mm. Therefore, the optimum fiber length was found to be 20 mm to achieve maximum peak stress value.
4. The peak shear stress of jute-sand mixes increased by nearly 25% with increasing water content from 0 to 20% whereas, the same is decreased significantly by 54% (approximately) when the fiber length increased from 20 to 40%.
5. It can be concluded that, for jute-sand mixes, by increasing D_r from 30 to 50%, the peak shear stress is increased by 2%, while it increased by nearly 4% with increasing D_r from 50 to 80%.
6. Increasing the amount of fiber contents (by weight) in sand, the shear strength parameters such as cohesion and angle of internal friction increases.

References

1. Aggarwal P, Sharma B (2010) Application of jute fiber in the improvement of subgrade characteristics. In: Proceedings of international conference on advances in civil engineering, Trabzon, Turkey, pp 27–30
2. Amritha VG, Soorya SR (2017) Study on strength behaviour of jute reinforced CH soil. IRJET 4(11)
3. Gray DH, Ohashi H (1983) Mechanics of fiber reinforcement in sand. J Geotech Eng 109(3):335–353
4. Kulhar KS, Raisinghani M (2017) Shear strength performance of sandy soil reinforced with jute. J Basic Appl Eng Res 4(7):624–629
5. Sharma AK, Shivapulliah (2011) Soil stabilization with waste materials based binder. In: Proceedings of Indian geotechnical conference, Paper no H-119
6. Umar J, Sonthwal KK, Duggal AK, Irfan Md (2015) Soil stabilization using shredded rubber tyre. IRJET 2(9)
7. Wang Y, Panpan G, Shengbiao S, Haiping Y, Binxiang Y (2016) Study on strength influence mechanism of fiber-reinforced expansive soil using jute. Geotech Geol Eng. https://doi.org/10.1007/s10706-016-0028-4
8. ASTM D6913/D6913M (2017) Standard test methods for particle-size distribution (gradation) of soils using sieve analysis. ASTM International, West Conshohocken, PA
9. ASTM D2487 (2011): Standard practice for classification of soils for engineering purposes (unified soil classification system). ASTM International, West Conshohocken, PA
10. ASTM D854 (2014) Standard test methods for specific gravity of soil solids by water pycnometer. ASTM International, West Conshohocken, PA
11. ASTM D4253 (2016) Standard test methods for maximum index density and unit weight of soils using a vibratory table. ASTM International, West Conshohocken, PA
12. ASTM D4254 (2016) Standard test methods for minimum index density and unit weight of soils and calculation of relative density. ASTM International, West Conshohocken, PA
13. Arunavathi S, Eithiraj RD, Veluraja K (2017) Physical and mechanical properties of jute fiber and jute fiber reinforced paper bag with tamarind seed gum as a binder—an eco-friendly material. AIP Conf Proc 1832:040026
14. Eldesouky HM, Morsy MM, Mansour MF (2016) Fiber-reinforced sand strength and dilation characteristic. Ain Shams Eng J 7:517–526
15. Maity J, Chattopadhyay BC, Mukherjee SP (2012) Behaviour of different types of sand randomly mixing with various natural fiber. J Inst Eng India Ser A 93(2):97–104. https://doi.org/10.1007/s40030-012-0014-7
16. ASTM D698 (2012) Standard test methods for laboratory compaction characteristics of soil using standard effort (12400 ft-lbf/ft^3 (600 kN-m/m^3)). ASTM International, West Conshohocken, PA
17. IS: 2720 (Part-XIII) (1986) Methods of test for soils: determination of friction angle and cohesion using direct shear test. Bureau of Indian Standards, New Delhi, India

Chapter 16
Numerical Studies on Effects of Embankment Layer Construction Period on Consolidation Settlements of Underlain Soft Soil

Rai Bahadur Reang, Sujit Kumar Pal, and Sanjay Paul

Introduction

Embankments constructions are one the most common geotechnical engineering structures encountered by site civil engineers. Especially, such types of constructions are more important in case of prefabricated vertical drain constructions [9, 10–13]. Furthermore, many researchers have conducted considerable research on the behavior of embankments and its staged loadings over soft clay treated with PVDs in terms of field behavior [1–8]. Sandy type of soils is mostly preferred for embankment constructions for railway, highway, or other heavy industrial structures, and hence, care must be taken of, especially, when the underlain ground is of clayey type. Embankment fills cannot be place over the ground in a single day or within a few days. They are constructed stage wise (layer by layer) up to a desired height. The completion periods are selected based on the design, but the height of each layer is selected randomly by the site engineers. Therefore, it is a desired addition to study the effect of time interval for each embankment layer construction on consolidation settlement of underlain soil.

Since a huge embankment structure cannot be modeled physically, a numerical model was selected to study the required objective. Since modeling of three dimensional embankment structures requires a modern computerized work station, a two dimensional numerical model was analyzed using ABAQUS software. By studying numerical models, a comprehensive idea can be made roughly how would be the consolidation settlement trend of both embankment fills and underlain soils.

Therefore, the main objective of this study is to analyze numerically the effects of embankment construction period on consolidation settlement of underlain soft soils.

R. B. Reang (✉) · S. K. Pal · S. Paul
National Institute of Technology Agartala, Agartala, Tripura, India
e-mail: raireang@gmail.com

K. Muthukkumaran et al. (eds.), *Ground Improvement Techniques*, Lecture Notes in Civil Engineering 297, https://doi.org/10.1007/978-981-19-6727-6_16

Modified Cam-Clay Model

Researchers at Cambridge University formulated the first critical-state model for describing the behavior of soft soils: the modified Cam-clay models. The model is capable of describing the stress–strain behavior of soils; in particular, the model can predict the pressure-dependent soil strength and the compression and dilatancy (volume change) caused by shearing. Because the model is based on critical-state theory, it predicts unlimited soil deformations without changes in stress or volume when the critical state is reached.

The parameter κ defines the elastic behavior of the soil in the Cam-clay model. The parameter κ is related to the swelling index through the equation $\kappa = C_s/2.3$. The parameter λ is related to the compression index through $\lambda = C_c/2.3$. The strength parameter M is related to the internal friction angle of the soil, φ', as follows:

$$M = \frac{6 \sin \varphi'}{3 - \sin \varphi'} \tag{16.1}$$

The other details of the model can be found in the book *applied soil mechanics with ABAQUS applications* by Sam [8].

Details of the Soil Profile and the Embankment

In this study, the height of each of layer of the embankment is 0.6 m giving a total height of 3.6 m. The top width of the embankment is 60 m, and the bottom is 110 m wide. The underlain soft soil is 8 m deep, and the bottom is extended up to an impervious rock layer. Only half-width of the embankment and soft soil is shown in Fig. 16.1. The engineering properties of the embankment and soft soils are different.

Numerical Modeling

Properties, Mesh, and Boundary Condition

Two dimensional numerical analyzes were conducted using ABAQUS software. Properties of the soil, meshing, and boundary conditions are assigned before analysis steps. The embankment material is linear elastic with bulk density $(\rho) = 1900\ \text{kg/m}^3$, Young's modulus $I = 2000$ kPa, Poisson's ratio $(\nu) = 0.3$, permeability $(k) = 0.08$ m/s, initial void ratio $(e_0) = 0.9$, and specific gravity $(G) = 2.67$. The underlain clay soil is elastoplastic, obeying the Cam-clay model (discussed in Sect. 2), and the properties are shown in Table 16.1. Some other geotechnical properties are liquid limit $(w_l) =$ 48.5%, plastic limit $(w_p) = 28.6\%$, specific gravity $(G_s) = 2.61$, MDD $= 1.62\ \text{kg/cm}^3$,

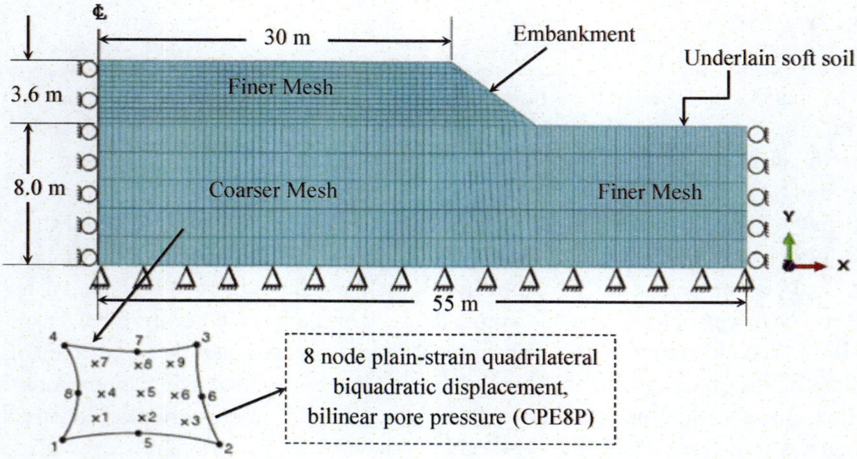

Fig. 16.1 Details of the embankment (half-width), boundary condition, and type of element used in the study

OMC = 25.5%, permeability (k_s) = 2.6 × 10^{-11} cm/sec. Finer and coarser mesh were applied for embankment and underlain soil, respectively, to simulate the sequential construction procedure of an embankment and to calculate the resulting consolidation settlement in the soft soil. The type of element for both soils is 8 noded plain-strain quadrilateral, coupled displacement-pore pressures, elements (CPE8P). Center line of the embankment is x-symmetry (U1 = UR2 = UR3 = 0), i.e., the movement is allowed only in vertical direction. The bottom surface of the underlain soil is restrained (U1 = U2 = 0), no movement is allowed in any direction, and it is connected with an impervious rock layer. The exterior boundary is allowed to move only in vertical direction (U1 = 0). Flow of pore water is allowed through the top surface of the underlain soil (Fig. 16.1), i.e., $u_0 = 0$.

Table 16.1 Cam-clay model parameters of the soft soil

General		Plasticity	
ρ (kg/m^3)	1980	Λ	0.186
k (m/s)	2.2 × 10^{-8}	Stress ration, M	1.5
γ_w (kN/m^3)	9.81	Initial yield surface size = $p_0'/2$ (kPa)	100
e_0	1.1	Wet yield surface size	1
Elasticity		Flow stress rate	1
κ	0.086		
ν	0.4		

Analysis Step

In the analysis part, firstly, the whole embankment is removed from the finite element mesh, i.e., only soft soil is considered, and then, the embankment layers are added, layer by layer, to the top surface of the soft soil in subsequent calculation steps. The first step is "geostatic" step. This command is invoked to make sure that equilibrium is satisfied within the soft soil body. The geostatic step makes sure that the initial stress condition in any element within the clay body falls within the initial yield surface of the Cam-clay model. In this step, the effective self-weight of the soil is applied using the "body-force" option. In other words, the type of the force is "body-force."

In the second step, the first embankment layer is constructed on top of the soft soil body. Just like step one, the force of the first layer is applied using "body-force" option, and the time period used is four days, i.e., the time interval to construct a layer is four days. Here, the permeable boundary of the underlain soil is made inactive (only at the interface between top surface of the soil and bottom surface of the embankment first layer) and only top surface of the first embankment layer and its sloping face are made permeable.

In the third step, the second embankment layer is constructed in the same manner described in second step. Now, the permeable layers are top surface of second embankment layer and its sloping face. The interface between the top surface of first layer and bottom surface of second layer is made impermeable. In the same manner, the other embankment layers are also constructed, up to six layers. Step 8 is a consolidation step with duration of 1000 days.

Results and Discussion

When a layer of an embankment is constructed, settlement occurs first at the top surface of underlain soil followed by that of the embankment. Settlement gradually increases as the number of construction of embankment layers is also increased. In the present analysis, each individual embankment layers are considered to be constructed at a same time interval (4 days for each layer) for a total embankment height of 3.6 m. Later on, several time intervals have also been selected such as 5, 6, 7 days up to 14 days to construct the same height of the embankment and the corresponding change in settlement, and pore pressures were determined. Figure 16.2 shows the deformation pattern of the underlain soil as well as the embankment due to stage wise construction of embankment layers. As soon as the first layer is constructed, settlement starts from the toe zone rather than from the middle section of the embankment. It is to be noted that greater settlement occurs just below the toe (Fig. 16.2). Dissipation pattern of pore pressure is shown in Fig. 16.3. It is observed that pore water get released first from the toe and then from the top center of the ground. This phenomenon propagates toward the exterior part of the ground as shown

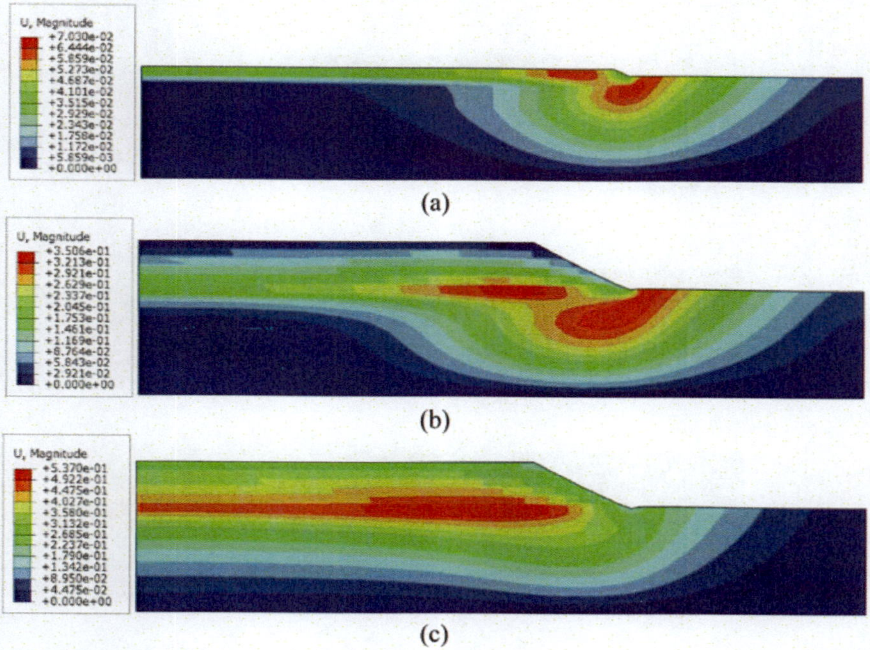

Fig. 16.2 Settlement of the underlain soil and the embankment (constructed at 5 days time interval) after **a** first layer construction, **b** sixth layer construction, **c** full consolidation period

by the figure. And, finally (after final consolidation period), maximum pore water gets release from underlain soil due to highly imposed surcharge pressures.

Effects on Settlement of Underlain Soft Soil

Final settlement was obtained after the complete consolidation period. Figure 16.4 shows the variations of consolidation settlements of soil for various time intervals. It can be seen that even though the settlement results are different at intermediate (during construction) period, the final consolidations are almost same for all time intervals. The results of consolidation settlement at the top surface are 457.41, 458.32, 459.14, 459.96, 460.53, 461.14, 461.70, 462.22, 462.71, 463.16, and 463.58 mm after 3000 days for 4, 5, 6, 7, 8, 9, 10, 11, 12, 13, and 14 days interval, respectively. It is to be noticed that after 300 days, settlement curves become parallel to x-axis, i.e., constant. The difference of settlement between 4 and 14 days interval is only about 6.17 mm, and hence, it is not a big concern to be considered. All these results were found to be increasing with an increase in vertical pressure. It was also observed that the results of consolidation settlement for all time intervals remain closer after

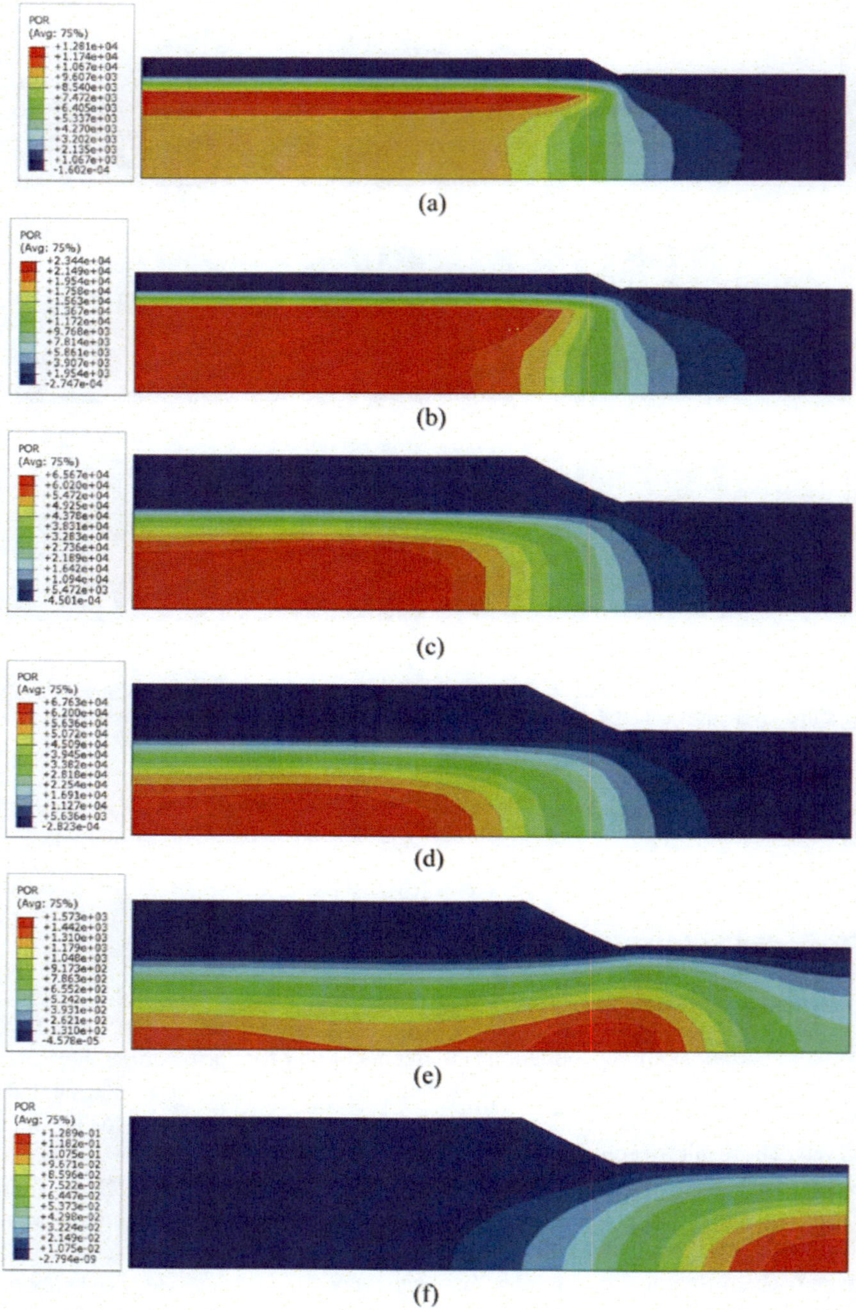

Fig. 16.3 Pore pressure distribution on underlain soil (5 days interval for a layer construction) after **a** first layer construction, **b** second layer construction, **c** sixth layer construction, **d** initial consolidation stage, **e** intermediate consolidation stage, **f** full consolidation period

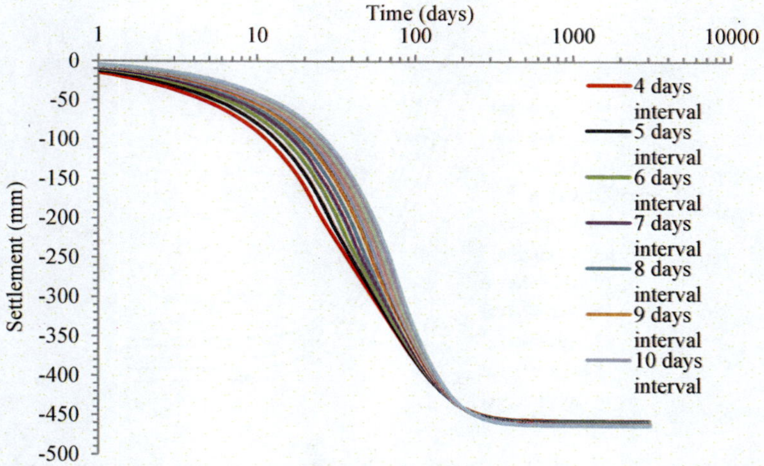

Fig. 16.4 Variation of settlement at top ground (underlain soil) surface with time at different intervals of construction periods

3000 days even after preloading pressures are changed. It indicates that settlement of underlain soil do not greatly depends on each layer construction period but on the magnitude of final surcharge loadings.

Effects on Settlement of Embankment

Figure 16.5 exhibits consolidation settlement of center top surface of the embankment fills. The final settlement results after 3000 days are 297.15, 285.00, 274.20, 263.19, 255.50, 247.23, 239.53, 232.30, 225.38, 218.92, and 212.79 mm for 4, 5, 6, 7, 8, 9, 10, 11, 12, 13, and 14 days interval, respectively. The figure exhibits that 4 days interval gives the largest settlement compare to others. And, settlement difference between 4 and 14 days interval is about 84.36 mm. This difference was observed to be fluctuating when preloading pressures change or when soil properties alter. Furthermore, it is to be noted that settlement does not occur all of sudden just after the construction of embankment layers. It takes few days to settle. First the underlain soil settles, then embankment. Most of the curves produce a constant result after 400 days. Therefore, from this point of view, it can be said that 4 days interval would produce the largest settlement and hence may be selected for the construction.

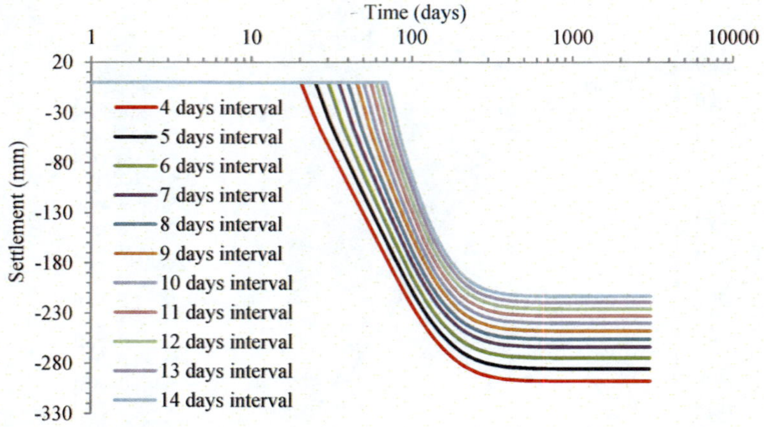

Fig. 16.5 Variation of settlement at top embankment surface with time at different intervals of construction periods

Effects on Pore Pressure Dissipation

When an embankment layer is constructed, pore water from underlain soil starts moving toward the top surface of the ground and then drains off through the embankment surface. Figure 16.6 shows the variations of pore pressure with time. For 4 days interval, it is seen from the figure that at the beginning, the magnitude of pore water pressure rises to 130 kPa and gradually decreases with time. The same trending was observed for all time intervals. The magnitude of these pressures gets decrease as the time interval increases. The minimum pressure is obtained from 10 days interval as 53 kPa. But, again, the pressure rises to 317, 219, and 283 kPa for 12, 13, and 14 days interval, respectively (Fig. 16.6). Due to this reason, the soil constituting the embankment may fail to restrain its shear strength. Therefore, these time intervals may not be suitable and hence should always select below 12 days intervals.

Effects on Lateral Movement of Embankment Fills

Lateral movement of soil mass of embankment layers occurs as the number of layer is increased. The results are illustrated in Fig. 16.7. In the numerical modeling, lateral movement of embankment fills just at the toe was analyzed. The rate of lateral movement is higher for time interval 4 days compare to others. During the consolidation period, the curves are found to be reversing giving a straight line after some days. It indicates that elastic rebound occurs tremendously on the soil mass leading to an increase in stress at this zone. Due to this stress, the soil mass becomes

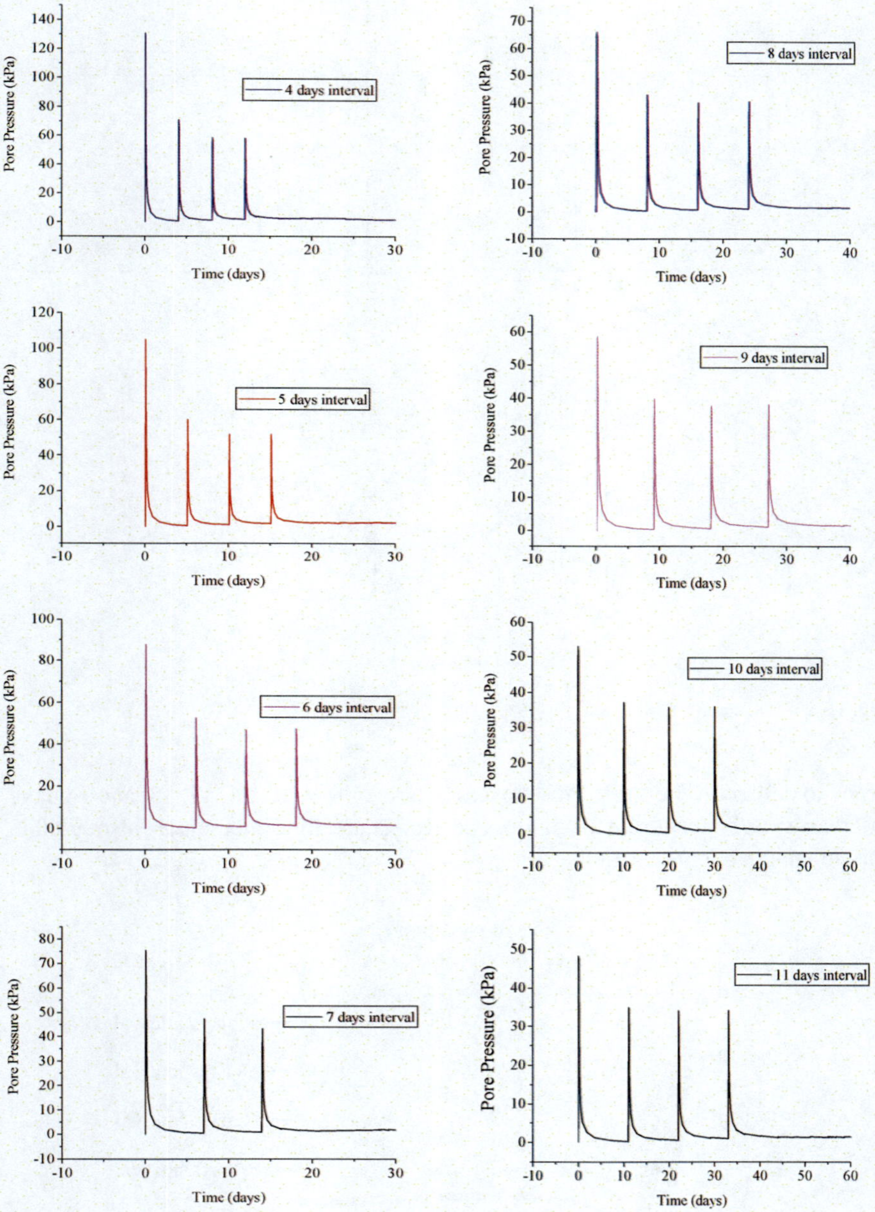

Fig. 16.6 Variations of pore water pressure with time at the top center of underlain soil

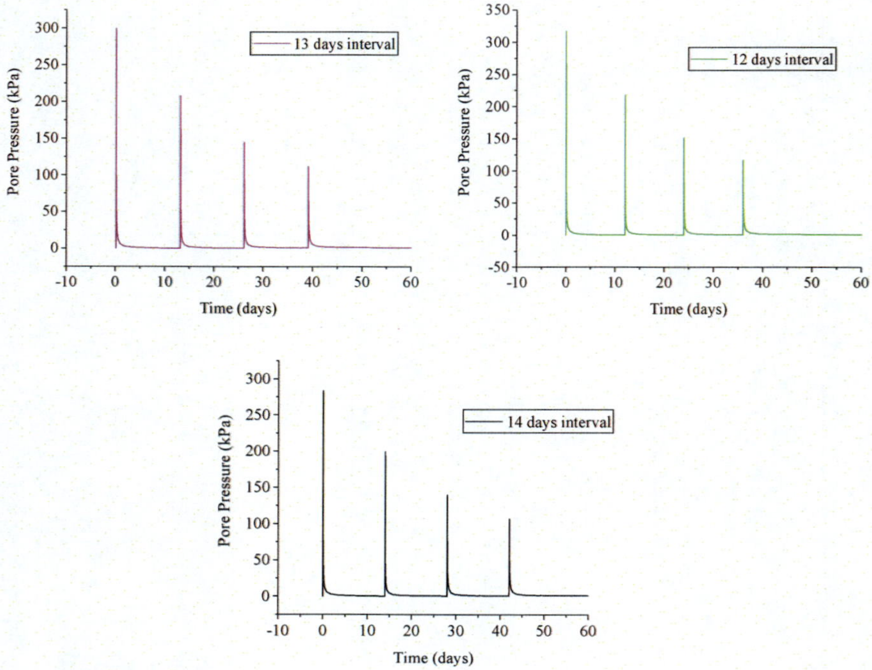

Fig. 16.6 (continued)

prone to collapse. The stress must be less in case of 8, 9, 10, 11, 12, 13, and 14 days interval as can be seen from the figure. Therefore, these time intervals are also suitable for embankment construction.

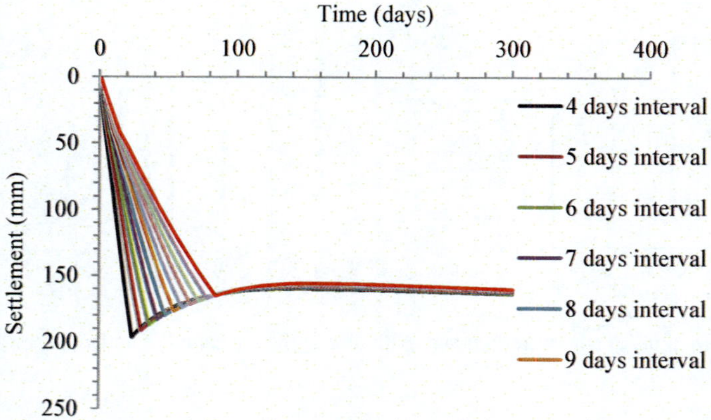

Fig. 16.7 Lateral movement of embankment soil with time

Conclusions

Based on the analysis made from consolidation settlement of underlain soil and embankment, dissipation of pore water and lateral movement of embankment fills it may be concluded that the time interval for the construction of each embankment layer do not greatly affect the overall stability of the embankment. But, it is observed that the maximum pore pressure is only 53 kPa for 11 days interval, which is the least among all other pore pressure results. However, the result may be different for other types of soils. Therefore, while an embankment is constructed in the field, the time interval should be kept in between 8 to 11 days which will produce considerable impacts to the embankment construction. More studies can be made in details using prefabricated vertical drains constructed with embankments.

References

1. Al-Soud MS (2016) Numerical analysis of prefabricated vertical drains improved soft soil beneath an embankment during staged construction. J Eng Dev 20(1):151–163
2. Bergado DT, Patawaran MAB (2000) Recent development of ground improvement with PVD on soft Bangkok clay. In: Proceeding of International seminar on geotechnics in Kochi, Japan
3. Bergado DT, Balasubramaniam AS, Jonathan Fannin RJ, Holtz RF (2002) Prefabricated vertical drains (PVDs) in soft Bangkok clay: a case study of the new Bangkok International Airport project. Can Geotech J 39:304–315
4. Crawford CB, Fannin RJ, Deboer LJ, Kern CB (1992) Experience with prefabricated (wick) drains at Vernon, B.C. Can Geotech J 29:67–79
5. Dahr AS, Saddique A, Fakrul Ameen S (2011) Ground improvement using preloading with prefabricated vertical drains. Int J Geoeng Case Hist 2(2):86–92
6. Davie JR, Young LW, Lewis MR (1988) Accelerated consolidation of soft clays using wick drains. In: Proceeding of second international conference on case histories in geotechnical engineering, pp. 1019–1024
7. David RC (1988) Performance of prefabricated drains in soft soils. In: Proceeding of second international conference on case histories in geotechnical engineering, pp 1069–1074
8. Helwany S (2007) Applied soil mechanics with ABAQUS applications. Wiley, United State of America
9. Indraratna B, Rujikiatkamjorn C, Ewers B, Adams M (2010) Class A prediction of the behavior of soft estuarine soil foundation stabilized by short vertical drains beneath a rail track. J Geotech Geoenviron Eng. https://doi.org/10.1061/(ASCE)GT.1943-5606.0000270,686-696
10. Karim MR, Lo SC (2015) Estimation of the hydraulic conductivity of soils improved with vertical drains. Comput Geotech 63:299–305
11. Karunaratne GP (2011) Prefabricated and electrical vertical drains for consolidation of soft clay. Geotext Geomembr 29(4):391–401
12. Tran-Nguyen HH, Edil TB, Schneider JA (2010) Effect of deformation of prefabricated vertical drains on discharge capacity. Geosynth Int 17(6):431–442. https://doi.org/10.1680/gein.2010.17.6.431
13. Watabe Y, Shinsha H, Yoneya H, Ko C (2014) Description of partial sandy layers of dredged clay deposit using penetration resistance in installation of prefabricated vertical drains. Soils Found 54(5):1006–1017

Chapter 17
Improvement of Soft Ground by Employing Granular Piles Below Raft

Dhanraj Nath, Plaban Deb, and Sujit Kumar Pal

Introduction

Granular piles are established or install into the engineering field to increase the bearing capacity and decrease the settlement of soft and weak soils. There are several techniques of ground improvement that has been established for structural construction of improper sites. Granular piles may be used to increase the bearing capacity in that sites which are having very poor ground conditions subjected to excessive settlement due to high compressibility and very low strength. This practice is assumed as one of the most cost effective and versatile systems for improvement of ground. The installation of granular piles involves removal of native soil by means of granular material. In granular piled raft foundation, the granular materials can be used in place of concrete material or steel piles because of their numerous additional advantages.

For analysis of granular piled raft foundation, similar approach may be used like as piled raft foundations. There are various numbers of possible methods to stabilize the expansive soils such as thermal, chemical and mechanical stabilization, soil replacement, cohesive non-swelling layer, and sand cohesion. But these techniques are costly and not very much effective for construction over large area is carried out such as construction of highways over the expansive soil belts. The approach of granular piles is employed to improve weak marine clays and has anticipation of being employed in improving the behavior of expansive soils in such regions.

Balaam and Booker [1] developed an analytic solution with the help of theory of elasticity for the foundation settlement and evaluating the expression for moment distributions across the foundation embedded in soft soils improved with granular piles. Zhou and Ooi [13] presented discrete element simulations of non-spherical

D. Nath (✉) · S. K. Pal
Department of Civil Engineering, National Institute of Technology, Agartala 799046, India
e-mail: dhanrajnath22@gmail.com

P. Deb
Department of Civil Engineering, Chandigarh University, Mohali, Punjab 140413, India

K. Muthukkumaran et al. (eds.), *Ground Improvement Techniques*, Lecture Notes in Civil Engineering 297, https://doi.org/10.1007/978-981-19-6727-6_17

and spherical materials of granular piles and investigated the distribution of base pressure beneath the granular piles. Madhav et al. [8] studied the distribution of load and the settlement of granular piled raft by using continuum approach. They found that due to increase in stiffness of granular pile or increase in relative length of granular pile the normal stresses at the raft–soil interface was decreased. Hasan and Samadhiya [7] carried out numerical analysis and conducted laboratory model tests on geosynthetic reinforced granular piles using unit cell concept under short-term loading condition. The effect of several factors such as encasement stiffness, shear strength of clay, diameter and length of granular piles has been deliberate in which the load is applied either over the area of granular piles or over the entire area of cylindrical tank. Elsawy and El-Garhy [6] analyzed numerically by using Plaxis 3D software to investigate the performance of granular piles under a uniformly loaded raft foundation. The various parameters such as lateral bulging and bending moment of granular piles, and the overall settlement of raft have been studied. Basu et al. [2] studied experimentally and developed a new granular pile, i.e., randomly mixed fiber granular pile instead of plain granular pile. Samanta and Bhownik [10] performed 3D numerical analysis to investigate the response by installing the stone column below the raft foundation. To make stronger the shallow soft soil strata, the stone columns were used, so that the overall behavior of piled raft foundation has been improved.

In previous years, so many experimental and numerical analyses have been carried out to understand the behavior of foundation reinforced with granular piles. Maximum of the research works are based on granular piles without providing a raft over the pile. There are limited studies on granular piled raft foundation or raft resting on granular piles. In this present study, the bearing capacity and decrease in settlement are analyzed by developing a granular piled raft model with and without providing cushion layer between granular piles and raft utilizing three-dimensional finite element software ABAQUS 6.12.

Numerical Modeling

Finite Element Model

The finite element software ABAQUS 6.12 was used for the simulation of granular piled raft foundation in this study. To model the interaction of the structure, the master–slave concept was used in which the smooth surface was considered as slave surface and the hard surface was considered as master surface and the soil was modeled as slip model. The interaction between the soils and piles, piles and raft, raft and soils was considered when cushion layer is absent. When cushion layer is provided between rafts and piles, the interaction between piles and cushion layer, cushion layer and soils, raft and cushion layer was considered. By using penalty method, the tangential and normal behaviors of contact between the surfaces were defined and the contact between the surfaces was assumed as hard contact with coefficient of friction 0.2. The Mohr–Coulomb model were used to define the behavior of cushion layer, soils and granular piles; however, the raft was modeled as linear elastic material.

Study of Model Configuration

In the present study, 2 × 2, 3 × 3 and 4 × 4 group of granular piled raft models, i.e., 4, 9, and 16 numbers of granular piled raft models has been developed by using three-dimensional finite element software ABAQUS 6.12. The length and diameter of the granular piles were considered as 15 m and 0.8 m, respectively. The c/c spacing between the granular piles is taken as the four times of the diameter of the granular piles. The soils, cushion layer and raft were modeled as an 8-nodded solid brick element, and the pile was modeled using tri angular prism element. Vertical load is applied throughout the area of raft to analyse the bearing capacity and settlement of raft over soft clayey soils. At first, the bearing capacity of soft clayey soils has been analyzed in 4, 9, and 16 numbers of granular piled raft models without providing cushion layers between raft and granular piles and after that same configurations were used to analyze the granular piled raft with the presence of different thickness of cushion layer. The one-quarter pattern of the piled raft was modeled due to reduce the computational time. The modeling of the numerical study is given in Table 17.1.

Table 17.1 Modeling of the numerical study

Model of the granular piled raft	Diameter and length of granular pile	Size of raft L × B	Size of cushion layer	Difference in thickness of cushion layer	Difference in thickness of raft
2 × 2	0.8 m and 15 m	7 m × 7 m	2 times of size of raft	Without cushion layer, 0.75 m, 1 m, 1.25 m	1 m
3 × 3	0.8 m and 15 m	14 m × 14 m	2 times of size of raft	Without cushion layer, 0.75 m, 1 m, 1.25 m	1 m, 1.4 m, 1.8 m
4 × 4	0.8 m and 15 m	14 m × 14 m	2 times of size of raft	Without cushion layer, 0.75 m, 1 m, 1.25 m	1 m, 1.4 m, 1.8 m

Table 17.2 Properties and characteristics of materials

Property	Granular piles	Soils	Cushion layer
Unit weight (kN/m^3)	20	15	19
Young's modulus (kPa)	30,000	1100	15,000
K_h (m/s)	1.16×10^{-4}	3.47×10^{-9}	1.2×10^{-4}
K_v (m/s)	1.16×10^{-4}	1.16×10^{-9}	1.2×10^{-4}
Poisson's ratio, υ	0.3	0.3	0.3
Cohesion, c´ (kPa)	5	1	3
Friction angle, ϕ (°)	40	20	28

Material Characteristics and Property

Three-dimensional (3D) model of granular piled raft has been developed by utilizing finite element-based software ABAQUS 6.12. The granular piles are embedded in clayey soil, and raft resting on both granular piles and cushion layers is carried to study the bearing capacity soils and settlement characteristics of raft. In this analysis, the material properties of the soft clay, cushion layer, and granular piles are partly taken from the data of previous published works by Yoo and Kim [12], Tan et al. [11], and Rajesh and Jain [9]. The properties of granular piles soils and cushion layer are provided in Table 17.2. Mohr–Coulomb model have been considered to model the behavior of soils, granular piles, and cushion layer. The raft is modeled as the linear elastic material having density 25 kN/m^3 and young modulus 2×10^7 kN/m^3.

Mesh Pattern and Boundary Condition

For creation of mesh, the soil continuum, raft and cushion layer are considered as 8-node brick element (C3D8R), and in the case of granular pile, it was considered as triangular prism element. As far as possible, all the surfaces were partitioned to developed a quality mesh and avoid non-convergence error. The mesh was created at all the surfaces in such a way that all the nodes are connected and matched up with each other. To develop more refined understanding of the contact surfaces in connection areas, the mesh was made smaller in size, whereas it was enlarged for other areas which were not as critical for this work. In respect of the boundary condition, the end surface of the soil was closed in all three directions; i.e., there were not any kind of translation and rotation and the top surface was free. For the vertical surface of the soil, the boundary condition for *y*–*z* plane was defined as XSYMM, and for the *x*–*y* plane, it was defined as ZSYMM [3–5]. The initial conditions of geostatic stress are well-defined for soil, and the gravity is applied to the whole model. To avoid or reduce the distribution of stress at the boundary, the horizontal boundary was taken as five times the width of raft (5B) and the vertical boundary was considered as three times the length of the granular pile (3L). The spacing between the piles is kept as 4 times of the diameter of the pile of c/c distance. The mesh pattern and its dimension of the granular piled raft element are shown in Fig. 17.1a, b.

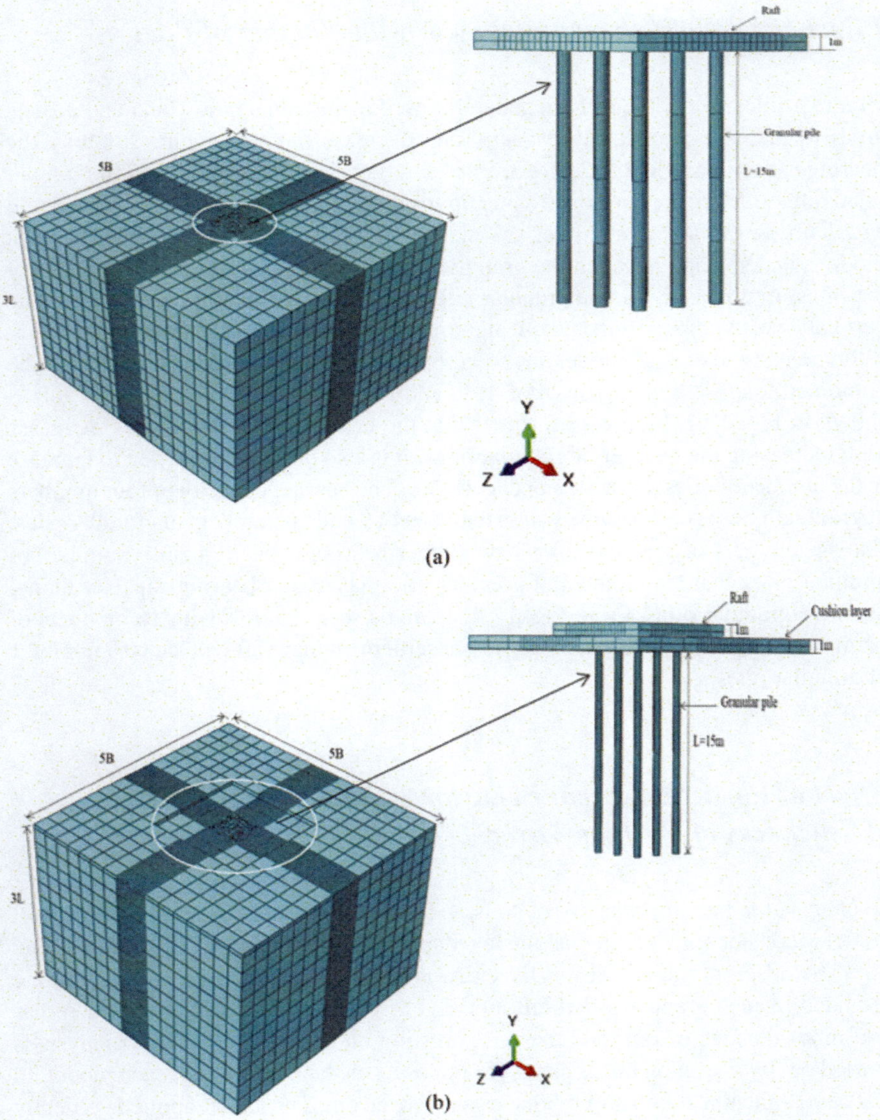

Fig. 17.1 Mesh pattern and each dimension: **a** without cushion layer and **b** with cushion layer

Results and Discussion

This section represents the study of results of tests of different numerical modeling through ABAQUS software. The results obtained from different models are compared and discusses in this section.

Effect of Load–Settlement Behavior with Number of Piles

Granular piles below the raft are generally used in soft soils in lowland and coastal areas to increase the bearing capacity and decrease the settlement. To study the bearing capacity of soil 2 × 2, 3 × 3, and 4 × 4, i.e., 4, 9, and 16 numbers of granular piled raft modeled, are developed using finite element software ABAQUS. The length and diameter of the granular piles are taken as 15 m and 0.8 m, respectively, in each models, and spacing between the granular piles is taken as 4 times the diameter of the piles. In this analysis, the granular piles are reinforced in soft clayey soils. The vertical load is applied throughout the area of the raft in each model, and load–settlement curve of each model has been found out through numerical analysis. The combined load–settlement curve of 4, 9, and 16 numbers of granular piled raft is shown in Fig. 17.2. Figure 17.2 shows that the bearing capacity of soils increases with increasing the number of granular piles. It is also observed that, due to increase in the number of piles the settlement decreases. The bearing capacity of four numbers of granular piles is comparatively less than that of 9 and 16 numbers of granular piles. The 16 numbers of granular piles have higher bearing capacity than 9 numbers of granular piles again 9 numbers of granular piles has higher bearing capacity than 4 numbers granular piles. From Fig. 17.2, it can be seen that raft with 16 numbers of granular piles provide comparatively lesser settlement than raft with 4 and 9 numbers of granular piles.

Effect of Load–Settlement Behavior with Variation of Thickness of Cushion Layer

To analyse the bearing capacity of soils, 4, 9, and 16 numbers of granular piled raft model is considered without cushion layer and with cushion layer between piles and raft. The different thicknesses of the cushion layer of 0.75 m, 1 m, and 1.25 m are provided in each granular piled raft model. The cushion layer size is considered as two times the size of raft, and the raft is resting on cushion layer. This analysis is carried out by assuming the thickness of raft as a constant, i.e., 1 m in each model. In this case, total nine number of model tests is carried out. The load is applied vertically throughout the area of the raft, and the load–settlement curve is plotted by using the results obtained from numerical analysis. The combined load–settlement curve in

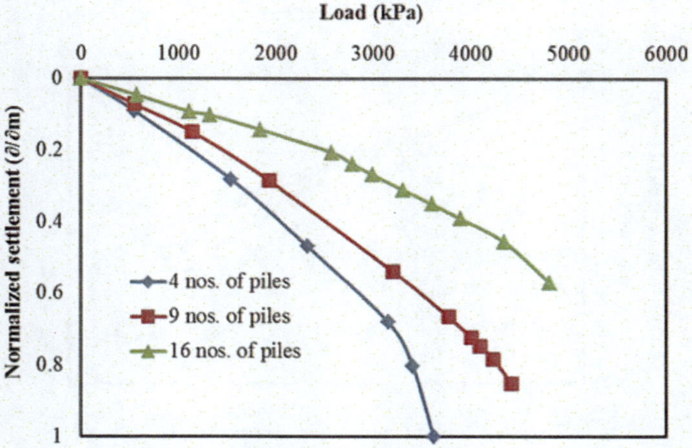

Fig. 17.2 Combined load–settlement curve for 4, 9, and 16 numbers of granular piled raft

each 4, 9, and 16 numbers of granular piled raft model is shown in Fig. 17.3a–c. Due to the presence of cushion layer between raft and piles, the bearing capacity increases in each model tests as compared to without cushion layer. From the figure, it is observed that in case of 1.25 m thick cushion layer, the bearing capacity increases about 18–24% as compared to without cushion layer. It can be also observed that the settlement is reduced about 24–30% by providing the 1.25 m thick cushion layer in between the piles and raft.

Effect of Thickness of Raft

To analyse the effects of different raft thickness on granular piled raft, 3 × 3 and 4 × 4, i.e., 9 and 16 numbers of granular piled raft modeled is developed in which the ratio of raft thickness to width (t_r/B) is taken as 0.072, 0.1, and 0.128; i.e., the thickness of raft is considered as 1 m, 1.4 m, and 1.8 m. The vertical load is applied throughout the area of raft, and load–settlement curve is plotted which is shown in Fig. 17.4a, b. From both the cases of granular piled raft model (i.e., Fig. 17.4a, b), it is observed that when the thickness of raft increases from 1 to 1.8 m the bearing capacity of soil increases about 19.5–23%. This is due to the increase of raft stiffness as the thickness of raft increases.

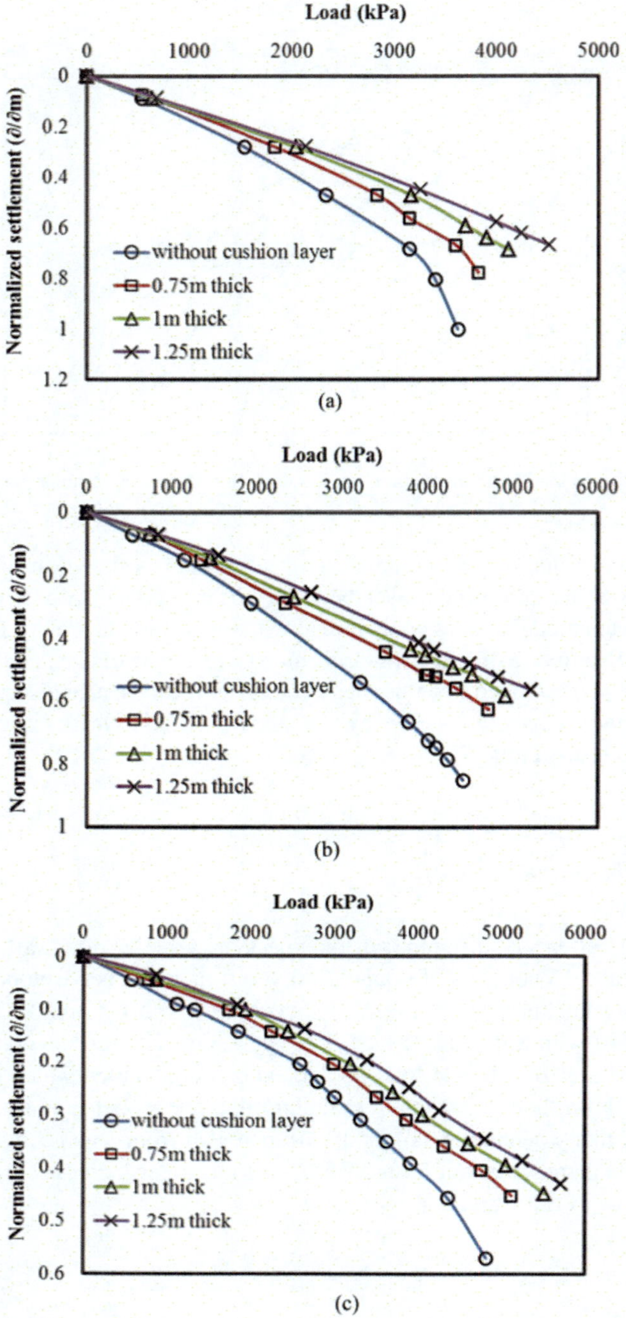

Fig. 17.3 Comparisons of load–settlement curve without cushion layer and variation of thickness of cushion layer for: **a** 4 numbers, **b** 9 numbers, and **c** 16 numbers of granular piled raft

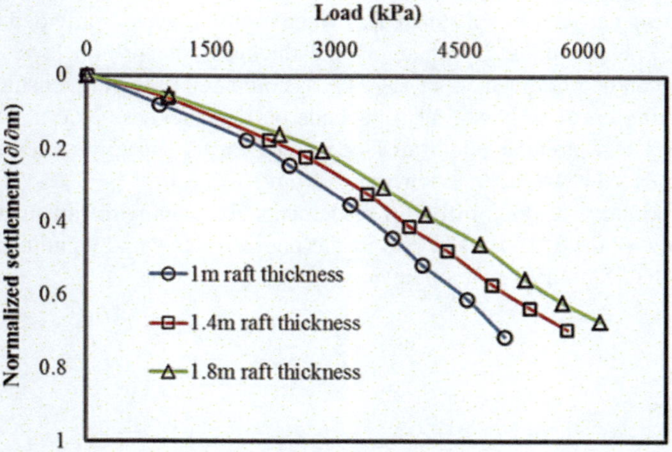

(a)

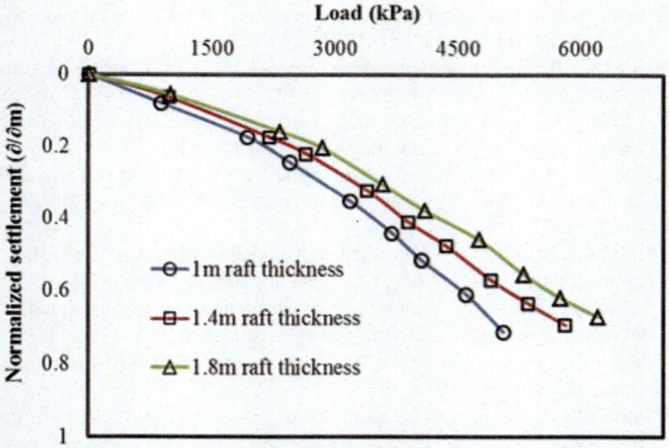

(b)

Fig. 17.4 Comparisons of load–displacement curve by variation of thickness of raft in **a** 9 numbers and **b** 16 numbers of granular piled raft

Conclusion

Based on the numerical analysis, the following conclusions can be drawn:

1. Granular piles lead an essential and effective role to increase in bearing capacity of soft soil and reducing the settlement.
2. As the number of granular piles increases, the bearing capacity increases comparatively. The 16 numbers of granular piles are having higher bearing capacity than 9 numbers of granular piles and 9 numbers of granular piles are having higher bearing capacity than 4 numbers of granular piles.

3. The bearing capacity of soils increases when cushion layer is provided between raft and granular piles. By changing the thickness of cushion layer, bearing capacity can be improved about 18–24% as compared to without cushion layer.
4. The bearing capacity of soil also depends on the thickness of raft. In both 3 × 3 and 4 × 4 granular piled raft model, the bearing capacity increases about 19.5–23% as the raft thickness increases from 1 to 1.8 m.
5. On performance of piled raft with settlements, 16 piles with raft thickness to width ratio of 0.1–0.13 is more efficient as compared to 4 and 9 numbers of pile in case of 1.25 m thick cushion layer.

References

1. Balaam NP, Booker JR (1981) Analysis of rigid raft supported by granular piles. Int J Numer Anal Methods Geomechan 5(4):379–403
2. Basu P, Samadhiya NK, De Dalal SS (2018) An experimental study on random fiber mixed granular pile. Int J Geotech Eng 12(1):1–12
3. Deb P, Pal SK (2019) Analysis of load sharing response and prediction of interaction behaviour in piled raft foundation. Arab J Sci Eng 44(10):8527–8543
4. Deb P, Pal SK (2020) Load-settlement and load sharing behaviour of piled raft foundation resting on layered soils. Acta Geotech Slovenica 2020(1):71–86
5. Deb P, Pal SK (2021) Structural and geotechnical aspects of piled raft foundation through numerical analysis. Mar Georesour Geotechnol. Published on 28th June 2021. https://doi.org/10.1080/1064119X.2021.1943083
6. Elsawy MB, El-Garhy B (2017) Performance of granular piles-improved soft ground under raft foundation: a numerical study. Int J Geosynth Ground Eng 3(4):1–13
7. Hasan M, Samadhiya NK (2017) Performance of geosynthetic-reinforced granular piles in soft clays: model tests and numerical analysis. Comput Geotech 87:178–187
8. Madhav MR, Sharma JK, Sivakumar V (2009) Settlement of and load distribution in a granular piled raft. Geomechan Eng 1(1):97–112
9. Rajesh S, Jain P (2015) Influence of permeability of soft clay on the efficiency of stone columns and geosynthetic-encased stone columns—a numerical study. Int J Geotech Eng 9(5):483–493
10. Samanta M, Bhowmik R (2019) 3D numerical analysis of piled raft foundation in stone column improved soft soil. Int J Geotech Eng 13(5):474–483
11. Tan SA, Tjahyono S, Oo KK (2008) Simplified plane-strain modelling of stone-column reinforced ground. J Geotech Geoenviron Eng 134(2):185–194
12. Yoo C, Kim SB (2009) Numerical modeling of geosynthetic encased stone column-reinforced ground. Geosynth Int 16(3):116–126
13. Zhou C, Ooi JY (2009) Numerical investigation of progressive development of granular pile with spherical and non-spherical particles. Mech Mater 41(6):707–714

Chapter 18
Multivariate Regression Model to Predict Geotechnical Properties of Fly Ash-Stabilized Clayey Soil

Niranjan Shekar and Sanku Konai

Introduction

Fly ash is the product of combustion of subbituminous coal in power plants and requires to be disposed safely in landfills. However, reuse of the fly ash is recommended by many countries in the interest of sustainable development. Hence, the use of fly ash as a stabilizer improves the engineering properties of soil as well as reduces the use of energy consumption and greenhouse gases. Fly ash improves the shear strength and bearing capacity of soils. It also reduces plasticity index and shrinkage limit, which has an impact on the properties of fine-grained soils.

India with 197 thermal power plants generates 226.13 million tons of fly ash of which only 83% is utilized remaining 17% of fly ash is dumped in landfills. But, the utilization has its own hurdles of lab testing the fly ash and soil that has to be improved. If a lot of data is collected about major soil types and fly ash types, then data can be processed to obtain empirical formulas which reduces the time required in testing the materials. This saves time, and the easy implementation yields in higher adoptability. Data collection is time exhaustive process, but once collected, the process becomes highly efficient.

From literatures available, fly ash as a stabilizing agent provides sufficient improvements in geotechnical properties of soil. For example, Karthik et al. [1] stabilized red soil with 6% of fly ash and achieved 3.5 times the bearing capacity when compared to virgin soil. Also, predictive models can be used to predict geotechnical properties of fly ash stabilized soils like CBR and compressive strength. This is highlighted by Trivedi et al. [2] who used a model based on genetic algorithm to predict the variation of CBR with the addition of fly ash. The input values used for this study were liquid limit, plasticity index, optimum moisture content, and percentage

N. Shekar (✉) · S. Konai
Department of Civil Engineering, National Institute of Technology, Rourkela 769008, India
e-mail: niranjan14shekar@gmail.com

K. Muthukkumaran et al. (eds.), *Ground Improvement Techniques*, Lecture Notes in Civil Engineering 297, https://doi.org/10.1007/978-981-19-6727-6_18

fly ash added; for their analysis, Evolver 5.7 an add-in software of excel was used to predict CBR value with 99% accuracy.

Another key finding was mixed regression models predict the values better than linear model. Obianyo et al. [3] proved this by performing multivariate regression to predict strength of bone stabilized soil and found that non-linear and mixed models performed better than linear models with highest R^2 value of 97%. Also, Microsoft Office Excel is a versatile tool in performing regression analysis; Risaldar et al. [4] performed regression analysis using Microsoft Excel 2010 by stabilizing black cotton soil with red mud and gypsum and found that 25% of red mud and 1% of gypsum replacement was optimum.

Other important findings are the existence of interrelation among various geotechnical properties highlighted in Experimental Study to Correlate the Test Results of PBT, UCS, and CBR with DCP on various soils in soaked condition (Patel et al. [5]). Hence, in the present study, fly ash is chosen as the stabilizing agent and clayey soil as the construction material which is one of the problematic soils; then, predictive CBR relations were established using regression analysis.

Materials and Methodology

Materials Used

Clayey Soil. The soil used in this study was gathered from Tennis Court renovation site at NIT, Rourkela. The soil was sieved separately through 20-mm sieve and 4.75-mm sieve and stored in sacks at room temperature in lab. Geotechnical properties of the soil were thoroughly studied in the laboratory, and it is classified as intermediate compressible clay as per IS 1498 (1970). Wet sieve analysis of clay was done as per IS: 2720 (Part 4): 1985. The soil was then tested for liquid limit, plastic limit, and specific gravity. The general properties of the soil obtained are tabulated in Table 18.1.

Table 18.1 Clay soil properties

S. No.	Properties	Value
1	Specific gravity	2.69
2	Natural moisture content	6–8%
3	Liquid limit	36%
4	Plastic limit	18%
5	Plasticity index	18%
6	Shrinkage limit	14.7%
7	Optimum moisture content (OMC)	12.75%
8	Maximum dry density (MDD)	1.985 g/cm^3
9	Free swell index	22%

Table 18.2 Fly ash properties

S. No.	Properties	Value
1	Specific gravity	2.23
2	Natural moisture content	0.55–0.60%
3	Maximum dry density	13.82 kN/m^3
4	Optimum moisture content	20%
5	Cohesion	2 kPa
6	Angle of internal friction	31.6°
7	Fly ash classification	Class-F

Fly Ash. The mineral residue collected from electro static precipitators (ESPs) installed in coal-fired power plant exhaust stacks. These ashes generally consist of alumina, silica, and iron. The fly ash is generally micro-sized spherical particles which is silty and non-plastic in nature. Depending on constituents of the coal used, fly ash can be both amorphous and crystalline in nature. The general properties of the fly ash used is in Table 18.2.

Methodology Adopted

Experiments. To evaluate the effectiveness of fly ash as a stabilizing additive in clayey soils, series of tests were conducted by varying the content of fly ash added to clayey soil in range of 4–32% (multiples of 4%) by dry weight of the soil taken. The tests performed and Indian Standard codes adhered to are as follows: Atterberg limits, modified Proctor test, free swell index test, California bearing ratio (CBR) Test, and direct shear test.

Regression Analysis. To perform regression analysis and derive empirical relations between various geotechnical properties, the following methodology is utilized:

- To establish positive and negative relation between each geotechnical property
- To derive empirical relation for parameters with strong relation
- To perform multivariate regression analysis on all possible combination of IVs
- Repeating the above steps with linear and non-linear models.

Results and Discussions

Modified Proctor Test

With increase in fly ash added, maximum dry density is decreasing; this is because less dense fly ash is replacing denser clay (Fig. 18.1).

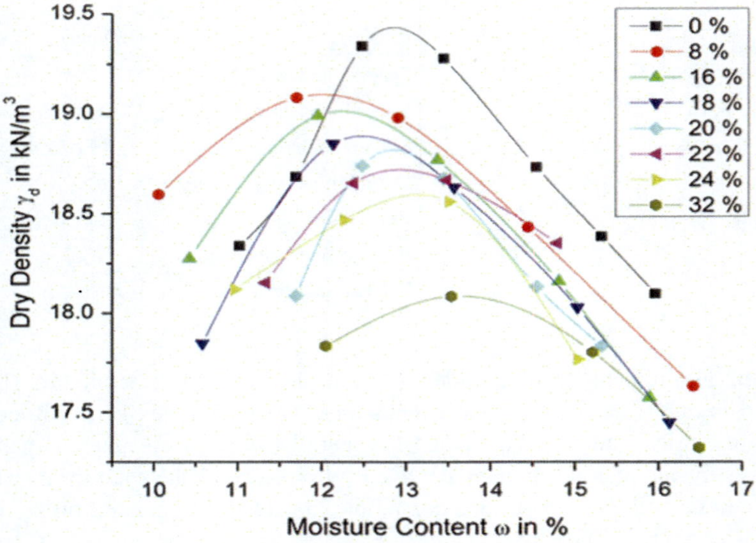

Fig. 18.1 Comparison graph of all modified Proctor test results

It can be observed that the variation of both MDD follows as trend of quadratic polynomial whereas OMC follows a cubic polynomial (Fig. 18.2).

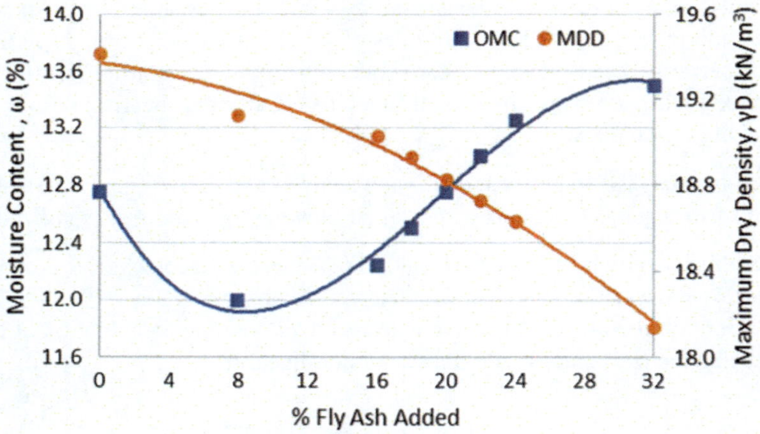

Fig. 18.2 Variation of OMC and MDD with increase in fly ash

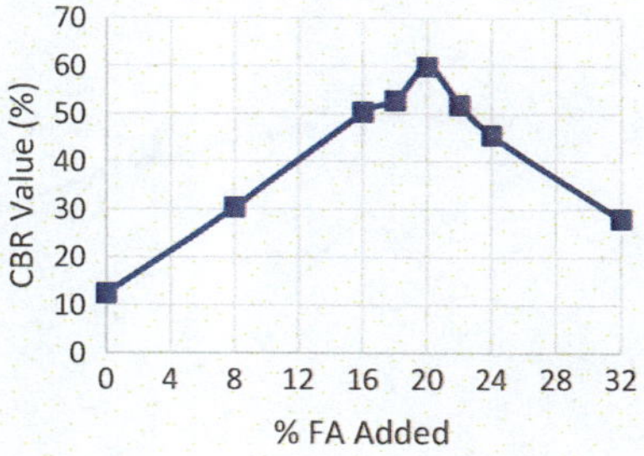

Fig. 18.3 Variation of unsoaked CBR value with increase in % FA

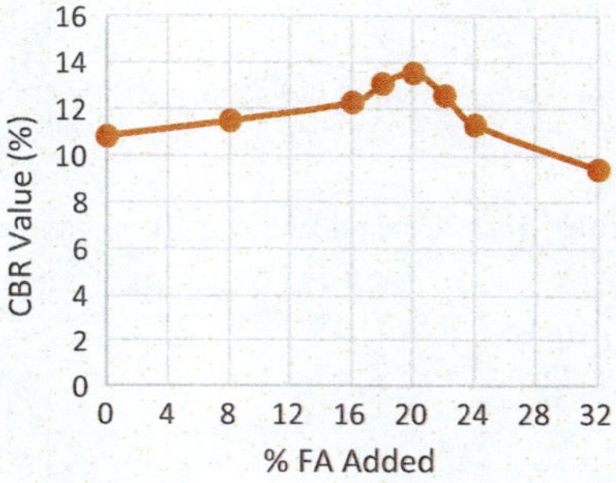

Fig. 18.4 Variation of soaked CBR value with increase in % FA

California Bearing Ratio Test

From Figs. 18.3 and 18.4, optimum fly ash content is 20% as both CBR (unsoaked) and CBR (soaked) are giving peak strength at 20% FA content.

Direct Shear Test

From Fig. 18.5, with increase in fly ash content, cohesion decreases, and angle of internal friction increases. This is in relation with properties of fly ash which is a non-cohesive, non-plastic material.

Regression Analysis

To reduce verbosity, CBR (soaked) will be abbreviated as CBRS, and CBR (unsoaked) will be abbreviated as CBRU only in equations. For ease of analysis, assumed increasing order of determination difficulty:

$$FA < G < SL < OMC < MDD < C < TPHI < CBRU < CBRS$$

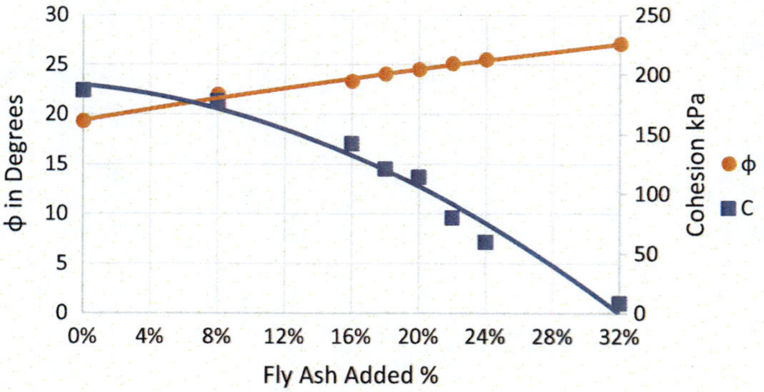

Fig. 18.5 Variation of C and φ with increase in fly ash

Table 18.3 Consolidated data for regression analysis

CBRU (%)	CBRS (%)	FA	OMC	MDD	G	SL (%)	C	TPHI
12.4	10.9	0	12.75	19.4	2.69	14.7	187	0.350
30.3	11.5	8	12.00	19.1	2.65	18.0	175	0.399
50.3	12.3	16	12.25	19.0	2.62	19.2	142	0.435
52.7	13.1	18	12.50	18.9	2.61	20.1	120	0.445
59.8	13.6	20	12.75	18.8	2.60	20.4	113	0.456
51.9	12.6	22	13.00	18.7	2.59	21.0	80	0.466

During prediction of various properties, the predicted equations have real benefits only when the dependent variable Y has higher order of determination difficulty than independent variable X. Below is the consolidated data considered for regression analysis (Table 18.3).

Univariate Linear Regression

Best Result.

Predicted Equation: $G = -0.005FA + 2.691$.

From the predicted equation, FA has a coefficient of −0.005 implying every 10% rise in FA content, G reduces by 0.05. For 100% replacement, the equation yields G = 2.231 which is the same value found experimentally in Table 18.2 tests on fly ash (Table 18.4; Fig. 18.6).

Other Result

See Table 18.5.

Univariate Non-Linear Regression

Best Result.

Table 18.4 Best linear regression result

Particular	Value
Dependent variable	G
Independent variable	FA
R	99.9%
R^2	99.7%
Standard Error	0.27%

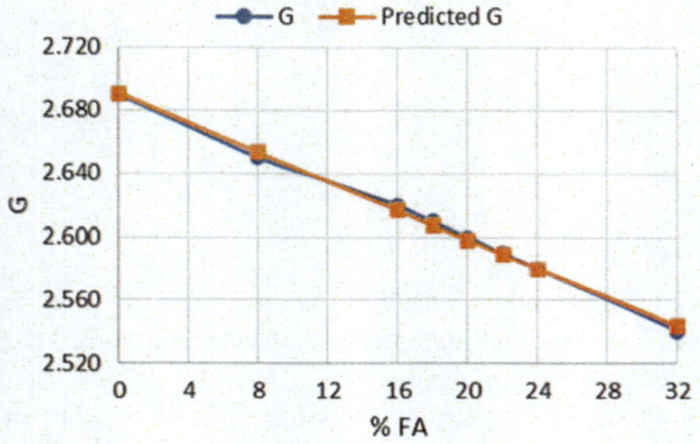

Fig. 18.6 G and Predicted G versus % FA

Table 18.5 Other univariate linear regression analysis results

Y	X	R (%)	R^2	Slope	Intercept	Equation
TPHI	FA	99.80	99.60%	0.005	0.355	TPHI = 0.005FA + 0.355
SL	FA	98.50	97.00%	0.003	0.154	SL = 0.003FA + 0.154
MDD	FA	−96.30	92.70%	−0.037	19.497	MDD = -0.037FA + 19.497
C	FA	−95.30	90.80%	−5.897	213.04	C = -5.897FA + 213.04

Predicted Equation: $MDD = -0.0008FA^2 - 0.0109FA + 19.372$.

From the graph, it can be observed the MDD decreases with increase in fly ash content (Table 18.6; Fig. 18.7).

Table 18.6 Best non-linear regression result

Particular	Value
Dependent variable	MDD
Independent variable	FA
R	96.69%
R^2	98.33%
Standard error	1.67%

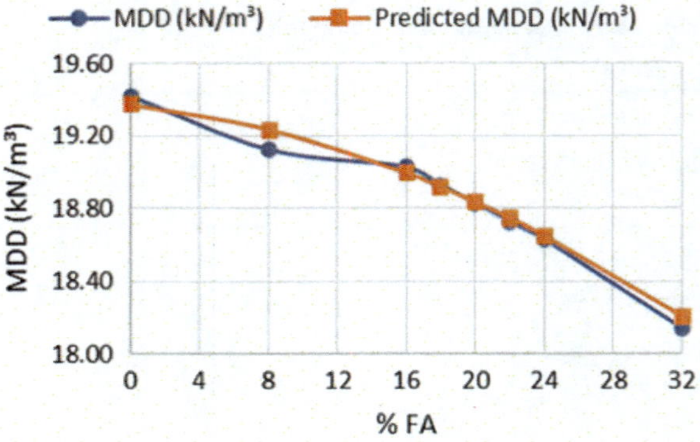

Fig. 18.7 MDD and predicted MDD versus % FA

Other Result.
See Table 18.7.

Multivariate Regression

The total number of independent variables (IV) available is 7, and number of test data available is 8. Considering 4 independent variable is assumed to be the best fit for the multivariate regression analysis.

Considering 4 IV out of 7 at a time, a total of 35 combinations of independent variables is considered for the analysis and prediction of both unsoaked CBR value and soaked CBR value.

Table 18.7 Other univariate linear regression analysis results

Y	X	R (%)	R^2 (%)	Equation	Linear R^2 (%)	Improvement (%)
MDD	FA	99.20	98.30	$MDD = -0.0008FA^2 - 0.0109FA + 19.372$	92.7	5.7
C	FA	98.70	97.50	$C = -0.1481FA^2 - 1.3232FA + 190.9$	90.8	6.7
TPHI	MDD	99.40	98.80	$TPHI = -0.0811MDD^2 + 2.9213MDD - 25.786$	90.5	8.3
SL	MDD	99.80	99.60	$SL = -0.0518MDD^2 + 1.882MDD - 16.873$	86.5	13.1
OMC	FA	99.30	98.50	$OMC = -0.0003FA^3 + 0.0168FA2 - 0.2232FA + 12.771$	44.6	53.9

Table 18.8 Best CBR (unsoaked) prediction result

Particular	Value
Dependent variable	CBRU
Independent variable	OMC, MDD, C, TPHI
R	99.5%
R^2	99.0%
Adjusted R^2	97.6%
Standard error	2.5%

Multivariate Regression CBR Un-Soaked

Best Result.

Predicted Equation: CBRU = −20.546 + 0.262OMC + 0.586MDD + 0.007C + 13TPHI.

From the graph, it can be observed the predicted CBR (unsoaked) value is very close to the experimental result obtained (Table 18.8; Fig. 18.8).

Other Result.

See Table 18.9.

Multivariate Regression CBR Soaked

Best Result

Predicted Equation: CBRS = −1.103+0.003FA+0.057OMC+0.001C+1.503SL.

From the graph, it can be observed the predicted CBR (soaked) value is very close to the experimental result obtained (Table 18.10; Fig. 18.9).

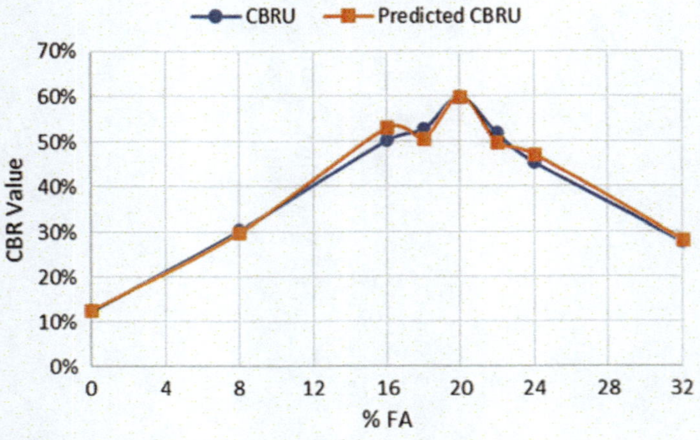

Fig. 18.8 CBRU and predicted CBRU versus % FA

Table 18.9 Other univariate linear regression analysis results

Independent Variables X_n				R^2 (%)	Adjusted R^2 (%)	Equation
OMC	G	C	TPHI	98.6	96.6	CBRU = −53.199 + 0.319OMC + 14.68G + 0.009C + 23.243TPHI
OMC	MDD	G	C	97.6	94.3	CBRU = 21.987 + 0.202OMC + 1.238MDD −18.463G + 0.006C
FA	G	C	TPHI	96.0	90.7	CBRU = −95.632 + 0.076FA + 32.979G + 0.002C + 19.032TPHI
OMC	C	SL	TPHI	96.0	90.6	CBRU = −13.389 + 0.493OMC + 0.013C + 7.668SL + 10.238TPHI

Table 18.10 Best CBRS prediction result

Particular	Value
Dependent variable	CBRS
Independent variable	FA, OMC, C, SL
R	96.3%
R^2	92.8%
Adjusted R^2	83.3%
Standard Error	0.5%

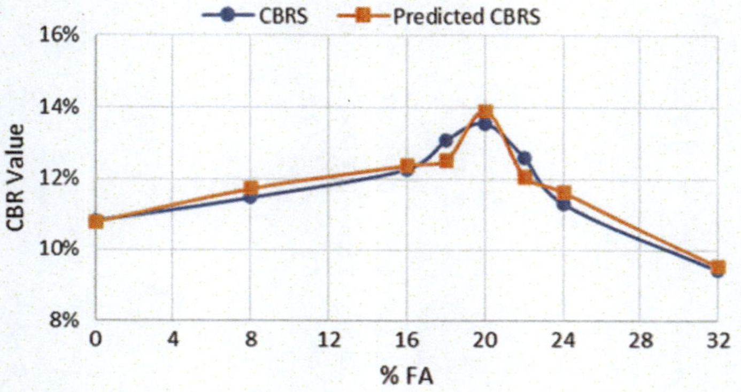

Fig. 18.9 CBRS and predicted CBRS versus % FA

Other Result

See Table 18.11.

Figure 18.10 showing the adjusted R^2 value of all combinations of MRA for predicting CBR (soaked) and CBR (unsoaked).

The selected IV are very good at predicting CBR (unsoaked) values whereas are not satisfactory in predicting soaked CBR (soaked) values.

Conclusion

Modified Proctor Test

- Maximum dry density reduced as the percentage fly ash added to soil increases
- Optimum moisture content reduces first till 8% fly ash and then increases linearly up to 32%.

Table 18.11 Other univariate linear regression analysis results

Independent variables X_n				R^2 (%)	Adjusted R^2 (%)	Equation
OMC	TPHI	C	SL	91.90%	81.20%	CBRS = −1.226 + 0.055OMC + 0.001C + 1.121SL + 0.632TPHI
OMC	TPHI	C	MDD	91.50%	80.10%	CBRS = −1.686 + 0.036OMC + 0.04MDD + 0.001C + 1.133TPHI
OMC	TPHI	C	G	91.50%	80.10%	CBRS = −4.044 + 0.039OMC + 1.047G + 0.001C + 1.859TPHI

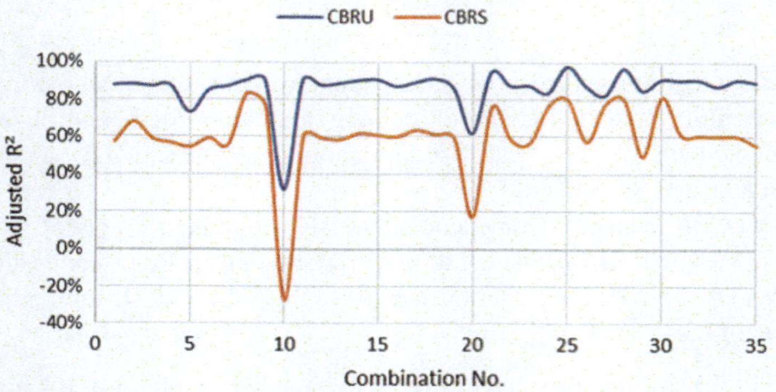

Fig. 18.10 All adjusted R^2 values of MRA for CBRU and CBRS

CBR Test

- Optimum fly ash content to be added to achieve good CBR value is 20% for both unsoaked CBR and soaked CBR test
- When optimum fly ash content of 20% is added by dry weight, the increase in CBR (unsoaked) value with respect virgin soil is 5 times, while the increase in CBR (soaked) value with respect virgin soil is 1.25 times
- From the above results, **20% fly ash** addition by dry weight of soil is recommended for clayey soil stabilization.

Direct Shear Test

- With increase in fly ash, cohesion of the soil is reducing quadratically up to 32% fly ash by dry weight of soil
- With increase in fly ash, angle of internal friction is increasing quadratically up to 32% fly ash by dry weight of soil
- This is in relation with properties of fly ash which is a non-cohesive, non-plastic material which has low cohesion and high angle of internal friction.

Regression Analysis

- Univariate Analysis
  - Linear: Out of 60 data pairs, 30 data pairs show strong correlation for univariate linear regression analysis with R^2 value greater than 80%
  - Non-Linear: Leaving the 30 data pairs showing strong linear correlation, 8 data pair shows strong correlation for univariate non-linear regression analysis with R^2 value greater than 95%.

- Multivariate Analysis
  - Linear CBR (unsoaked): 12 independent variable combinations out of 35 successfully predicted CBR (unsoaked) value with coefficient of regression value above 90%. The best IV combination was OMC, MDD, C, and TPHI with adjusted R^2 value of 97%
  - Linear CBR (soaked): 4 independent variable combinations out of 35 successfully predicted CBR (soaked) value with coefficient of Regression value above 80%. The best IV combination was FA, OMC, C, and SL with adjusted R^2 value of 83%.
- C and TPHI was found to be having the most impact in predicting CBR (unsoaked) and CBR (soaked) values
- The selected IV are very good at predicting CBR (unsoaked) values whereas are not satisfactory in predicting soaked CBR (soaked) values. This general trend was observed from Fig. 18.10.

References

1. IS 2720 Methods of test for soil
2. Karthik S, Ashok EK, Gowtham P, Elango G, Gokul D, Thangaraj S (2014) Soil stabilization by using fly ash. J Mech Civ Eng 10(6):20–26
3. Obianyo II, Anosike-Francis EN, Ihekweme GO, Geng Y, Jin R, Onwualu AP, Soboyejo ABO (2020) Multivariate regression models for predicting the compressive strength of bone ash stabilized lateritic soil for sustainable building. Constr Build Mater 263:120677
4. Patel MA, Patel HS (2012) Experimental study to correlate the test results of PBT, UCS, and CBR with DCP on various soils in soaked condition. Int J Eng (IJE) 6(5):244
5. Risaldar N, Rajashekhar MS, Patel M (2017) Stabilization of black cotton soil with red mud and formulation of linear regression between properties of the mixes

Chapter 19
A Study on the Strength Aspects of Alkali-Activated Red Mud-Crusher Dust-Blended Geopolymer

Subham Jena

Introduction

Geopolymers are created by dissolving alumina-silicate sources in a highly alkaline medium, releasing silica and alumina species, followed by diffusion, exchange, and oligomerization between the dissolved species, resulting in the formation of small coagulated structures, and finally hardening to form a solid. Geopolymers are inorganic alumina-silicate polymers having the chemical formula $\mathbf{M_n\text{-}(Si\text{-}O_2)_z\text{-}(\text{-}Al\text{-}O)_n}$, where M is the alkaline earth metals. They are generally stable up to 1250 °C and possess excellent binding properties comparable to that of cement. A survey shows that every 1 kg of cement emits nearly 1 kg of carbon dioxide into the atmosphere, which leads to global warming. These are low-carbon materials and do not produce so much carbon dioxide as compared to cement.

On the other hand, the safe disposal of red mud from alumina refineries has been challenging because of its high alkaline nature as it contaminates groundwater and surface water. It thus creates a hazardous risk to humans and animals. Crusher dust is also an industrial waste from the crusher plant, rich in silica, and acts as a filler material. Thus, disposing of these wastes is a significant burden to the engineers, aesthesis, and scientists.

Badanoiu et al. [1] used glass cullet and red mud to synthesize foamed geopolymers by thermal treatment of the wastes at a temperature between 600 °C and 800 °C and reported the strength of geopolymer quite appreciable that can be used for practical purposes. Kourti et al. [2] used plasma treatment of air pollution control residues for the synthesis of geopolymers by the use of direct current reporting to have a geopolymer of the strength of nearly (130 MPa) and density of about (2300 kg/m^3) with porosity of (5.5%) which clearly shows higher compressive strength than

S. Jena (✉)
Department of Civil Engineering, Indian Institute of Technology (Indian School of Mines), Dhanbad, Jharkhand 826004, India
e-mail: subhamjena.iitism@gmail.com

K. Muthukkumaran et al. (eds.), *Ground Improvement Techniques*, Lecture Notes in Civil Engineering 297, https://doi.org/10.1007/978-981-19-6727-6_19

Metakaolin or ground-granulated blast furnace slag geopolymers. Phummiphan et al. [3] used two by-products like fly ash and residue of calcium carbide for developing binders for stabilizing marginal lateritic soil as a geopolymer for sustainable pavement base with the use of alkaline activator NaOH + Na_2SiO_3 to 10 M. The seven days strength of marginal lateritic soil with the help of geopolymers made from fly ash and residue of calcium carbide with the different ratios of the above-used activators meets the strength requirement for the pavement used by both light and heavy vehicles for long term also. Yaghoubi et al. [4] had taken various proportions of fly ash and slag mixed with silt investigated their strength and did their microstructural analysis. The microstructural results confirmed that the calcium, mainly available in the slag, attributed more to strength development than the silica and alumina present in the fly ash. The change in the stiffness also added to the growth in strength. Based on results, 30% sodium hydroxide and 70% sodium silicate achieved the highest strength with 15% slag and 5% fly ash and can be used as sustainable binders in projects with significant benefits. Overall, an increase in stiffness was observed from the SEM and XRD analyses.

Bilondi et al. [5] used glass powder for geopolymer synthesis and confirmed that the compressive strength of the specimens treated using geopolymer was increased from 3 to 8 Mpa compared to untreated sample resulting in the maximum strength at 90 days of curing. The strength and the frictional strain values of all specimens treated with the geopolymer were increased compared to the original sample. However, the studies based on the stabilization of red mud and crusher dust using geopolymer were found to be limited. The current study uses two different types of industrial wastes, namely red mud and crusher dust. Both have been mixed in a fixed proportion and activated using an alkali activator, sodium hydroxide, synthesizing geopolymers. The alkali solution sodium hydroxide (97% purity) was used in different concentrations.

Materials Used and Methodology

Bauxite tailing, also known as bauxite residue or red mud or red sludge, is an industrial by-product from an alumina refinery. After being extracted from the sludge that is left over, the bauxite is highly rich in alumina is known as red mud. It is also known as alumina refinery residues (ARR) and is very alkaline and mainly composed of iron oxide and alumina. Every year globally, nearly about, 120 million tons of red mud are produced. Its storage causes a significant problem as it pollutes the ground and surface water by percolation. Crusher dust, a by-product of mining and crushing industries, is generally treated as waste material. However, it has many applications around and on construction sites. It is chemically rich in silica. It can also be used as an aggregate to create a substitute for concrete. The costs of crusher dust are comparatively low as compared to other building materials. It is fire, heat resistant, and non-plastic in nature. It also has some applications in horticulture as a natural fertilizer. It is an ideal material to stop mineral leaching in soils, thus reducing waterlogging. It can also be used as a compacting material. Odisha is a land of mines. This study brought red

Fig. 19.1 Red mud

mud from Vedanta Alumina Refinery, Langijarh, Raygada district, Odisha, India. The crusher dust used in the present study was collected from Gotidhara near Kalunga, Odisha, India. The alkali used was sodium hydroxide, originally in pellets with a molecular weight of 40 gm/mole and a specific gravity of 2.13 at 20 °C and 97% purity. The sodium hydroxide was brought from Rourkela, India. This was used based on the number of similar studies that have analyzed the effect of the alkaline activator composition on varying ratios.

After collecting the red mud, it was oven-dried at 105 °C for 24 h before its laboratory use. Then, the grain size analysis of various combinations of red mud and crusher dust was tested. The variety which comes to be well graded was taken into consideration for compaction purposes. The different combinations of red mud and crusher dust are taken in this study are 90% RM and 10% CD, 85% RM and 15% CD, 80% RM and 20% CD, 70% RM and 30% CD, 60% RM and 40% CD, 55% RM and 45% CD, 50% RM and 50% CD. Particle size distribution of the above combinations was done, and the one of well graded is taken for compaction with water. Then, compaction with sodium hydroxide is done as alkaline activation (Figs. 19.1 and 19.2).

Results and Discussion

Basic Geotechnical Properties

Fundamental geotechnical properties of red mud and crusher dust are conducted as per Indian standards. Their values are summarized in Tables 19.1 and 19.2.

Fig. 19.2 Crusher dust

Table 19.1 Basic properties of red mud

S. No.	Properties	Value
1	Specific gravity (G)	2.97
2	Liquid limit (LL)	38.80%
3	Plastic limit (PL)	25.65%
4	Shrinkage limit (SL)	23.12%
5	Plasticity index (PI)	13.15%
6	pH	9.75
7	Curvature coefficient (C_c)	1.00
8	Uniformity coefficient (C_u)	1.03

Table 19.2 Basic properties of crusher dust

S. No.	Properties	Value
1	% Clay	0.5%
2	% Silt	22.5%
3	% Sand	75%
4	% Gravel	2%
5	G	2.5
6	pH	7.77
7	Curvature coefficient (C_c)	25
8	Uniformity coefficient (C_u)	0.64

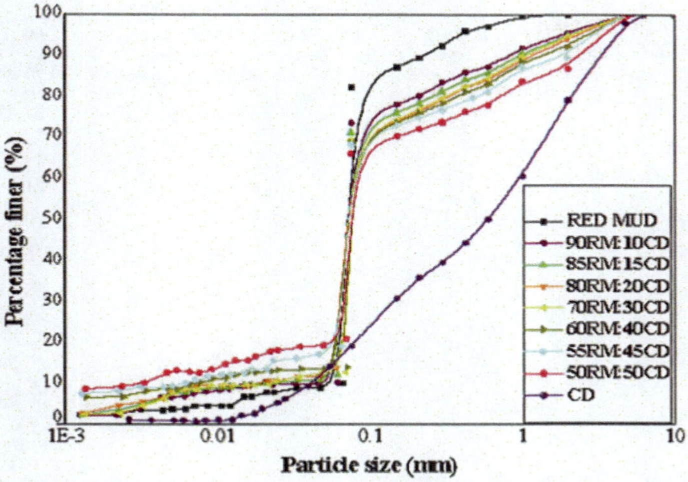

Fig. 19.3 Particle size distribution curves of RM, CD, RM-CD mixes

Grain Size Analysis and Specific Gravity

The grain size analysis of RM, CD, and RM-CD mixes (Fig. 19.3) is experimented with as per ASTM D7928. RM is observed to have uniformity coefficient (C_u) and curvature coefficient (C_c) values of 1.03 and 1.00, respectively, and that of CD are 0.64 and 25, respectively, which are, therefore, poorly graded materials. However, the C_u value of all RM-CD mixes is found to be greater than 4. The mixture of 90% RM with 10% CD and 85% RM with 15% CD comes to be well graded with C_c of 2.98 and 2.99, respectively. The C_c value of the remaining mixes ranges from 3.08 at 20% CD to 20.91 at 50% CD, but the combination remains poorly graded. The median diameter of red mud and crusher dust (d_{50}) was 0.07 mm and 0.8 mm, respectively. The specific gravity (Gs) value was studied as per ASTM D854. RM is observed to have particular gravity (Gs) value of 2.97 which is greater than that of the CD of value 2.5 due to the presence of a higher percentage of iron content in RM as indicated by Table 19.3. Vick [6] also reported the specific gravity value lies in between 2.8 and 3.3. The specific gravity of RM mix with 10, 15, 20, 30, 40, 45, and 50% CD is found to decrease with an increase in the percentage of CD.

Compaction Characteristics

The standard Proctor test/light compaction test (as per IS: 2720 (Part 7)–1980) was performed for the RM mixed with 10, 15, 16, 18, 20, 22, 24, 26, 28, 30% CD to determine the optimum moisture content (OMC) and the maximum dry density (MDD) which are presented in Fig. 19.4. Light compaction test is adopted because

Table 19.3 MDD and OMC of various RM: CD mixes

RM (%)	CD (%)	MDD (kN/m^3)	OMC (%)
90	10	17	22.41
85	15	17.4	21.82
84	16	17.49	21.39
82	18	17.61	21.2
80	20	17.67	20.95
78	22	17.78	20.71
76	24	17.81	20.48
74	26	17.78	20.51
72	28	17.67	20.54
70	30	17.64	20.63

the mixture is sandy. It is tested that by mixing 10, 15, 16, 18, 20, 22, 24% CD with RM, the MDD increases by 1.75, 0.52, 0.69, 0.34, 0.62, 0.17%, respectively, and the OMC decreases by 2.63, 1.97, 1.00, 1.18, 1.15, 1.12%, respectively. This is because of reduction in the percentage of clay, which reduces the resistance toward the movement of the soil grains at the time of compaction. It is also observed that the maximum degree of flocculation/optimum combination in terms of strength is achieved at the combination of 76%RM with 24% CD. Hereafter, the curves start moving downward with MDD decreasing. The graph of MDD, OMC vs. ratio of RM to CD is shown in Fig. 19.6 as it is observed that the optimum combination in terms of strength is achieved at 76%RM with 24% CD. Hence, with the above combination, alkali treatment with sodium hydroxide is done. The graph of MDD, OMC with sodium hydroxide as an alkaline activator is shown in Fig. 19.7. It is seen that the MDD and OMC are nearly the same when it is treated with a 2 M solution of sodium hydroxide solution as compared to water. As seen, the shrinkage limit of red mud is higher than that of OMC of various percentages because the geopolymer will develop cracks on moisture reduction (Fig. 19.5).

It is thus observed that the optimum maximum dry density (MDD) of 17.81 kN/m^3 and optimum moisture content (OMC) of 20.48% with water occurs at 76% RM + 24% CD combination, as seen from Fig. 19.6. However, on mixing with sodium hydroxide as an alkaline activator, the MDD and OMC come to be nearly the same as that with water, as observed from Fig. 19.7 because of the alkaline nature of red mud used as raw material (Table 19.4).

Microstructure Analysis

Microstructural images of the raw materials RM, CD, and the optimum combination of the RM-CD mix obtained using a scanning electron microscope (SEM) are shown

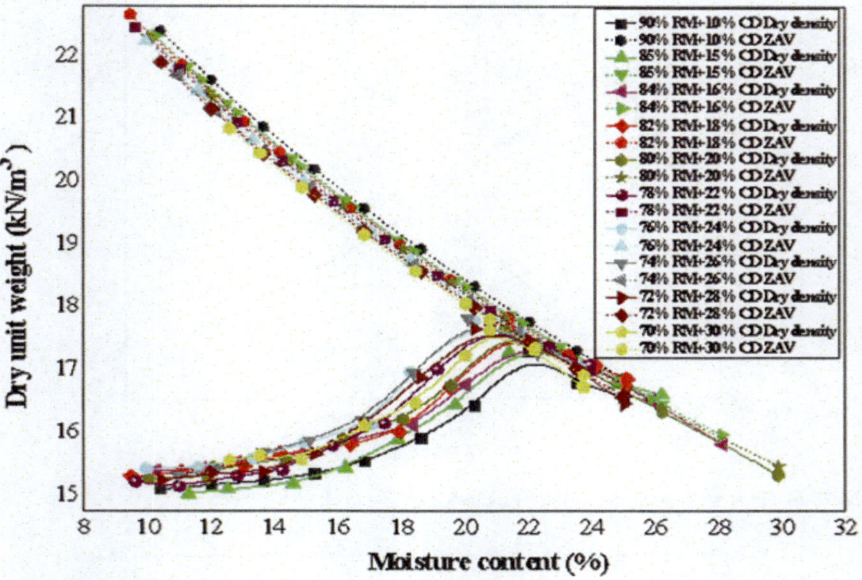

Fig. 19.4 Compaction curves with water

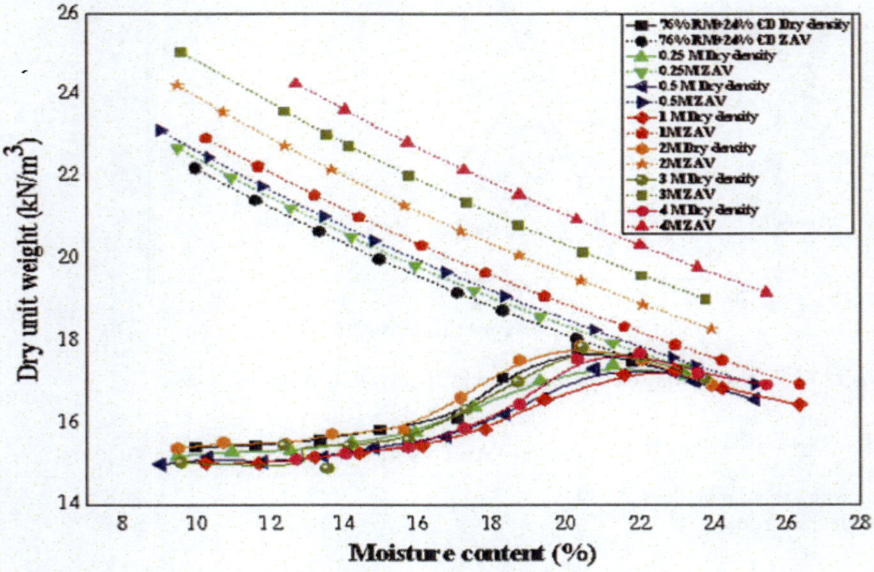

Fig. 19.5 Compaction curves with sodium hydroxide of different molarities

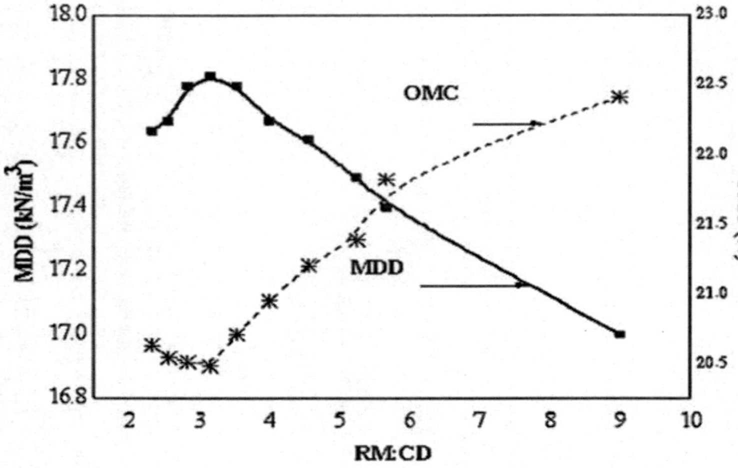

Fig. 19.6 Graphs showing the ratio of mix to MDD and OMC

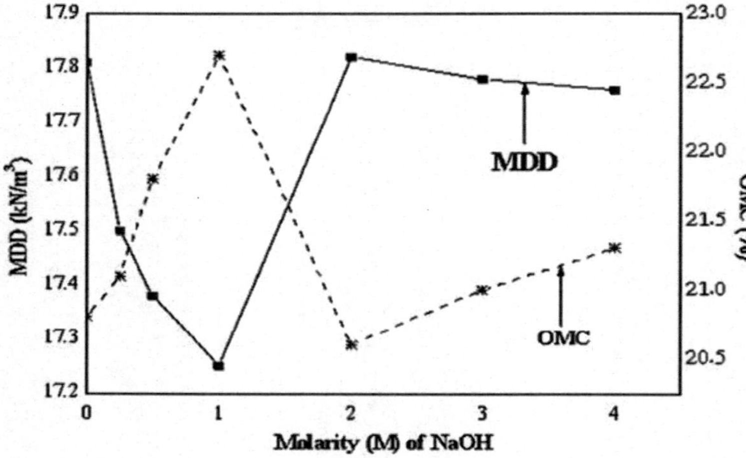

Fig. 19.7 Graphs showing the ratio of mix to MDD and OMC versus molarities

Table 19.4 MDD and OMC with 76% RM + 24% CD mix with various molarities of sodium hydroxide

Molarity (M) of NaOH	MDD (kN/m^3)	OMC (%)
0	17.81	20.8
0.25	17.5	21.1
0.5	17.38	21.8
1	17.25	22.7
2	17.82	20.6
3	17.78	21.0
4	17.76	21.3

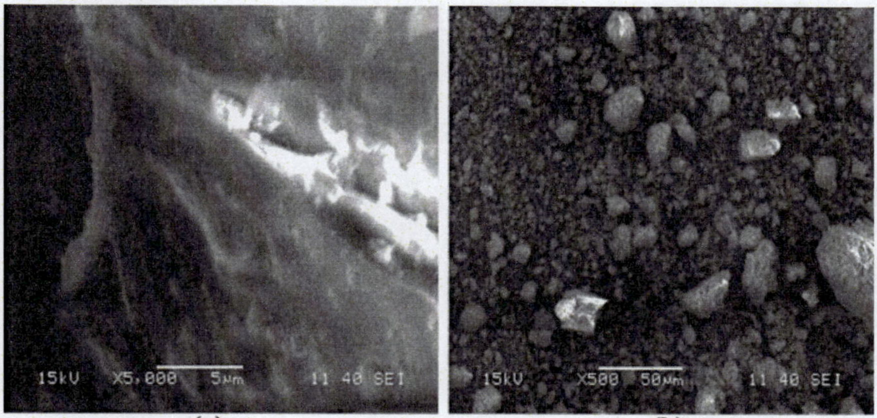

(a) (b)

Fig. 19.8 SEM image of red mud at **a** 5000X, **b** 500X magnification

below. The microstructural images of red mud and crusher dust at various magnifications are shown in Figs. 19.8 and 19.9, respectively. It can be inferred that the red mud is a fine-grained material, and crusher dust is coarse-grained. In contrast, on mixing 76% RM + 24% CD, it comes to be uniformly grained, which can be best suited for the synthesis of geopolymer as uniformly grained structures are highly dense, clearly shown in Fig. 19.10. The energy-dispersive X-ray spectroscopy (EDS) of red mud and crusher dust and 76% RM + 24% CD and their elemental composition are shown in Figs. 19.14, 19.15, 19.16, and Tables 19.5, 19.6, 19.7, 19.8, 19.9 and 19.10. The XRD graphs of RM, CD, and 76%RM + 24%CD are shown in Figs. 19.20, 19.21, 19.22, respectively. Figure 19.20 presents XRD of red mud, which shows the presence of bornite, burnt ochre, calcite ($CaCO_3$), green cinnabar, hematite (Fe_2O_3), praseodymium, quartz (SiO_2), Uranium oxide. Broad humps at 30° in 76% RM + 24% CD mix indicate the presence of a significant amorphous stage, which leads to the development of pozzolanic activity [7] (Figs. 19.11, 19.12, 19.13, 19.17, 19.18 and 19.19).

Unified Compressive Strength

The unified compressive strength values for 76% red mud and 24% crusher dust with 2 M NaOH solution are shown below. As shown from the graph below, the strength of the geopolymer increases exponentially with increase in curing days, and the ratio slightly approaches less than 1 (Fig. 19.23).

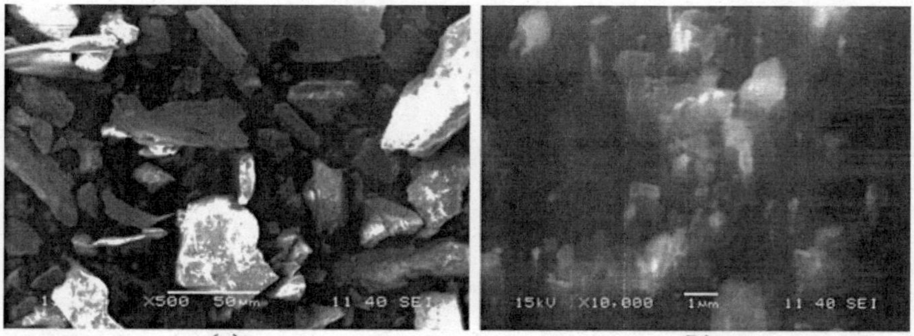

(a) (b)

Fig. 19.9 SEM image of crusher dust at **a** 500X, **b** 10000X magnification

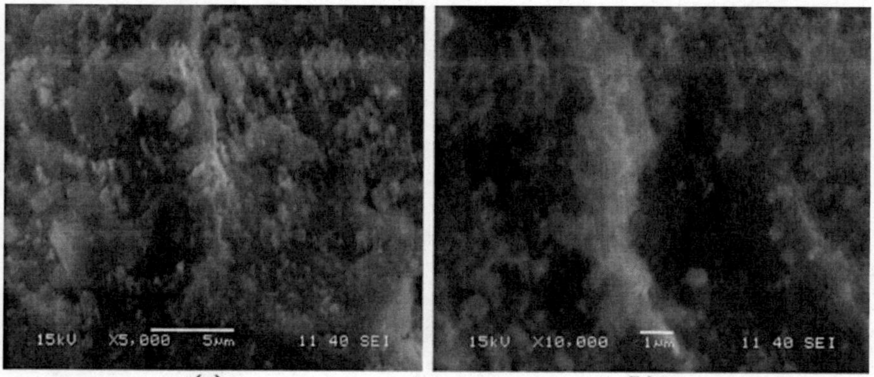

(a) (b)

Fig. 19.10 SEM image of 76% RM + 24% CD at **a** 5000X, **b** 10000X magnification

Table 19.5 Elemental composition of red mud

Element	Weight%	Atomic%
OK	31.33	59.54
Na K	1.06	1.40
Al K	1.49	1.68
Si K	0.87	0.95
Ca K	1.07	0.81
TiK	7.46	4.73
Mn K	0.45	0.25
Fe K	56.27	30.64
Totals	100.00	

Table 19.6 Elemental composition of crusher dust

Element	Weight (%)	Atomic (%)
O K	58.49	70.35
Na K	7.70	6.45
Mg K	0.45	0.35
Al K	8.43	6.02
Si K	23.46	16.08
P K	0.64	0.40
Ca K	0.58	0.28
Mn K	0.15	0.05
Zn K	0.10	0.03
Totals	100.00	

Table 19.7 Elemental composition of 76% RM + 24% CD

Element	Weight%	Atomic%
O K	52.06	65.89
Na K	0.92	0.81
Mg K	0.67	0.55
Al K	4.65	3.49
Si K	38.03	27.42
P K	1.58	1.03
K K	0.13	0.07
Ca K	0.13	0.07
Fe K	1.83	0.67
Totals	100.00	

Table 19.8 Elemental composition of 76% RM + 24% CD activated at 2 M NaOH

Element	Weight (%)	Atomic (%)
O K	66.07	78.68
Na K	4.74	3.93
Mg K	0.36	0.28
Al K	10.95	7.73
Si K	8.71	5.91
K K	0.92	0.45
Ca K	0.49	0.23
Ti K	2.08	0.83
Fe K	6.26	2.13
Cu L	-0.58	-0.17
Totals	100.00	

Table 19.9 Elemental composition of 76% RM + 24% CD with water

Element	Weight%	Atomic%
O K	60.38	74.86
Na K	5.60	4.83
Mg K	0.22	0.18
Al K	12.34	9.08
Si K	9.07	6.40
Ca K	0.91	0.45
Ti K	2.14	0.88
Mn K	0.13	0.05
Fe K	9.01	3.20
Co K	0.21	0.07
Totals	100.00	

Table 19.10 Elemental composition of 76% RM + 24% CD activated at 0.5 M NaOH

Element	Weight%	Atomic%
O K	62.44	75.88
Na K	4.39	3.72
Mg K	0.34	0.27
Al K	10.92	7.87
Si K	12.36	8.56
K K	2.18	1.09
Ti K	1.54	0.63
Mn K	0.08	0.03
Fe K	4.89	1.70
Co K	0.03	0.01
Cu L	0.82	0.25
Totals	100.00	

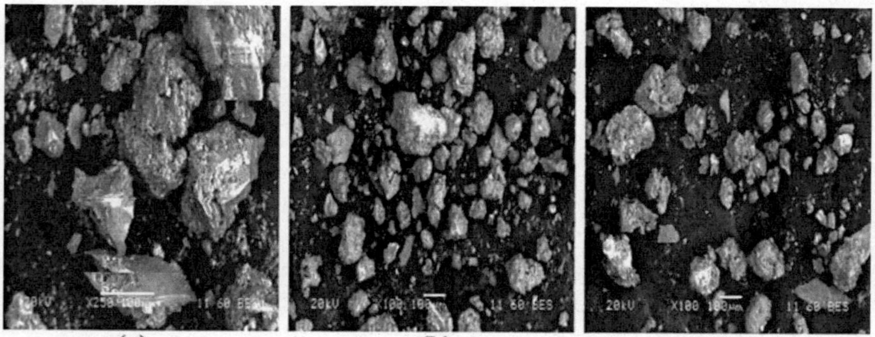
(a) (b) (c)

Fig. 19.11 SEM image of 76% RM + 24% CD activated at 0.5 M sodium hydroxide

Fig. 19.12 SEM image of 76% RM + 24% CD activated at 2 M sodium hydroxide

Fig. 19.13 FE-SEM image of 76% RM + 24% CD with water

Conclusions

According to the author's best understanding, developing a geopolymer using stabilized red mud with alkali-activated crusher dust and discussions of its compaction characteristics is the initial move in this direction.

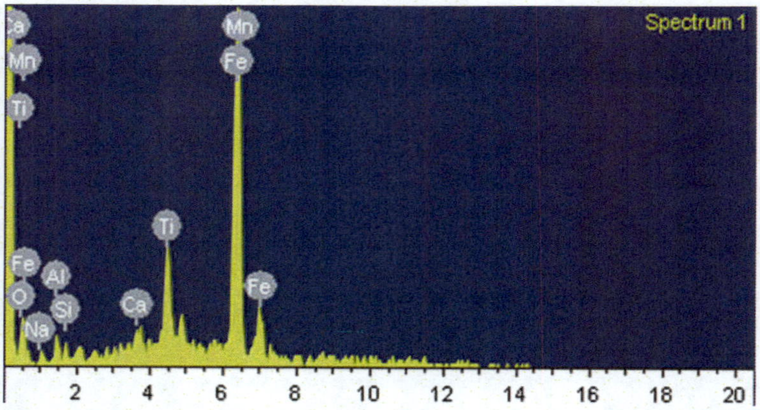

Fig. 19.14 EDX for red mud

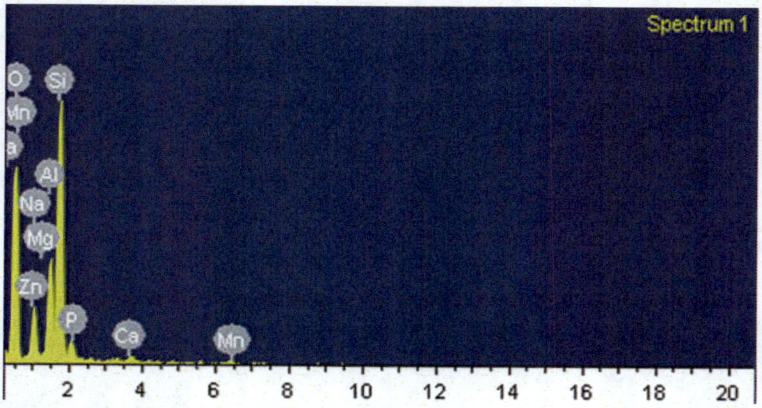

Fig. 19.15 EDX for crusher dust

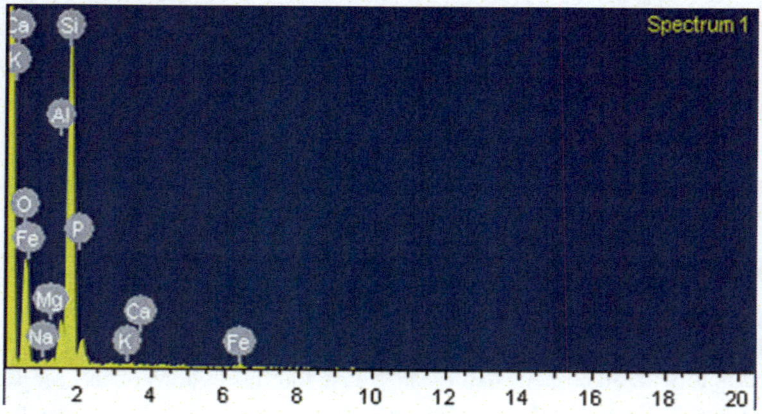

Fig. 19.16 EDX for 76% RM + 24% CD

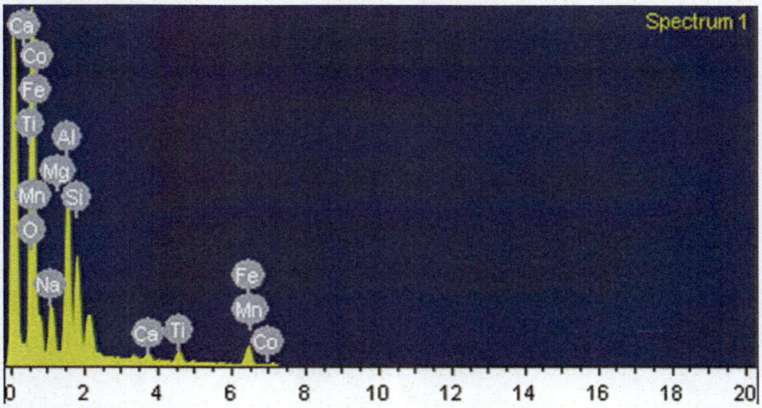

Fig. 19.17 EDX for 76% RM + 24% CD activated at 2 M NaOH

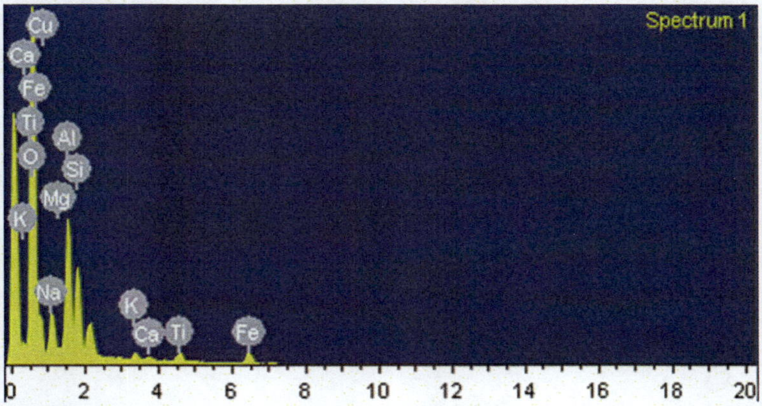

Fig. 19.18 EDX for 76% RM + 24% CD with water

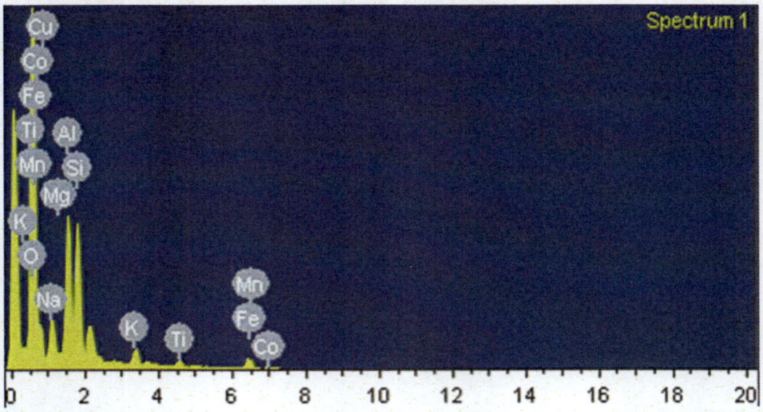

Fig. 19.19 EDX for 76% RM + 24% CD activated at 0.5 M NaOH

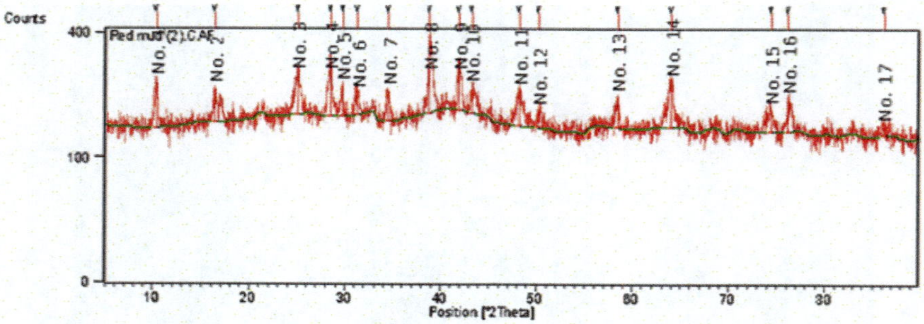

Fig. 19.20 XRD of RM. 1 Bornite, 2 Burnt ochre, 3 Calcite, 4 Chromium, 5 Green cinnabar, 6 Hematite, 7 Praseodymium, 8 Quartz, 9 Silica, 10 Uranium oxide

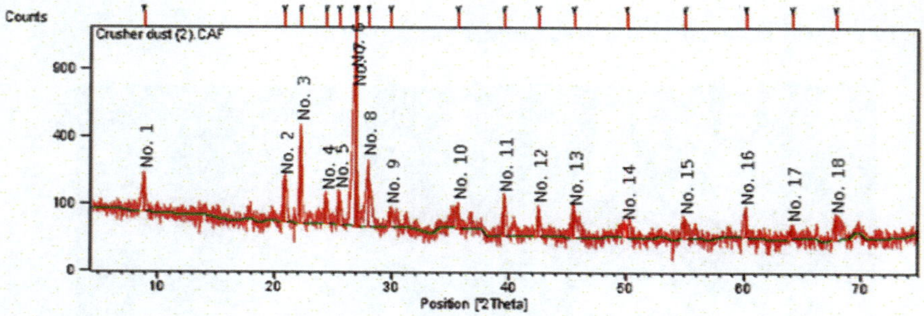

Fig. 19.21 XRD of CD. 1 Silica, 2 Quartz, 3 Fluorite, 4 Hematite, 5 Burnt ochre, 6 Bornite, 7 Calcite

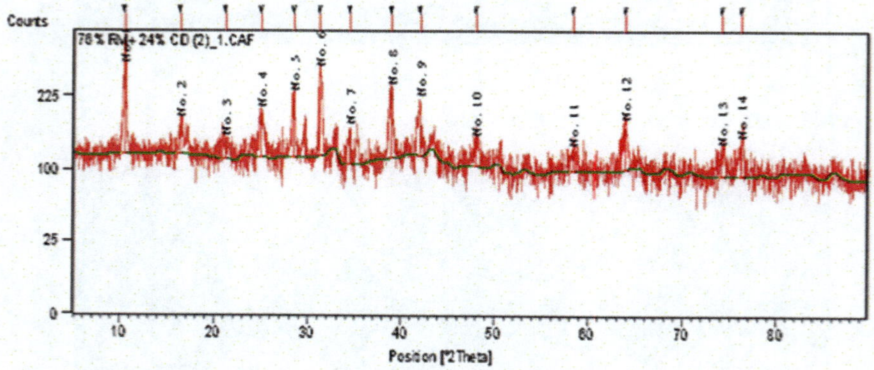

Fig. 19.22 XRD of 76% RM + 24% CD. 1 Burnt ochre, 2 Hematite, 3 Silica, 4 Quartz, 5 Calcite

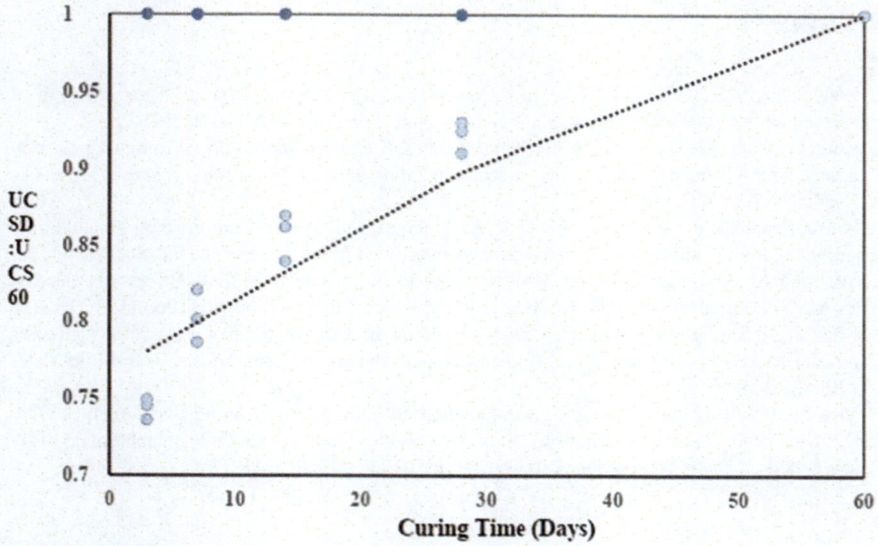

Fig. 19.23 UCS results of the sample with curing period

Relying on red mud and crusher dust laboratory investigations with sodium hydroxide as an alkali activator based on the study, the following inferences are drawn.

1. The mixture of 90% RM with 10% CD and 85% RM with 15% CD comes to be well graded with C_c of 2.98 and 2.99, respectively, and ($C_u > 4$).
2. RM is seen to have specific gravity (Gs) value of 2.97, more significant than that of the CD of value 2.5 due to more percentage of iron content in RM.
3. The optimum combination of red mud and crusher dust is 76% red mud and 24% crusher dust for geopolymer formation as obtained by compaction using water.
4. The optimum molarities of sodium hydroxide solution are 2 M for the geopolymer formation as obtained by compaction using sodium hydroxide as an alkaline activator.
5. It is observed that the MDD and OMC are nearly identical when treated with a 2 M solution of sodium hydroxide solution as compared to that of water with the above optimum combination.
6. The strength aspects of red mud and crusher dust geopolymer and sodium hydroxide as an alkali activator are not found to have so much strength with curing days.

References

1. Badanou AL, Saddi THAL, Stoleriu S, Voicu G (2015) Preparation and characterization of foamed geopolymers from waste glass and red mud. Constr Build Mater 84:284–293
2. Kourti I, Rani AD, Boccaccini AR, Cheeseman CR (2011) Geopolymers from DC Plasma-treated air pollution control residues, metakaolin, and granulated blast furnace slag. J Mater Civ Eng 23(6):735–740
3. Phummiphanane G, Amu OO, Fajobi AB, Afekhuai SO (2015) Stabilisation of marginal lateritic soil using high calcium fly ash-based geopolymer. Road Mater Pavement Des 25(4)
4. Yaghoubi M, Arulrajah A, Disfani MM, Horpibulsuk S, Darmawan S (2018) Effects of industrial by-products based geopolymers on the strength development of soft soil. Soils Found 58:716–728
5. Bilondi MP, Toufigh MM, Toufigh V (2018) Experimental investigation of using a recycled glass powder-based geopolymer to improve the mechanical behavior of clay soils. Constr Build Mater 170:302–313
6. Vick SG (1990) Planning and design and analysis of Tailings Dams. Wiley, New York
7. Das SK, Yudhbir (2006) A simplified model for prediction of a pozzolanic characteristic of fly ash, based on chemical composition. Cem Concr Res 36(10):1827–1832

Chapter 20
A State-of-the-Art Review on Electro-osmotic Treatment for Stabilization of Soft Soils

B. K. Pandey, S. Rajesh, and S. Chandra

Introduction

The infrastructure on soft ground is expanding considerably around the world. Infrastructure projects are critically hindered by the nature and behavior of the encountered soil. Once soft soils are encountered during infrastructure projects, it requires improvement methods to decrease water content, reduce compressibility, fast-track the consolidation rate and increasing the undrained shear strength and bearing capacity [15, 19, 31]. The selected soil modification methods should be able to provide solution which is economically feasible and structurally stable. Moisture content in soft ground plays a decisive role in the stability of soft ground infrastructure. The strength is also governed by the rate of dissipation of excess pore water pressure. The soft soil consolidation using stone columns, preloading, vertical drains or encased stone columns [29, 33, 34, 36, 37] helps in dissipating the developed pore pressure. Nonetheless, in numerous projects related to reclamation, infrastructure on soft clay, dredging and mining, the dredge or fill materials have lower undrained shear strength and extremely higher moisture content. Therefore, these kinds of extremely challenging sites need to be improved and employing traditional soil improvement techniques will not yield desired outcome. In such scenario the application of electro-osmotic treatment may be handy [10, 35].

The electrokinetic phenomena work on the assumption that by using a small electrical gradient between the electrodes positioned in the soil to be improved, counterions travel to anode and cathode. Thus, substantial physicochemical movement happens inside the soil, causing elimination of pore water, inorganic, heavy metals and organic contaminates existing in the soils, along with significant alteration in the

B. K. Pandey (✉) · S. Rajesh · S. Chandra
Indian Institute of Technology Kanpur, Kanpur, Uttar Pradesh 208016, India
e-mail: balbirpandey123@gmail.com

B. K. Pandey
Shri Mata Vaishno Devi University, Kakryal Katra, Jammu and Kashmir 182320, India

K. Muthukkumaran et al. (eds.), *Ground Improvement Techniques*, Lecture Notes in Civil Engineering 297, https://doi.org/10.1007/978-981-19-6727-6_20

soil composition, fabric orientation, and pH of the soft clay [2, 9, 20, 36]. The electrokinetic phenomena like electro-migration, electroosmosis and electrophoresis are commonly applied to improve the soils [30]. Extensive research has been conducted to recognize the details of electrokinetic phenomena for soft ground modification [10]. The literature have emphasized the benefits of in-situ electroosmosis modification in pile bearing capacity, bearing capacity, undrained shear strength, reduction of liquefaction, drainage, and changes in physicochemical parameters of soft ground; concluded that this modification method is faster, cheaper, and more permanent [1, 5, 32, 39, 42]. These studies have made substantial understanding relating to the process of electroosmosis. There is some evidence that the chemical reactions that occur during electroosmosis lead to changes in the fundamental index, physicochemical and engineering properties of soft ground, but a comprehensive review of these aspects is limited [3, 7, 13, 23, 24, 40].

The primary aim of this review is to systematically examine the critical features that affects the efficacy of electro-osmotic technique and summarizing the outcomes of several studies to understand changes in soil moisture content, consolidation properties and undrained shear strength due to electro-osmotic treatment.

Fundamental Principle

In the soil system, the liquid, heat, chemicals or electrical flow rate is proportional to the gradient. The gradient may be result of electrical and chemical concentrations, hydraulic head or temperature [26]. Interestingly, a gradient (P_j) of one kind led to a movement of another (T_j) as per Eq. (20.1).

$$T_j = \alpha_{ij} P_j \tag{20.1}$$

where α_{ij} is coupling coefficient. If the electric gradient dictates the movement of pore water in soft soil, then phenomena is called electroosmosis and treatment is called electro-osmotic treatment. This technique of modification is mainly appropriate for soft clays with lower permeability, which needs strengthening [14, 17]. Figure 20.1 illustrates the representation of the different phenomena occurring during the electrokinetic modification of the soil. Specifics of these processes may be found from numerous supplementary publications [22].

Electroosmosis is one of numerous electrokinetic processes that performs similar to electrochemical cell and causes water flow when a voltage gradient is applied to a soil water system [12, 17, 22]. After applying the voltage gradient, the cations are attracted to the cathode from the outer diffusion layer. During the process, they drag surrounding free and bound water along with it, causing a pore water movement toward the cathode as shown in Fig. 20.1. The transport of pore water decreases the moisture content, thus increasing the soil shear strength and effective stress. The elements that govern the amount of moisture movement are (i) presence of counterions and (ii) cation existence in soil water [3, 23]. However, the rate of transport of

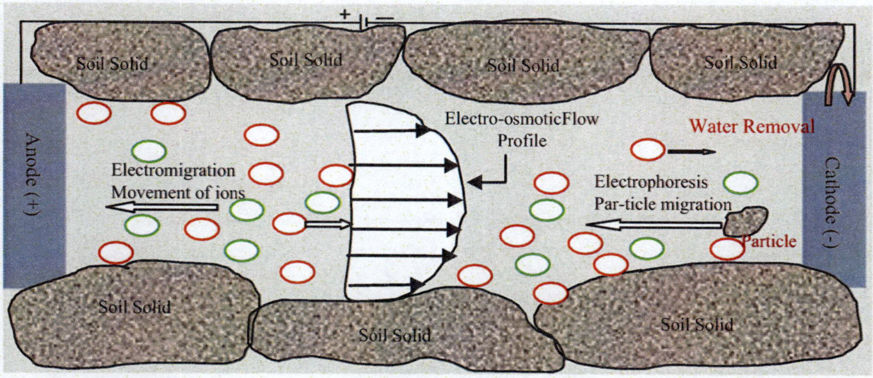

Fig. 20.1 Illustration of different phenomena occurring during the electrokinetic modification

pore water and resultant direction is assessed after the transmission of momentum by counterions existing in soil water, as shown in Fig. 20.1.

Factors Affecting the Efficiency

The electrokinetic modification for in-situ implementation is accompanied by the electrochemical alterations [26]. These alterations are extremely reliant on the medium and the chemical constituents of soil and is very location specific. Figure 20.2 illustrates factors influencing the efficacy of electrokinetic phenomena in soft clay. The important factors can be broadly classified into soil, electrode, and operational properties. The electrokinetic phenomena in the soil cause altercation of ions, electrolysis, heat generation, modification in mass composition and pH of soil. Further, it also causes deterioration of electrode materials causing alteration in mineral phase. The electrochemical reactions at the anode and cathode [23] are indicated by Eqs. (20.2)–(20.4):

At anode

$$2H_2O - 4e^- \rightarrow O_2 \uparrow +4H^+, \quad E_o = 1.23\ V \tag{20.2}$$

At cathode

$$2H_2O + 4e^- \rightarrow 2H_2 \uparrow +4OH^-, \quad E_o = -0.83\ V \tag{20.3}$$

$$2H_3O^+ + 2e^- \rightarrow 2H_2O + H_2 \uparrow, \quad ,E_o = 0\ V \tag{20.4}$$

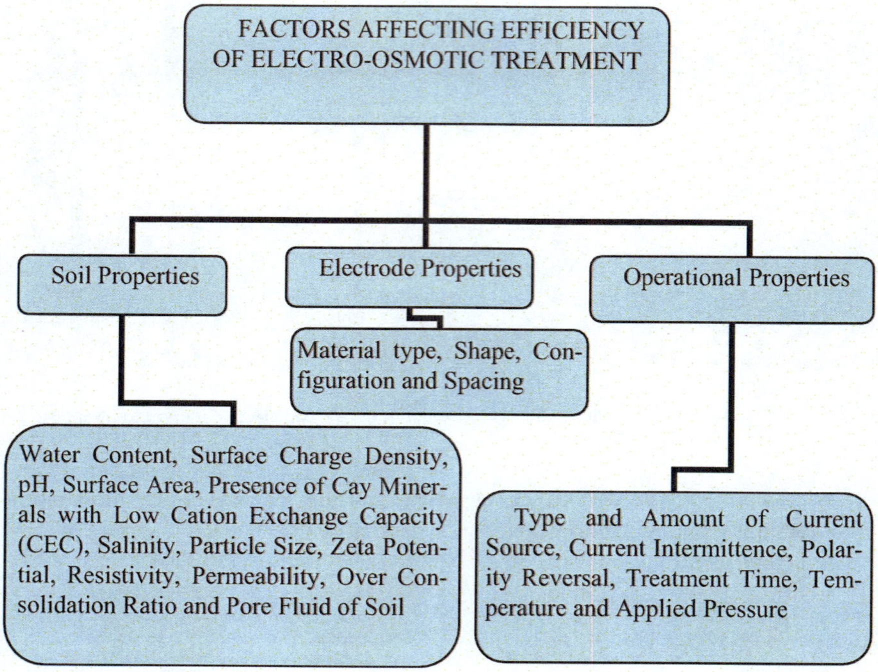

Fig. 20.2 Factors influencing the efficacy of electrokinetic phenomena in soft clay

The term E_0 is standard electrode potential measured at 298 K. Due to formation of OH^- and H^+ at the cathode and anode, local conductivity of soil is affected. At the same time, it alters the pH of the soil in the region of electrode.

Observations on Modification of Soil Properties

Moisture Content

This modification method is predominantly used to decrease the moisture content of encountered soil [4, 5, 11, 38, 41]. Thus, to evaluate the effectiveness of electro-osmotic technique, the degree of modification in moisture content of the soft soil at the end of the modification is evaluated. The modification in moisture content is stated as the ratio of the change in moisture content to that of the original moisture content of the soft soil. Figure 20.3 shows the variation of water content between the electrodes. The test detail and its duration of cited literature is stated in the alphameric data format, for example, Micic et al. [25] [$EO_t = 144$ h, test-4 K-EK] specifying 144 h of treatment time (EO_t) equivalent to a test legend 4 K-EK as described in Micic et al. [25]. Tajudin [41] reported higher reductions of 5.94% and 0.33% in the vicinity of anode and cathode at termination of test (i.e., 14 days). Further, Bjerrum et al. [5] used mild steel bar as an anode and cathode for in-situ reduction of moisture content and reported reduction of approximately 24.9% in the vicinity of anode after 120 days of treatment. However, close to cathode, alteration in the moisture content was inconsequential (1.17%). This behavior is perhaps due to the greater electrode spacing (i.e., 2 m). To overcome this issue of lesser decrease in moisture content close to cathode Micic et al. [25] varied the voltage gradient and adopted current intermittence. The decrease in moisture content at the treatment time was 15.59% close to anode and 4.93% near cathode. Although the moisture content of the soil near the cathode decreased due to the current interruption, the difference was much smaller than the moisture content of the soil next to the anode, resulting in irregularity in the moisture content of the soft soil being treated.

The polarity reversal helps in achieves uniformity in moisture content reductions. Bergado et al. [4] studied the modification soft clay by coupling PVD with electroosmosis and utilizing the reversal of polarity during the test. It was observed that the polarity reversal helps in uniform reduction in moisture content irrespective of testing condition. Comparable outcomes were observed by Rittirong et al. [38]. Further, the heterogeneity in moisture content reduction can also be decreased by coupling electroosmosis with other soil improvement methods. Comparable observations have been made by Liu et al. [21], at the end of 120 h, the decrease in moisture content close to anode was 33.15% and close to cathode it was 40.87%. Fu et al. [11] stated that the use of vacuum preload with electrodes at different distances achieved a 46.46% reduction nearby anode and 49.73% close to cathode. Meanwhile, Kaniraj and Yee [18] chose changed pump intervals to achieve uniform moisture content within the treatment zone.

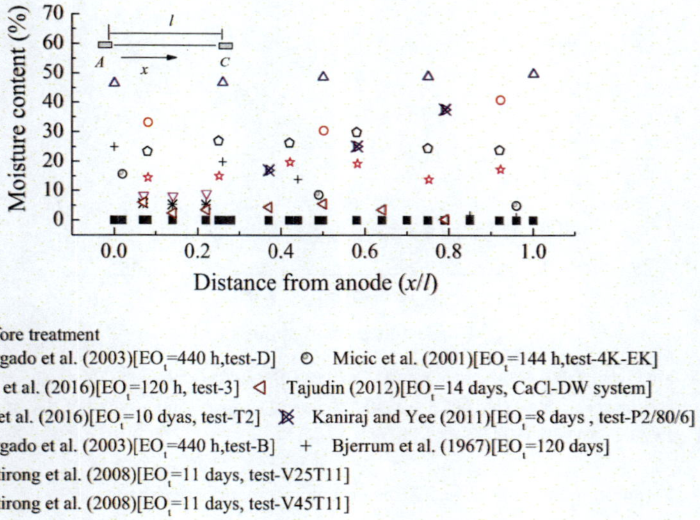

Fig. 20.3 Variation in reduction of moisture content between electrodes

The above discussions confirms that significant consistency in the moisture content reductions can be accomplished by (a) decreasing pumping time, (b) current intermittence (c) polarity reversal (d) scheming suitable layout of electrodes, and (e) coupling the electro-osmotic treatment with other improvement technique. The reasoning behind decreasing moisture content is due to increased electrical gradients, alterations in improvement time, initial moisture content, type of soil and their parameters.

Undrained Shear Strength

The alteration in the undrained shear strength of soft soil is observed between the electrodes after the treatment. The efficacy of the process in improving the undrained shear strength is articulated as the ratio of alteration in the undrained shear strength at the termination of the test to that of original strength of soft soil. Figure 20.4 shows the variation in the undrained shear strength between electrodes, acquired from experimental observations of Micic et al. [25], Chew et al. [8], and others. Burnotte et al. [6] observed 130.9% increase in undrained shear strength close to anode and roughly 100% in the vicinity of cathode. A comparable pattern was witnessed by scientists like Bergado et al. [4]. Tajudin [41] witnessed a maximum increase happening at the mid of the electrodes and the increase is extremely reliant on treatment duration. Remarkably, if the reversal of polarity was implemented, a uniform enhancement in the soil strength was detected between the anode and cathode. The enhancement was due to electrochemical reactions, cementation and increase in the soil density that ensues after treatment.

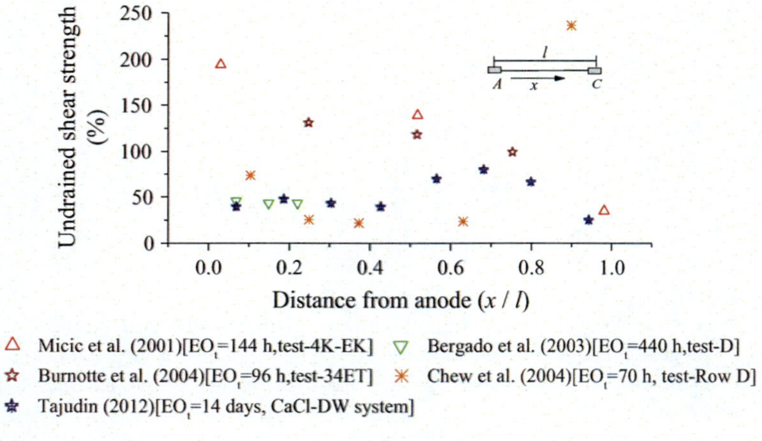

Fig. 20.4 Variation in undrained shear strength enhancement between the electrodes

Consolidation and Dewatering Characteristics

Figure 20.5 illustrates the representation of dewatering practice normally employed in-situ. The free water existing in the soil voids may be removed by generating a hydraulic head as can be seen in Fig. 20.5a, or by incorporating materials having high permeability. The physical dewatering methods like filtration, gravimetric settling, centrifugation, surcharge loading, and others have greater efficacy in removing part of interstitial water along with free water situated outside diffused double layer as noticed in Fig. 20.5b. Though, the discharge rate reliant on the field situations, the electro-osmotic dewatering is useful in dissipating a substantial part of vicinal and interstitial water. This may lead to alteration in the soil microfabric [23, 32]. Though, the electro-osmotic dewatering is not effective in removing chemically bound water. For supplementary data on electro-osmotic dewatering readers may refer to Fourie et al. [10] and Glendinning et al. [14].

It is apparent of findings mentioned above that the electro-osmotic modification of soil is highly efficient in expediting removal of pore water and the impacts are permanent. The electro-osmotic flow causes formation of negative pore pressure inside the soft soil, in the vicinity of cathode [16] causing enhancement in effective stress and soil consolidation. The frequently encountered in-situ field condition is permitted drainage from the cathode without allowing re-circulation of water through anode. This condition results in the formation of negative pore pressure as discussed above [27, 28] and evaluated using Eq. (20.5):

$$u_e(x) = -\eta\left({}^{k_e}/_{k_h}\right)\gamma_w E x = -\Delta\sigma' \tag{20.5}$$

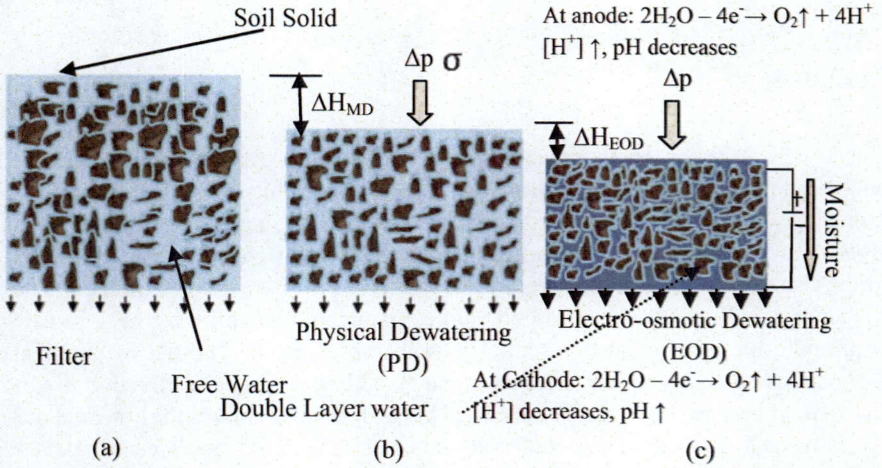

Fig. 20.5 Schematic of dewatering techniques employed in the field (modified after Pandey and Rajesh [32])

where $\Delta\sigma'$ = increment in effective stress (kPa), x = measured distance, η = empirical factor following Mohamedelhassan and Shang [28] and E = applied voltage gradient (V/m). It is observed in Eq. 20.5 that the electrokinetic consolidation is chiefly ruled by the hydraulic conductivity ratio of the soil. The coupling the electroosmosis process with surcharge loading has higher efficacy improving the soft soil consolidation as witnessed Shang [39]. Supplementary material on electrokinetic consolidation can be obtained following Malekzadeh et al. [24].

Practical Applicability

Numerous case studies based on in-situ and experimental works disclose that the applying of electro-osmotic technique for improving the soft soils is beneficial and encouraging. The probable applications of electro-osmotic treatment and its benefits are accelerated rate of consolidation settlement, increase in the secondary compression, and enhancement in factor of safety, helpful in removing different kinds of organic, inorganic or radioactive contaminants from the soil and reduction in the carbon footprint during treatment and highly economical. This method provides an essential alternative to sustainable soil improvement techniques, as it poses less threat to regional biodiversity and habitat changes than other methods. There are always concerns about the cost of electroosmosis treatment. Most of the cost is related to the manufacture and setting up of electrodes and electrical bills. The economy this method can be significantly minimized and compared to other conventional ground stabilization methods by choosing the appropriate electrodes and suitably developing reversal of polarity and current intermittence scheme.

Conclusion

This article describes detailed of soil modification using electro-osmotic process. It can be concluded that this method provides a great opportunity for soil improvement. An inclusive review was conducted, including in-situ and laboratory observations, to assess the viability of using electro-osmotic method to enrich soil parameters. This article also describes the several factors that influences the improvement process. The effects of electro-osmotic method on various parameters of soil have been proven to promote its prospective application in soil improvement. The list of potential applications based on in-situ and experimental studies exhibits the effectiveness of this method over conventional treatments. This review also explores the potential use of electro-osmotic method in geotechnical engineering with the development of state-of-the-art technology and new electrode materials that develop appropriate schemes for polarity reversal and current interruption. However, to be able to transform the data into a design requires further investigation to be required.

References

1. Adamson LG, Chilingar GV, Beeson CM, Armstrong RA (1966) Electrokinetic dewatering, consolidation and stabilization of soils. Eng Geol 1(4):291–304
2. Alshawabkeh AN, Acar YB (1996) Electrokinetic remediation II: theoretical model. J Geotech Eng 122(3):186–196
3. Barker JE, Rogers CDF, Boardman DI, Peterson J (2004) Electrokinetic stabilization: an overview and case study. Proc ICE - Ground Improv 8(2):47–58
4. Bergado DT, Sasanakul I, Horpibulsuk S (2003) Electro-osmotic consolidation of soft Bangkok clay using copper and carbon electrodes with PVD. Geotech Test J 26(3):277–288
5. Bjerrum L, Moum J, Eide O (1967) Application of electro-osmosis to a foundation problem in a Norwegian quick clay. Géotechnique 17(3):214–235
6. Burnotte F, Lefebvre G, Grondin G (2004) A case record of electroosmotic consolidation of soft clay with improved soil–electrode contact. Can Geotech J 41(6):1038–1053
7. Casagrande L (1949) Electro-osmosis in soils. Geotechnique 1(3):159–177
8. Chew SH, Karunaratne GP, Kuma VM et al (2004) A field trial for soft clay consolidation using electric vertical drains. Geotext Geomembr 22:17–35
9. Chien S, Ou C, Wang M (2009) Injection of saline solutions to improve the electro-osmotic pressure and consolidation of foundation soil. Appl Clay Sci 44:218–224
10. Fourie AB, Johns DG, Jones CJFP (2007) Dewatering of mine tailings using electro-kinetic geosynthetics. Can Geotech J 172:160–172
11. Fu H, Wang J, Wang P (2017) Experimental study on the combined application of vacuum preloading—variable-spacing electro-osmosis to soft ground improvement. Geosynth Int 24(1):72–81
12. Gargano S, Lirer S, Liguori B, Flora A (2020) Effect of the pore fluid salinities on the behaviour of an electrokinetic treated soft clayey soil. Soils Found 60:898–910
13. Gill RT, Harbottle MJ, Smith JWN, Thornton SF (2014) Electrokinetic-enhanced bioremediation of organic contaminants: a review of processes and environmental applications. Chemosphere 107:31–42
14. Glendinning S, Lamont-Black J, Jones CJFP (2007) Treatment of sewage sludge using electrokinetic geosynthetics. J Hazard Mater 139(3):491–499
15. Han J (2015) Principles and practice of ground improvement. Wiley, Hoboken, New Jersey
16. Hansbo S (2008) Soil improvement by means of electro-osmosis. In: Proceedings of the 6th International conference on case histories in geotechnical engineering. Arlington, VA, pp 1–14
17. Jones CJFP, John L-B, Glendinning S, Bergado DT, Eng T, Fourie A, Liming H, Pugh RC, Romantshuk M, Simpanen S, Yan-Feng Z (2008) Recent research and application in the use of electrokinetic geosynthetics. In: EuroGeo4 conference, pp 1–30
18. Kaniraj SR, Yee JHS (2011) Electro-osmotic consolidation experiments on an organic soil. Geotech Geol Eng 29(4):505–518
19. Kumar S, Solanki CH, Pandey BK (2015) Behaviour of prestressed geotextile-reinforced fine sand bed supporting an embedded square footing. Int J Geomate 8:1257–1262
20. Li Z, Xue Q, Katsumi T, Inui T (2014) Electric–hydraulic–chemical coupled modeling of solute transport through land fill clay liners. Appl Clay Sci 101:541–552
21. Liu Z, Yang J, Li Q (2016) Soil improvement by electro-osmosis with vacuum drainage in cathode. Electron J Geotech Eng 21(15):1–13
22. Lykelma J (1995) Fundamentals of interface and colloid science, vol II. Elsevier, Amsterdam
23. Mahmoud A, Olivier J, Vaxelarie J, Hoadley AFA (2010) Electrical field: a historical review of its application and contributions in wastewater sludge dewatering. Water Resource 44:2381–2407
24. Malekzadeh M, Lovisa J, Sivakugan N (2016) An overview of electrokinetic consolidation of soils. Geotech Geol Eng 34(3):759–776
25. Micic S, Shang JQ, Lo Ky et al (2001) Electrokinetic strengthening of a marine sediment using intermittent current. Can Geotechn J 38(2):287–302

26. Mitchell JK, Soga K (2005) Fundamental of soil behaviour. Wiley, New York
27. Mohamedelhassan E (2009) Electrokinetic strengthening of soft clay. Proc ICE - Ground Improv 162(GI4):157–166
28. Mohamedelhassan E, Shang JQ (2002) Feasibility assessment of electro-osmotic consolidation on marine sediment. Proc ICE - Ground Improv 6(4):145–152
29. Murugesan S, Rajagopal K (2007) Model tests on geosynthetic-encased stone columns. Geosynth Int 14(6):346–354
30. Otsuki N, Yodsudjai W, Nishida T (2007) Feasibility study on soil improvement using electrochemical technique. Constr Build Mater 21:1046–1051
31. Pandey BK, Bajaj K, Singh AP (2013) Soil stabilization using pozzolanic material and jute fibre. In: Proceedings of Indian geotechnical conference. Roorkee, India, pp 1–6
32. Pandey BK, Rajesh S (2019) Enhanced engineering characteristics of soils by electro-osmotic treatment: an overview. Geotech Geol Eng 37(6):4649–4673
33. Pandey BK, Rajesh S, Chandra S (2020) Numerical evaluation of geogrid-encased stone columns in soft soil under embankment loading. Geotech Spec Publ 315:543–551
34. Pandey BK, Rajesh S, Chandra S (2020) 3-D finite element study of embankment resting on soft soil reinforced with encased stone column. In: Latha Gali M, Raghuveer Rao P (eds) Problematic soils and geoenvironmental concerns. Lecture notes in civil engineering, vol 88. https://doi.org/10.1007/978-981-15-6237-2_38
35. Pandey BK, Rajesh S, Chandra S (2021) Performance enhancement of encased stone column with conductive natural geotextile under k0 stress condition. Geotext Geomembr. https://doi.org/10.1016/j.geotexmem.2021.03.004
36. Pandey BK, Das SK, Kumar S (2021) Study of prestressed geotextile reinforced sand Supporting an embedded square footing. In: Patel S, Solanki CH, Reddy KR, Shukla SK (eds) Proceedings of Indian geotechnical conference 2019. Lecture notes in civil engineering, vol 137. https://doi.org/10.1007/978-981-33-6466-0_20
37. Rajesh S, Jain P (2015) Influence of permeability of soft clay on the efficiency of stone columns and geosynthetic encased stone columns-a numerical study. Int J Geotech Eng 9(5):483–493
38. Rittirong A, Douglas RS, Shang JQ, Lee EC (2008) Electrokinetic improvement of soft clay using electrical vertical drains. Geosynth Int 15(5):369–381
39. Shang JQ (1998) Electroosmosis-enhanced preloading consolidation via vertical drains. Can Geotech J 35(3):491–499
40. Shang JQ (2012) Electrokinetics: engineering applications and recent development. Geotech Spec Publ 217:1–8
41. Tajudin SAA (2012) Electrokinetic stabilization of soft clay. Dissertation, University of Birmingham, Birmingham (UK)
42. Zhang L, Wang NW, Jing LP et al (2017) Electro-osmotic chemical treatment for marine clayey soils: a laboratory experiment and a field study. Geotech Test J 40(1):72–83

Chapter 21
Study on Time-Viscosity Characteristics of Microfine Slag Grout with Hydrated Lime Activator

Amit Patel, Nazimali Chinwala, N. H. Joshi, and Mansi Parmar

Introduction

The rheological properties of grout are very important in many applications including the treatment of ground, cracks filling in concrete, decreasing the permeability of soil/rocks, etc. The pumpability and ability of grout to penetrate inside voids are dependent on the rheological properties of grout in particular yield stress and viscosity [1]. Reference [2] worked on the development of grout by using slag, cement, and different activators and found that with addition of the slag, cement mixture with activator gives satisfactory performance of grout. Reference [3] worked on the development of grout by using cement, fly ash and GGBS and found that with addition of GGBS and fly ash, the basic grouting properties of cement mixture are improved. Reference [4] evaluated the effect of slag, fly ash, and lime stone mix using grouting and found that fine limestone powder has a positive effect on rheology, strength, and shrinkage properties of grout. Reference [5] worked on the properties of cement grout modified with ultra-fine slag and found that the viscosity properties are improved by adding ultra-fine slag. Reference [6] worked on the rheological properties of microfine cement-based grouts by adding microfine fly ash, nanosilica, and superplasticizer and found that the fluidity and spreading ability of fresh grouts are improved and with addition of 0.5% nanosilica decreased the apparent viscosity of fresh grouts. Reference [7] studied the fluidity and rheological properties of cement-based grout mixed with rice husk ash and reported that addition of RHA to cement base grouts at different high w/b ratios significantly modifies the rheological properties of grouts.

A. Patel (✉) · N. H. Joshi · M. Parmar
Department of Applied Mechanics, Faculty of Technology & Engineering, The Maharaja Sayajirao University of Baroda, Vadodara 390001, India
e-mail: amitpatel22a@gmail.com

N. Chinwala
Department of Civil Engineering, Sardar Vallabhbhai National Institute of Technology, Surat 395007, India

K. Muthukkumaran et al. (eds.), *Ground Improvement Techniques*, Lecture Notes in Civil Engineering 297, https://doi.org/10.1007/978-981-19-6727-6_21

The plastic viscosity of grouts increases with increasing of RHA content regardless of the w/b ratio. On the other hand, the increase in w/b ratio reduces the viscosity of grout for any given RHA content. Reference [8] presented the effects of a polycarboxylate and polynaphthalene superplasticizer on the rheological properties of cement grouts, and the results show that grouts with polycarboxylate superplasticizer had higher viscosity compared with the polynaphthalene superplasticizer.

However, to the best knowledge of the authors, there is no systematic study about the rheological properties of microfine slag grouts containing hydrated lime as additive. Hence, it is important to quantify the effects of hydrated lime on the behavior of grout. So, the main objective of this study is to evaluate the effect of hydrated lime on the rheological properties of grout by using 0.5% of Sulfonated Melamine Formaldehyde Resin superplasticizer.

Methodology

Materials

Microfine Slag

Microfine slag (Alccofine 1203) was procured from Counto Microfine Products Pvt. Ltd., Goa. The physical and chemical composition as provided by the manufacturer are given in Table 21.1.

Hydrated Lime

Hydrated lime was procured from Suvidhinath laboratories, Vadodara.

Table 21.1 Physical and chemical analysis of microfine slag

Fineness (D_{10})	<2.5 μ
Specific gravity	2.86 ± 0.02
CaO	>30%
SiO_2	>30%
Al_2O_3	>20%
Manganese oxide (MnO), % by mass	1.24
Magnesium oxide (MgO), % by mass	6.1
Sulfide sulfur (S), % by mass	0.48
Sulfate (SO_3), % by mass	0.07
Insoluble residue %	0.73
Chloride contents, % by mass	0.020
Glass content,% by mass	>85
Moisture content, % by mass	0.1

Table 21.2 Properties of sulfonated melamine formaldehyde resin superplasticizer

Dosage	0.5% up to 4% as per by wt. of cement
Appearance	Clear to slightly hazy
Stability	12 month in closed container
Specific gravity	1.2–1.28
pH	Minimum 7.0
Water reduction	13–20%

Table 21.3 Time of afflux (sec) for microfine slag + hydrated lime (3%, 5%, 7%) grout at different w/ms ratio (0.8 and 2) and superplasticizer (0.50%)

Hydrated lime (%)	w/ms = 0.8	w/ms = 2
3	NP	30.5
5	NP	31.4
7	NP	32.9

Sulfonated Melamine Formaldehyde Resin Superplasticizer
SMFR superplasticizer was procured from Krishna Conchem Pvt. Ltd., Vadodara. The properties as provided by the manufacturer are given in Table 21.2.

Time Viscosity Properties of Grout

Fluidity Test

Fluidity is a measure of the easiness in pumpability of a grout. For slag/cement, fluidity is measured by marsh cone. In marsh cone test, about 2000 cc of grout is to be poured in the marsh cone and time of afflux is noted to pour 1000 cc of grout in measuring cylinder. The results of Time of Afflux (sec) with different w/ms ratios obtained are given in Table 21.3.

Viscosity Test

Viscosity is the very important parameter for defining a grout as it determines the groutability of grout in a given permeable porous medium. The viscosity of a grout in the present study is measured using Brookfield viscometer (RVT model). Figures 21.1 and 21.2 show the yield stress versus time, Figs. 21.3 and 21.4 shows the plastic viscosity versus time, and Figs. 21.5 and 21.6 shows the apparent viscosity

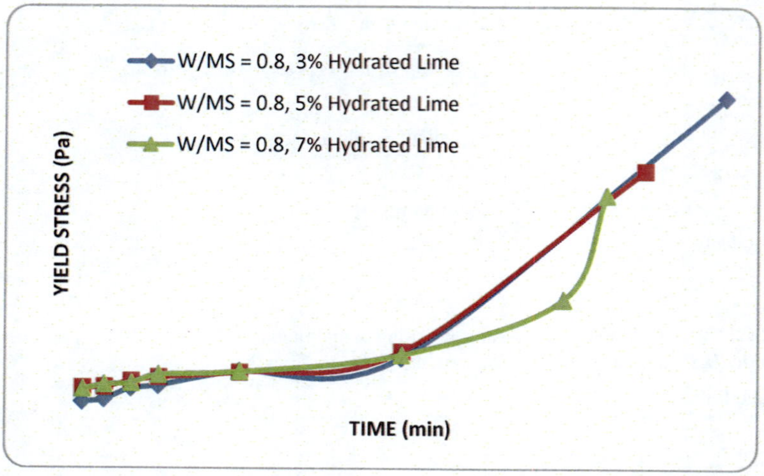

Fig. 21.1 Yield stress versus time for grout at different hydrated lime % for w/ms ratio 0.8

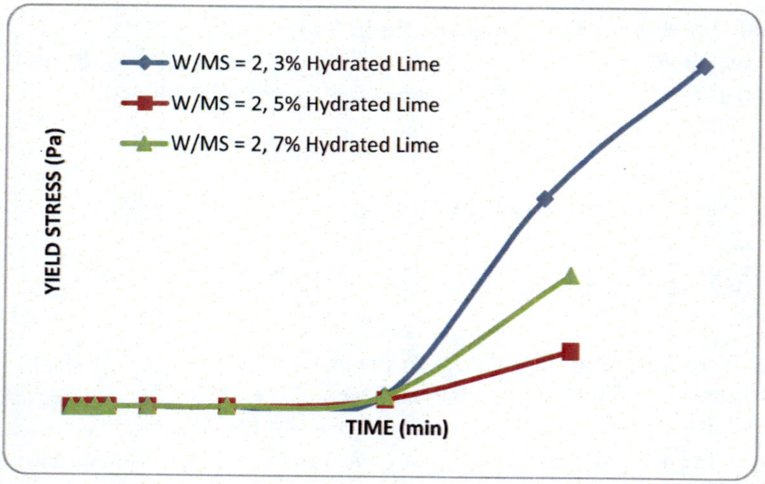

Fig. 21.2 Yield stress versus time for grout at different hydrated lime % for w/ms ratio 2

versus time of microfine slag grout at different hydrated lime % and 0.5% of SMFR superplasticizer for w/ms ratio 0.8 and 2, respectively.

Results and Discussions

From the marsh cone test, for w/ms ratio of 0.8, the grout was very thick so it was not possible to perform the test, while in the case of w/ms ratio 2, the time of afflux

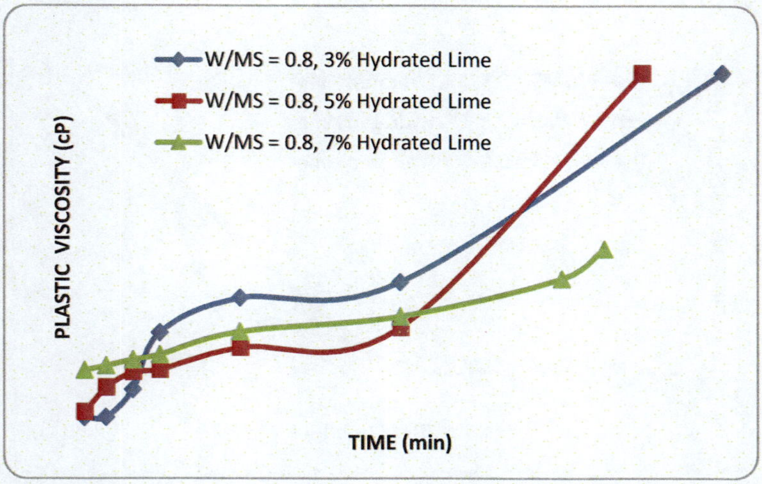

Fig. 21.3 Plastic viscosity versus time for grout at different hydrated lime % for w/ms ratio 0.8

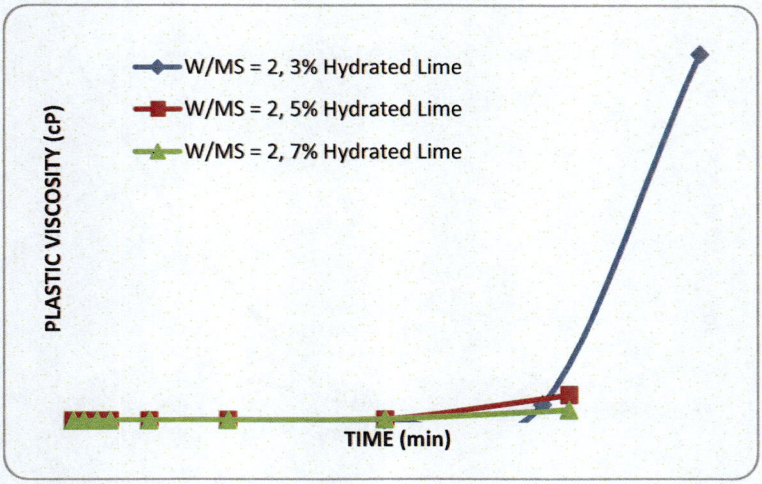

Fig. 21.4 Plastic viscosity versus time for grout at different hydrated lime % for w/ms ratio 2

increases as the hydrated lime content is increasing as the grout becomes thick with addition of hydrated lime. From viscosity test, it has been observed from the graphs of yield stress versus time, apparent viscosity versus time, and plastic viscosity versus time that for both the w/ms ratios, the initial viscosity remains low up to the limiting injection time with addition of 3% of hydrated lime and the viscosity increases with the addition of 5 and 7% of hydrated lime reaching to gellification time. So from viscosity point of view 3% of hydrated lime, activator is found to be optimum.

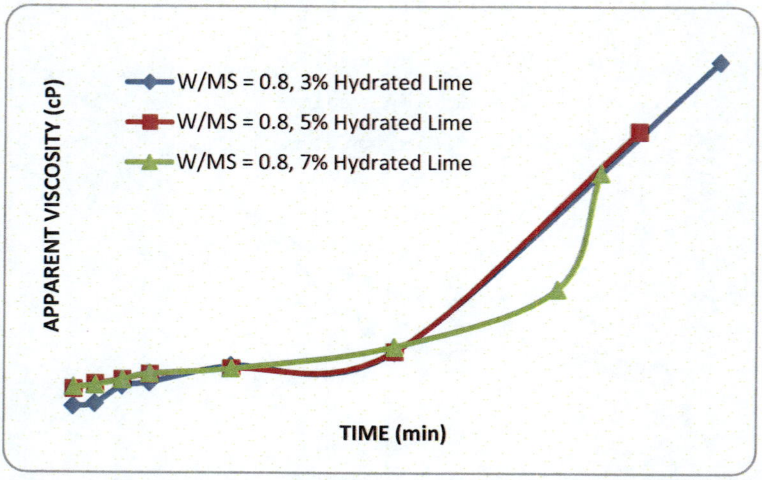

Fig. 21.5 Apparent viscosity versus time for grout at different hydrated lime % for w/ms ratio 0.8

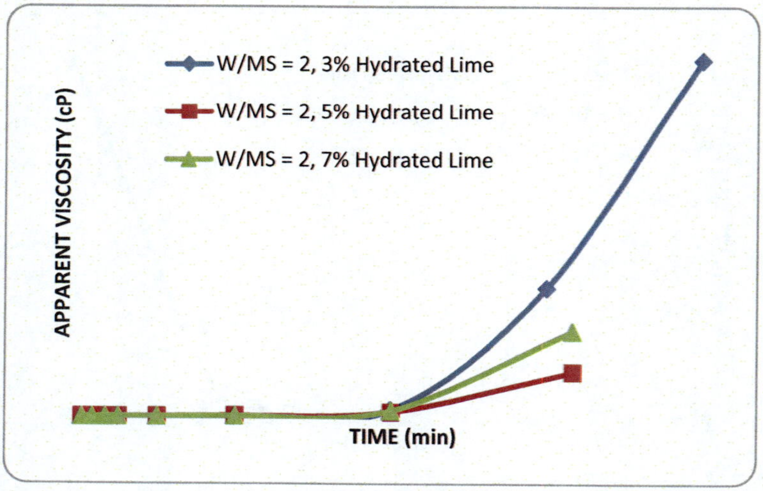

Fig. 21.6 Apparent viscosity versus time for grout at different hydrated lime % for w/ms ratio 2

References

1. Ferraris CF, de Larrard F (1998) Testing and modelling of fresh concrete rheology, p 61
2. Patel GN, Sinroja JM, Joshi NH (2010) Chemically activated microfine slag cement grouts. In: Indian geotechnical conference GEOtrendz, pp 567–568
3. Li S, Sha F, Liu R, Li W, Li Z, Wang G (2017) Properties of cement-based grouts with high amounts of ground granulated blast-furnace slag and fly ash. J Mater Civ Eng 29(11):04017219
4. Xiang J, Liu L, Cui X, He Y, Zheng G, Shi C (2018) Effect of limestone on rheological, shrinkage and mechanical properties of alkali—Activated slag/fly ash grouting materials. Constr Build Mater 191:1285–1292

5. Gopinathan S, Anand KB (2018) Properties of cement grout modified with ultra-fine slag. Front Struct Civ Eng 12(1):58–66
6. Zhang S, Qiao WG, Chen PC, Xi K (2019) Rheological and mechanical properties of microfine-cement-based grouts mixed with microfine fly ash, colloidal nanosilica and superplasticizer. Constr Build Mater 212:10–18
7. Celik F, Canakci H (2015) An investigation of rheological properties of cement-based grout mixed with rice husk ash (RHA). Constr Build Mater 91:187–194
8. Anagnostopoulos CA (2014) Effect of different superplasticisers on the physical and mechanical properties of cement grouts. Constr Build Mater 50:162–168

Chapter 22
Influence of Intensity and Position of Surcharge on the Performance of Soil Nail System

H. R. Krupa and S. K. Prasad

Introduction

Soil nailing is one of the popular methods of deep excavation and slope stability. In this method, the nails are installed into the ground at certain inclination to provide stability to the slope or to the deep excavated vertical cut. The reinforcing nail along with the grout develops tensile forces caused due to the deformation of soil toward the cut surface. The maximum tensile force developed is the function of deformation, bond strength, and the diameter of the grouted nail. Various applications of soil nailing include roadway cuts, road widening under existing abutments, tunnel portals, etc., [1]. The behavior of soil nail system depends on a number of factors that include soil properties like C-Φ parameters, surcharge, back slope inclination, etc., nail properties such as length, bond resistance, inclination of the nail, spacing between the nails, and to some extent, on facing element properties. Each of these properties is interdependent of the other and their combined effect on the overall performance of the soil is a complex phenomenon. Thus the effect of individual elements on the performance of the soil nail system is studied. Some of the researchers have done some detailed study on the mechanism of the soil nail system. The FHWA [2] reference manual gives the graph for back slope inclination on the soil nail wall. But the importance given to the surcharge intensity/pressure is less. In this paper, efforts are made to analyze the mechanism of the soil nail system under the influence of surcharge and its location. The main parameters that are considered for this study are horizontal displacement at wall face, vertical settlement of the 2 m strip footing carrying the pressure from superstructure, axial force in the nails and maximum axial

H. R. Krupa (✉)
JSS Science and Technology University (Formerly SJCE), Mysuru, India
e-mail: krupa.gangadikar@gmail.com

S. K. Prasad
Civil Engineering, Vidyavardhaka College of Engineering, Mysuru, India
e-mail: skprasad@vvce.ac.in

K. Muthukkumaran et al. (eds.), *Ground Improvement Techniques*, Lecture Notes in Civil Engineering 297, https://doi.org/10.1007/978-981-19-6727-6_22

force. The bending moment and the shear force of the soil nails contribute very less to the design and performance of the soil nail system [3] and thus are ignored in this paper. PLAXIS 2D is used for numerical simulations, thus considering the surcharge load to act on a strip footing plate.

Methodology

PLAXIS 2D is used for numerical simulations of soil nail wall. The input data parameters are calculated based on the preliminary design recommended by FHWA [2] reference manual. A two-dimensional plane strain model with 15 nodded CST elements is used for the finite element simulation. The model properties are taken from already published information for validation [4] and are as mentioned in Table 22.1.

Conventional Design Procedure

Conventional design process is a 2 step procedure involving preliminary design and final design. The normalized length and normalized tensile force are calculated as a part of preliminary design. The factors of safety against various internal failure modes (nail pullout failure and nail tensile strength failure) and external failure modes (facing flexural and punching shear failures) are calculated based on the maximum

Table 22.1 Model/material parameters

Parameter	Value
Wall height, H, m	10 m
Type of soil	Dense silty sand
Cohesion (kPa)	5
Angle of friction, degrees	31.5
Unit weight of the soil, kN/m^3	17
Modulus of elasticity of soil, MPa	20
Poisson's ratio	0.3
Method of nail installation	Rotary drilled and grouted
Grade of steel	Fe415
Modulus of elasticity of nail, GPa	200
Nail spacing, m × m	1 × 1
Nail inclination, degrees	15
Drill hole diameter, mm	100
Nail diameter, mm	20

Table 22.2 Conventional design summary

Description	Symbol (unit)	Value
Bond resistance	μ	0.29
Normalized SN length	L/H	0.65
Normalized T_{max}	t_{max}	0.29
Corrected length factor	L/H *	0.63
Corrected T_{max} factor	t_{max}*	0.28
Length of the nail	L (m)	**7.00**
Maximum tension in the nail	T_{max} (kN)	**47.60**
Tension at face	T_o (kN)	**28.56**
Pullout resistance per length	r_{po} (kN/m)	31.42
Pullout length	L_p (m)	2.45
Pullout resistance	R_{po} (kN)	76.98
FS against pullout	FS_{PO}*	**1.62**
Cross sectional area	A_t (mm^2)	206.46
Tensile resistance	R_T (kN)	85.68
FS for tensile strength	FS_T*	**1.80**
Flexure capacity	R_{FF} (kN)	80.07
Punching shear capacity	R_{FP} (kN)	155.69
FS against flexure failure	FS_{FF}	**2.80**
FS against punching shear	FS_{FP}	**5.45**

*Differentiate the general values with the normalised/corrected values

normalized tensile force and tensile force at the nail-face junction, are obtained in the final design step. Table 22.2 shows the conventional design summary for the considered model and material properties. It was observed that the factors of safety obtained are higher than the manual recommended limits and thus considering the system safe.

Numerical Simulations

The length of the nail for the model is decided based on the preliminary design. As mentioned earlier, PLAXIS 2D is used to simulate the soil nail model using 15 nodded CST elements with plane strain idealization. Mesh boundaries are defined far enough so as to minimize its influence on the performance of the system. Medium density of mesh is adopted globally but refined to fine mesh density in the vicinity of the soil nail wall. The nails, facing and the footing are considered as plates with respective equivalent properties. The soil is considered homogenous and satisfies Mohr- Coulomb criteria. The results of numerical simulation without any surcharge load are as shown in Table 22.3. Typical soil nail system with footing located on it is as shown in Fig. 22.1.

Table 22.3 Results from numerical simulation

Parameter	Symbol	Value
Horizontal displacement of the system	H (mm)	21.52
Maximum tension in the nail	T_{max} (kN)	56.21
Error from theoretical value	%	+15.32
Tension at face	T_o (kN)	51.41
Maximum shear force	Q (kN)	19.63
Maximum bending moment	M (kN m)	2.74
Pullout length	L_p (m)	6.53
FS for pullout of nail	FS_{PO}	**3.65**
FS for tensile strength of nail	FS_T	**2.31**
FS against flexure failure of face (R_{FF} = 157.91 kN/m^2)	FS_{FF}	**3.07**
FS against punching shear of face (R_{FP} = 204.62 kN/m^2)	FS_{FP}	**3.98**

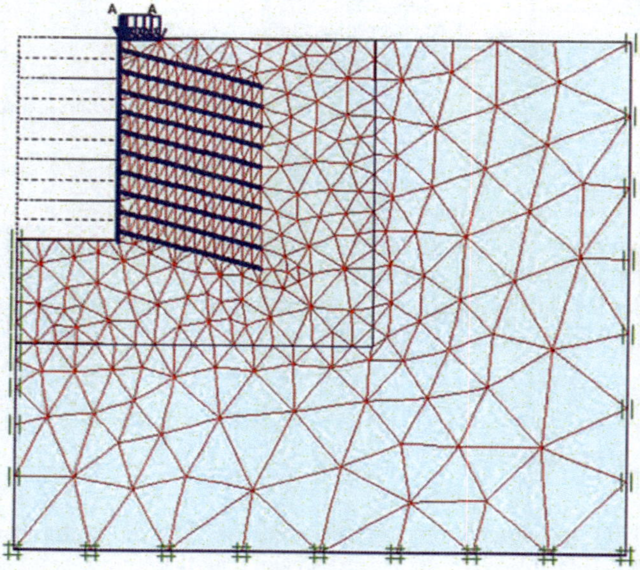

Fig. 22.1 Typical finite element model of soil nail system

Results and Discussion

The strip footing is placed at different locations and the surcharge pressure at which the system fails is recorded and is as shown in Fig. 22.2.

It was observed that the load at failure when the footing is at 9.5 m away from the wall edge is the lowest, i.e., 189 kN/m^2. Thus, 188 kN/m^2 is considered as the safe load for all locations in the analysis here after. At the edge, the load at failure was 450 kN/m^2 and moving away from the edge the load carrying capacity increases and then suddenly decreases. It can be observed that after a distance of 6.5 m away from the edge, the load carrying capacity reduces drastically up to 14 m from the edge and after 15 m, the load carrying capacity increases and remains almost constant.

Horizontal Displacement at Wall Face

For a constant load of 188 kN/m^2, the analysis was done where the footing locations are varied and the horizontal displacement patterns are observed. For simplicity, the entire system is divided into 3 zones—**Zone 1, Zone 2, and Zone 3**. The maximum displacements at wall face are plotted for different locations of footing as shown in Fig. 22.3.

The Zone 1 consists of nails and the footing is placed starting from the edge of wall face. It is observed that the horizontal displacement at wall face is about 0.75%

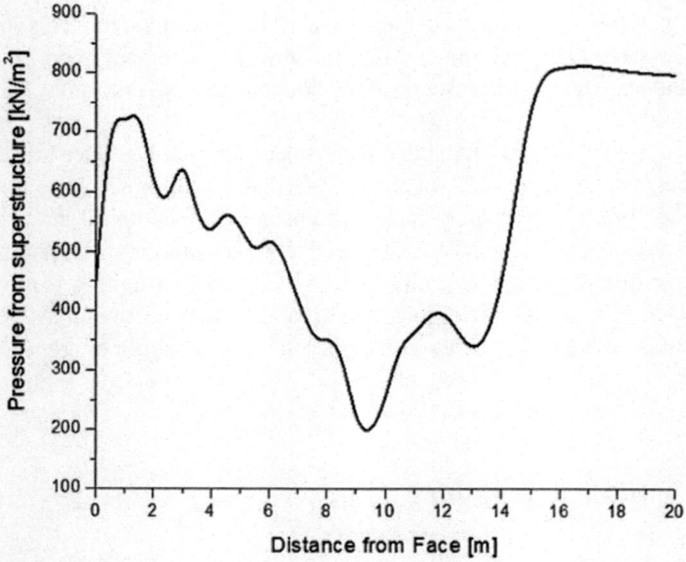

Fig. 22.2 Load carrying capacity of the soil nail system for different footing locations

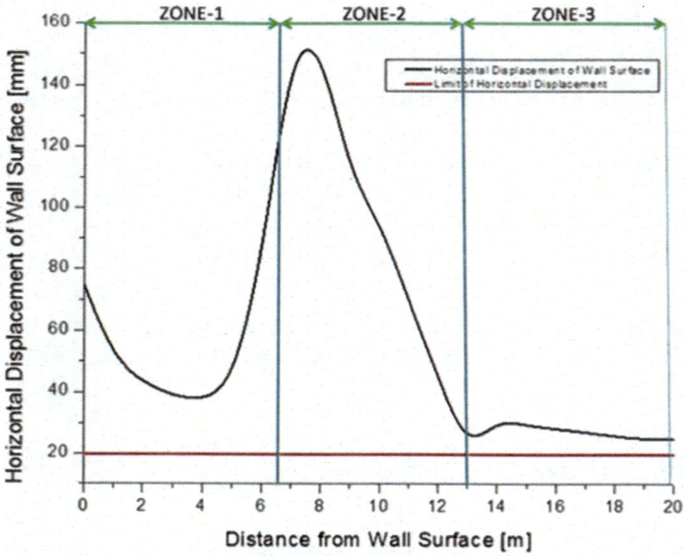

Fig. 22.3 Maximum horizontal displacement of wall for footing at various locations (for a constant Load of 188 kN/m^2)

of wall height when the footing is placed at the edge and it gradually decreases as the footing is moved away from the edge in the Zone 1. At the last location of Zone 1, i.e., at 6.5 m away from edge, the displacement is about 1.3% of wall height which is also greater than the FHWA recommended limits. In Zone 2, it is observed that the displacements increase and are more than 1.5% of wall height. This is because of vertical settlements of footing causing the outward movement of soil toward the wall face and also due to the presence of the boundary of two dissimilar materials at zone junctions.

Finally in Zone 3, it is observed that the displacements at wall face are decreased and vary between 0.2 and 0.3% of the wall height which is nearer to the FHWA recommended limits. The displacement patterns at wall face for all three zones are as shown in Figs. 22.4, 22.5, and 22.6, respectively. Precautions are to be taken when the footing is placed at edge and mostly be avoided. When footing is close to the edge, displacement of top surface of wall is a matter of concern and as the footing moves away from edge, the displacement of wall at lower surfaces are also matters of concern.

Vertical Settlement of Footing

The settlement profile from edge to edge of footing is insignificant. Thus average settlement of footing is considered at various locations on the soil nail system as

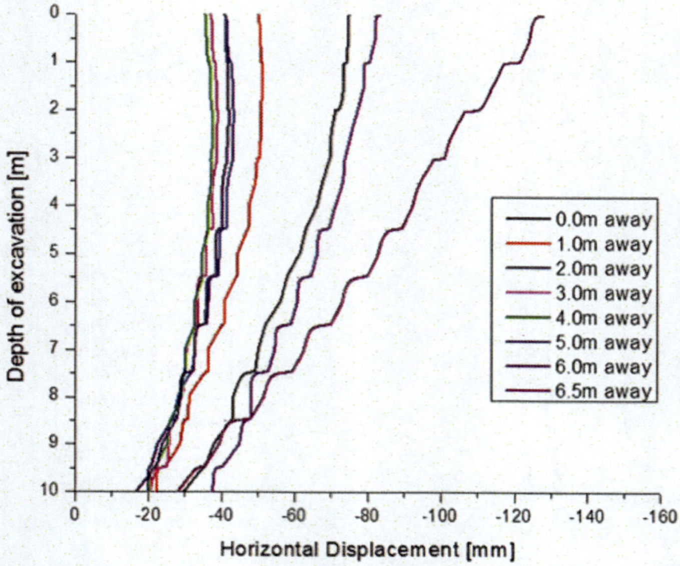

Fig. 22.4 Horizontal displacement patterns at wall face for Zone 1 (for a constant Load of 188 kN/m^2)

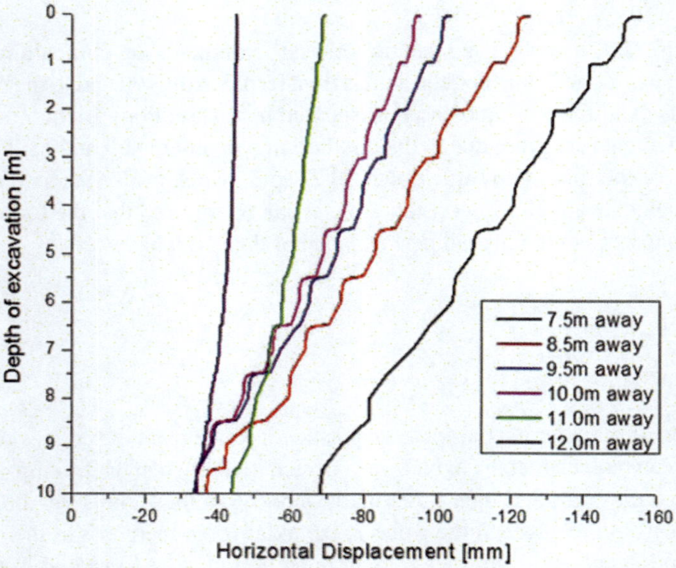

Fig. 22.5 Horizontal displacement patterns at wall face for Zone 2 (for a constant Load of 188 kN/m^2)

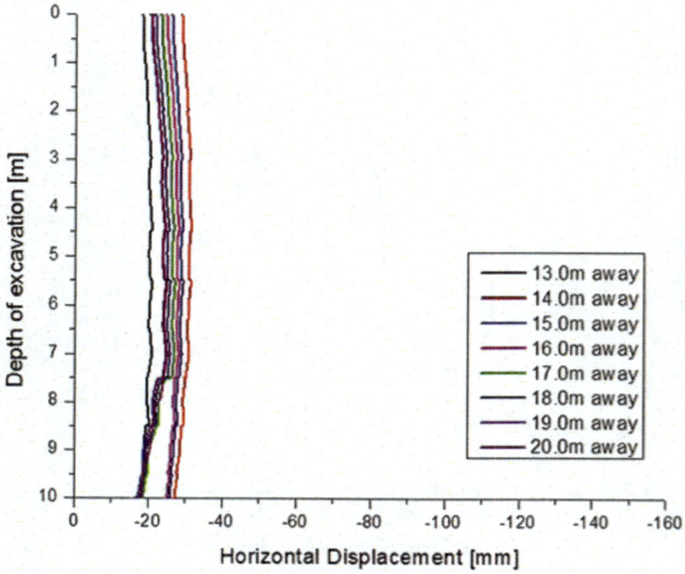

Fig. 22.6 Horizontal displacement patterns at wall face for Zone 3 (for a constant Load of 188 kN/m^2)

shown in Fig. 22.7. It is observed that the vertical settlements are comparatively less in Zone 1 than in Zone 2. Further, the load is transferred to the soil and to the nails and thus the nails axial force has increased (discussed in next section). In the Zone 2, the vertical settlements are more due to the outward movement of soil and nails causing more space for the footing to settle down. In Zone 3, which is also the safe zone the vertical settlements are almost constant, thus it can be inferred that the footing does not significantly affect the overall performance of the soil nail system.

Axial Force in Nails

The variation of maximum axial force versus the distance from the edge is shown in Fig. 22.8. It can be seen that the axial force is much higher when the footing is placed at the edge of the wall, and then it gradually reduces in the Zone 1, as the footing moves away from the edge. In the Zone 2, the axial force increases at the junction of Zone 1 and 2 and again shows a gradual reduction in axial force and is almost a constant in Zone 3. Also the maximum axial force developed in Zone 3 is almost equal to the maximum axial force developed in a soil nail system without surcharge on it.

Observing the axial force in nail 1 (Fig. 22.9), it can be seen that the axial force is more in Zone 1 and less in the other two zones. This is because the nail under the footing takes the load onto it and thus gets more stressed. Also, because of

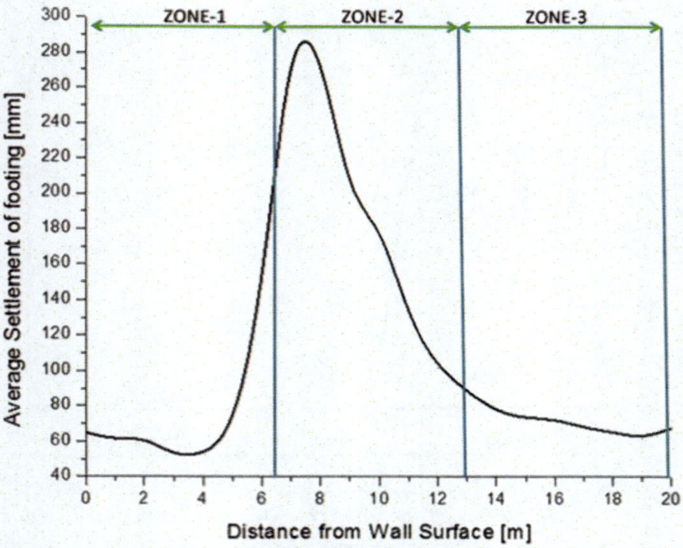

Fig. 22.7 Average settlement of footing at various locations (for a constant Load of 188 kN/m^2)

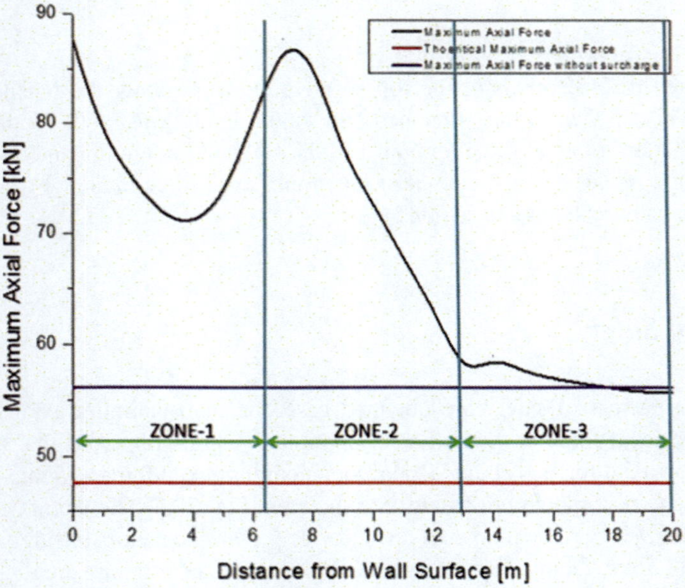

Fig. 22.8 Maximum axial force in the soil nail system with footings at various locations on top (for a constant load of 188 kN/m^2)

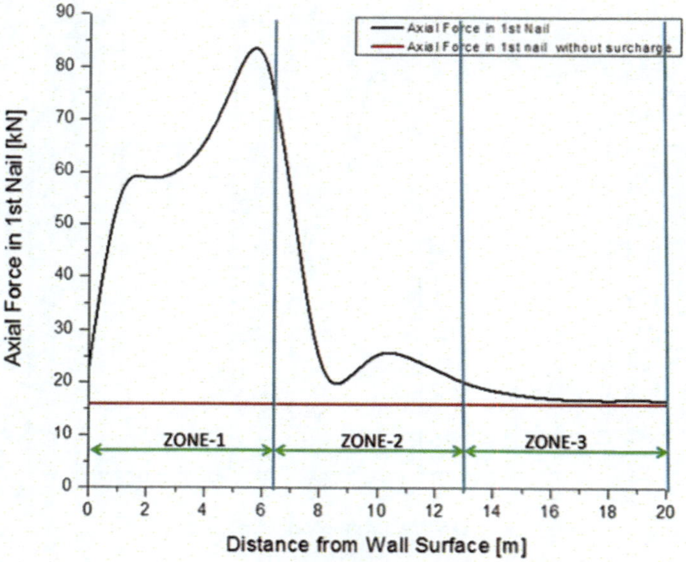

Fig. 22.9 Maximum axial force in the Nail 1 with footings at various locations on top (for a constant load of 188 kN/m^2)

the increased displacements at the top surface due to loading, the tensile forces increase. This applies to the other nails under the loading as well. Figure 22.10 shows the pattern of axial force for Nail 1 of the soil nail system. It can be seen that as the footing moves away, the point of maximum axial force in Nail 1 also moves along/stays below the loading region.

Factors of Safety

The various factors of safety are obtained based on the maximum axial force of the nail and axial force at the nail-face junction. It can be observed that the nail-face junction axial force is about 90% of the maximum axial force and thus follows the same pattern of maximum axial force shown in Fig. 22.8. The factors of safety increase with the decrease in axial force and vice versa. The factors of safety against nail pullout failure, facing flexural and facing punching shear failure are higher than recommended limits and almost nearer to the respective results of the soil nail system without surcharge. But the factor of safety against nail tensile strength failure is lesser than both the recommended limit and without surcharge model (Fig. 22.11).

Fig. 22.10 Pattern of axial force developed in Nail 1 for **Zone 1**

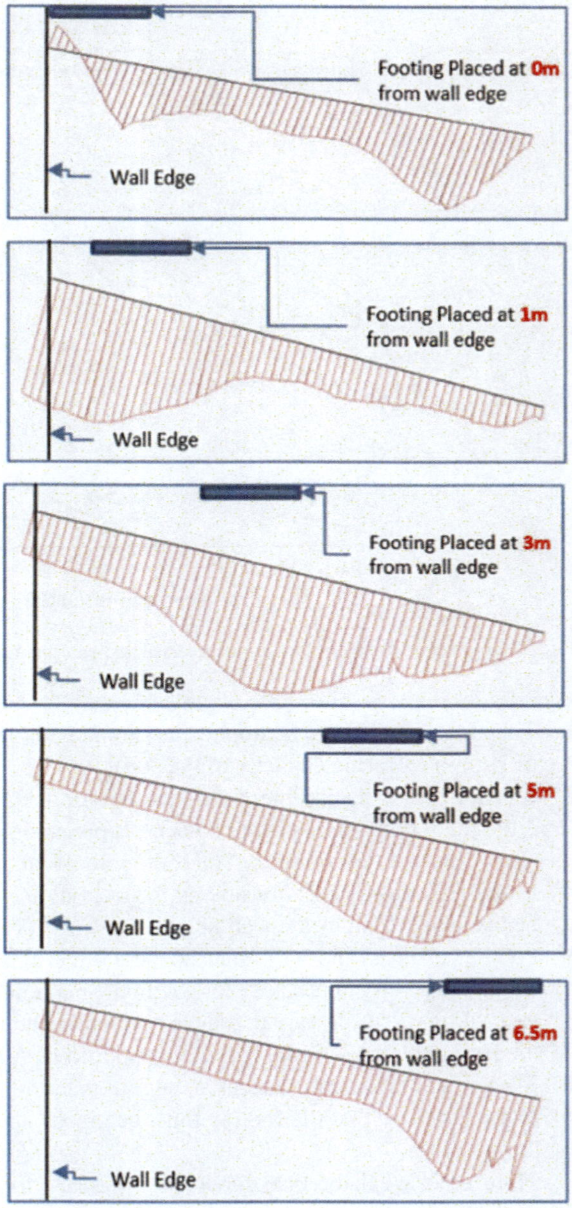

Concluding Remarks

The following conclusions were drawn from the observations made in the present analysis.

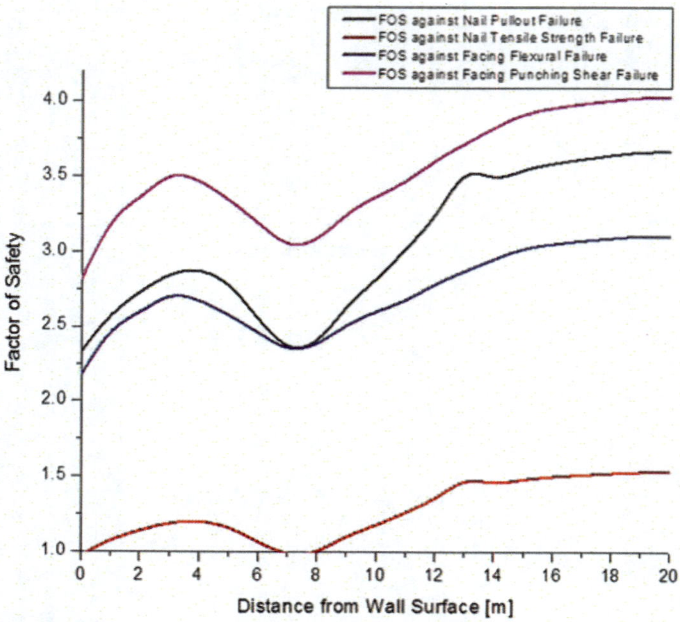

Fig. 22.11 Factors of safety for various internal and external failures

1. The maximum load to be applied as surcharge is 188 kN/m^2 only. (30% to 35% of Ultimate bearing capacity of the soil).
2. The horizontal displacements at Zone 1 and Zone 2 are much higher. For considering the footing to be placed in Zone 1, precautions are to be taken at wall face to reduce wall deformation. The footing location at Zone 2 must be completely avoided. Zone 3 is best considered for locating the footing, and the distance can be taken as 1.5 times the wall height.
3. Footing taken at the nail zone (Zone 1), the axial force developed in Nail 1 should be given importance along with the maximum axial force. The additional precautions are necessary to reduce horizontal deformation of wall by increasing the facing thickness and/or providing additional reinforcement for facing.
4. The vertical settlement patterns, axial force variation patterns and factors of safety plots also show that the footing must be placed approximately 1.5H away from the wall face.
5. More précised theoretical model is necessary for dealing with the surcharge applied soil nail system.

References

1. Briaud JL, Lim Y (1997) Soil-nailed wall under piled bridge abutment: simulation and guidelines. J Geotech Geo-Environ Eng, ASCE 123(11):1043–1050
2. FHWA (2003) Soil nail walls, Geotechnical Engineering Circular No. 7, Report No. FHWA0-IF-03-017, Federal Highway Administration
3. Elias V, Juran I (1991) Soil nailing for stabilization of highway slopes and excavations, Publication FHWA-RD-89-198, Federal Highway Administration, Washington, DC
4. Shivakumara Babu GL, Murthy BRS, Srinivas A (2002) Analysis of construction factors influencing the behaviour of soil-nailed Earth retaining walls. Gr Improv 6(3):137–143
5. Joshi B (2003) Behavior of calculated nail head strength in soil-nailed structures. J Geotech Geo-Environ Eng 129(9):819–828
6. Jewell RA, Pedley MJ (1992) Analysis for soil reinforcement with bending stiffness. J Geotech Eng 118(10):1505–1528
7. Juran (1985) Reinforced soil systems—Application In retaining structures'. Geotech Eng 16:39–81
8. Krupa HR, Prasad SK (2020) Influence of excavation phase on the performance of soil nail system, Theme 10. In: Proceeding of Indian geotechnical conference, 2020, Vishakapatnam Chapter, pp 35–46
9. Krupa HR, Prasad SK (2020) Influence of nail fixity to facing on the performance of soil nail system. Int J Eng Res Technol (IJERT) 09(08):805–810, ISSN: 2278-0181
10. Shivakumara Babu GL, Singh VP (2009) Appraisal for soil nailing. Indian Geotech J 39(1):81–95
11. Shivakumara Babu GL, Singh VP (2009) Simulation of soil nail structures using PLAXIS 2D, Spring Issue. PLAXIS Bull, pp 16–21

Chapter 23
Effect of Biopolymer Inclusion and Curing Conditions on the Failure Strain and Elastic Modulus of Cohesive Soil

Kopparthi Venkata Vydehi and Arif Ali Baig Moghal

Introduction

Under applied loads, the soil mass does not fail completely but exhibits movement/deformation in possible directions. To evaluate the permissible/allowable movement of soil, it is necessary to understand the stress–strain relationship and deformability parameters under static loads [1]. Despite different moduli from the stress–strain curve, the actual deformation or behavior of material at the selected deviator stress (generally 1/2 or 1/3 of the peak deviator stress) is represented by secant modulus [2]. Determining the secant modulus from the stress–strain curve by conducting unconfined compressive strength (UCS) test is considered to be simple method as the confining pressure is zero [3].

For the stabilized soil mass (e.g.: lime, cement, flyash), the deformability parameters depend on the interaction between the soil grains and stabilizer used which imparts ductile or brittle nature to the soil [4, 5]. Recent research concentrated on bio inspired materials such as EICP, MICP, and biopolymers instead of conventional stabilizers in view of reducing adverse impact on the environment [6, 7]. Biopolymer treated soils showed notable increase in strength and resistance to deformation under applied loads [8]. Among the various types of biopolymers, xanthan gum (G1) and guar gum (G2) are widely used for soil stabilization due to their effectiveness to improve properties [7, 8]. Based on the test conditions adopted in the laboratory, researchers evaluated the deformability properties such as elastic moduli (tangent modulus, secant modulus, Young's modulus) and failure strain from the obtained stress–strain curve of biopolymer treated soil. In general, the modulus values and failure strain increased with increase in dosage of biopolymer. Increase in elastic

K. Venkata Vydehi (✉) · A. A. B. Moghal
Department of Civil Engineering, National Institute of Technology, Warangal 506004, India
e-mail: vydehi56@student.nitw.ac.in; kvvydehi252@gmail.com

A. A. B. Moghal
e-mail: baig@nitw.ac.in; reach2arif@gmail.com

K. Muthukkumaran et al. (eds.), *Ground Improvement Techniques*, Lecture Notes in Civil Engineering 297, https://doi.org/10.1007/978-981-19-6727-6_23

modulus value by 5 times is observed for clay with 1% addition of xanthan gum compared to untreated soil [9]. The secant moduli of xanthan gum treated high plastic clay increased with increase in curing period and concentration of biopolymer and failure strain increased with increase in curing period [10]. Addition of 1% xanthan gum to the well graded sand with silt showed increase in Young's modulus value [11].

The present paper determines the effect of xanthan gum (G1) and guar gum (G2) addition on the mechanical properties of cohesive soil. Xanthan gum and guar gum are identified as non-toxic and does not impose negative impact on the environment. The secant modulus (E_s) and failure strain values were evaluated from stress–strain curve of UCS test. The variations in gum dosage and curing period were considered under different curing conditions. Mechanisms responsible for changes in the behavior of gum treated soil were examined by scanning electron microscopic (SEM) analysis.

Materials

Soil

Natural soil excavated at a depth of 1.2 m from the ground surface was used in the present study (location: National Institute of Technology, Warangal, India). Soil was oven dried, pulverized, sieved and was kept in air tight containers for the experimental work. The specific gravity of soil is found as 2.66. The liquid and plastic limits of soil are 47% and 22%, respectively. The optimum moisture content (OMC) and maximum dry density (MDD) are determined as 13.47% and 18.23 kN/m^3, respectively. All the properties were determined according to relevant ASTM standards. The selected soil has been classified as low plastic clay (CL) as per USCS (unified soil classification system) classification.

Biopolymers

Xanthan gum (G1) and guar gum (G2) procured from SRL Chemicals, Mumbai, India were used as stabilizers in the present study. The selected biopolymers are non-toxic and are proved as eco-friendly materials [12]. The dosages of the two biopolymers were kept at 0.5, 1, 2, and 4% by dry weight of soil based on the findings of the previous literature [8].

Methodology

Specimen Preparation and Curing

Soil was mixed with two selected biopolymers (G1 and G2) at various dosages (0.5, 1, 2, and 4%) in dry state. Water content equal to OMC of soil (13.47%) is added and mixed thoroughly to get homogenous mixture. The prepared sample is kept in desiccator for 24 h to attain moisture equilibrium. Cylindrical specimens (d = 38 mm; l = 76 mm) were prepared in triplicates for each mix proportion and curing period using static compaction technique.

Specimens were subjected to two different curing conditions, such as controlled and natural curing. In controlled curing, specimens were cured in desiccators under controlled temperature and relative humidity; whereas, in natural curing, specimens were subjected to daily variations in outside temperature and relative humidity values. The curing periods were maintained at 7, 28, and 60 days.

Unconfined Compressive Strength Test

Cured specimens were axially loaded in compression testing machine and sheared till failure state at a strain rate of 0.6 mm/min following the ASTM D2166 [13] standard. Readings were noted to evaluate the deformation of specimen at corresponding axial load. Stress and axial strain values are calculated to obtain the stress–strain curves of biopolymer treated soil.

Secant modulus (E_s) is defined as the slope of the line joining origin and the point at 50% peak axial stress as seen in Fig. 23.1. In addition, failure strain (%) values are evaluated for the biopolymer treated specimens. The pseudo-plastic behavior of biopolymer leads to bulging of soil specimen even at greater strain values without sample failure. Therefore, the strain values were calculated at peak stress for each gum dosage under applied static loads.

Microlevel Studies

Scanning electron microscopic (SEM) analysis is carried out on the tested samples to confirm the mechanism responsible for changes in mechanical properties of biopolymer treated soil.

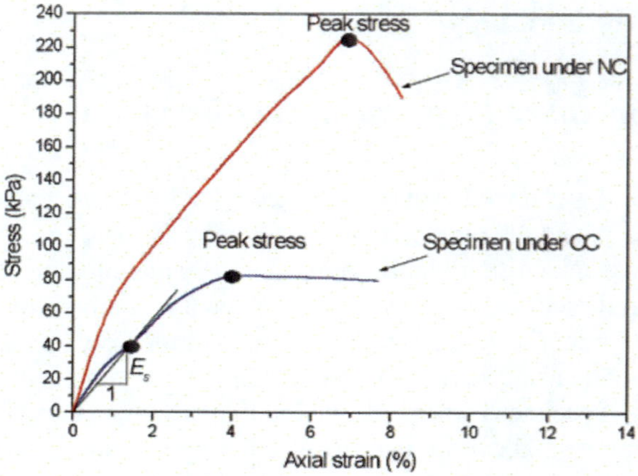

Fig. 23.1 Typical stress–strain curves for tested specimens

Results and Discussion

The typical stress–strain curves of biopolymer (G1/G2) treated soil at various curing conditions are presented in Fig. 23.1. For the specimens under controlled curing, at all the curing periods, the stress value continued to increase with increase in strain till it reaches peak value due to the binding of soil grains with biopolymer gel which offers more resistance to the applied loads. After reaching the peak stress, the biopolymer treated soil continued to bulge at its mid length even at higher strain values. At this stage, the stress value decreased slightly with increase in strain value resulting in softening behavior for biopolymer (G1/G2) treated soil specimens. In case of natural curing, during hardening process, the glassy structure of gum gel imparts brittle nature to the soil [14]. Therefore, sudden collapse of specimen is observed after experiencing peak stress.

Based on the previous findings from Vydehi and Moghal [8], significant increase in strength value is observed up to 28 days curing periods for all dosages of biopolymer (G1/G2). In addition, beyond the optimum gum dosage (4% for G1 and 2% for G2), marginal or no improvement in strength is observed. The results corresponding to these optimum conditions are presented in the following sections.

Variation in Secant Modulus (E_s) Values

The variation in secant modulus values of gum (G1/G2) treated soil at 28 days curing period under various curing conditions is presented in Fig. 23.2. In controlled curing condition, the E_s value increased with increase in gum dosage till 1% for both G1 and G2. At higher dosages (at 2 and 4%), marginal reduction in E_s value is observed. Beyond the optimum dosage, higher dosage of gum does not fetch any improvement in the targeted properties [11]. Under natural curing conditions, E_s value of gum

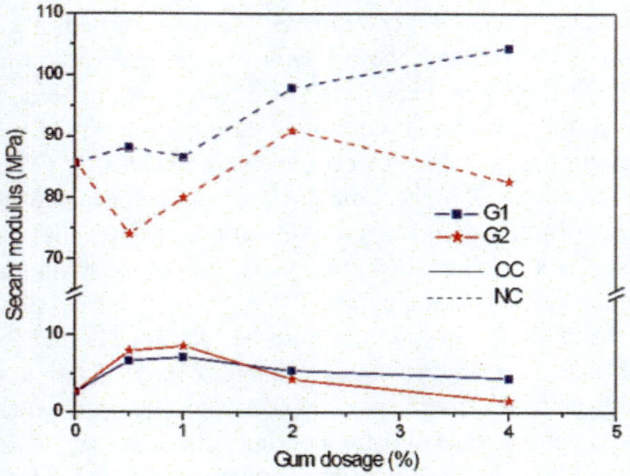

Fig. 23.2 Secant modulus values of gum treated soil at 28 days curing period

Table 23.1 Secant modulus (E_S) (MPa) values at various curing periods

Curing period (days)	E_S (MPa) at 50% peak stress			
	Controlled curing		Natural curing	
	G1 (at 4%)	G2 (at 2%)	G1 (at 4%)	G2 (at 2%)
7	4.52	5.23	73.17	78.43
28	4.43	4.33	104.43	91.02
60	4.72	3.56	126.77	109.44

treated specimens increased with increase in dosage (Fig. 23.2). The hardened gum binds the soil grains tightly, which are able to withstand greater stress values at lower axial strain.

Table 23.1 represents the variation in E_s values with increase in curing period at optimum dosages of gum (4% for G1 and 2% for G2) under different curing conditions. In controlled environment, the gum solution will be in gelatinous form and exhibits pseudo-plastic nature under applied loads [15, 16]. Therefore, no significant variation is observed in the E_s values with increase in curing period. Whereas under natural curing, the hardened gum gel offers greater resistance to applied loads and exhibits lower deformation values with higher stress values. Therefore, the E_s values increased significantly for both gums (G1/G2) with increase in curing period [9].

Variation in Failure Strain (%)

The variations in failure strain (%) of gum treated (G1/G2) and untreated soil at various curing conditions are presented in Fig. 23.3 and Table 23.2. Under controlled curing, the strain at peak stress is determined; whereas, for natural curing, strain at failure is determined to evaluate the behavior of gum treated soil as described in Fig. 23.1. In Fig. 23.3, for gum treated soil, the strain values increased with increase in dosage for both the gums under controlled curing representing the more ductile nature of soil-gum mixture [10]. However, the strain values for G2 treated soil are less than the untreated case at higher gum dosage. This is due to beyond the optimum dosage, addition of further gum leads to poor workability results in reduced stress and corresponding strain values. In similar lines, the failure strain values increased along with gum dosage for natural curing.

The failure strain values at optimum dosages (4% for G1 and 2% for G2) are presented in Table 23.2. At any gum dosage, notable change in failure strain values is not observed with increase in curing period under controlled curing due to gelatinous nature of hydrated gum. In case of natural curing with increase in curing period, the dehydration of moisture from soil-biopolymer mixture leads to hardening of mix and resulting in decrease in failure strain values. Similar results have been observed by Muguda et al. [14].

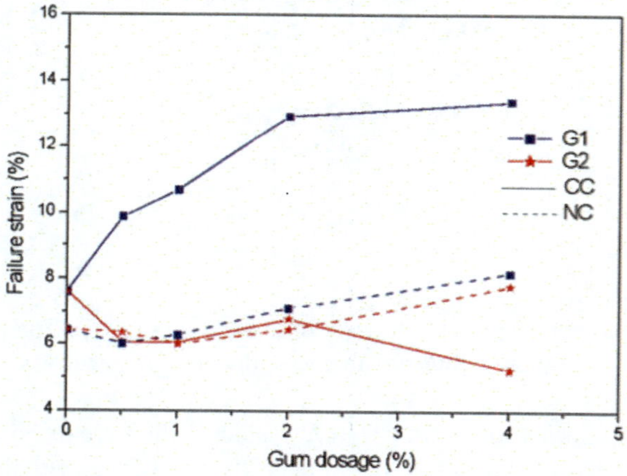

Fig. 23.3 Failure strain values of gum treated soil at 28 days curing period

Table 23.2 Failure strain (%) values at different curing periods

Curing period (days)	Failure strain (%)			
	Controlled curing		Natural curing	
	G1 (at 4%)	G2 (at 2%)	G1 (at 4%)	G2 (at 2%)
7	12.51	8.25	11.51	9.78
28	13.4	5.96	8.18	6.46
60	13.6	7.06	7.69	6.83

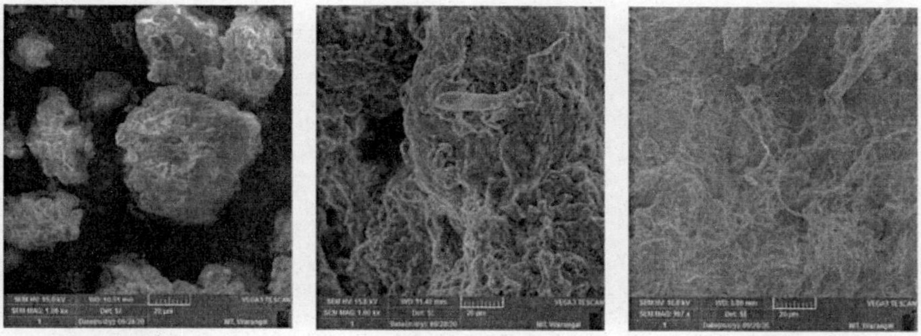

Soil Soil + 4% G1 Soil + 2% G2

Fig. 23.4 Scanning electron micrographs of gum treated soil

Surface Morphological Characteristics

The surface morphological images of xanthan gum (G1) and guar gum (G2) treated soil (at optimum dosages) are presented in Fig. 23.4. The binding of soil grains with hydrated/hardened biopolymer film causes improvement in the mechanical properties of soil such as secant modulus (E_S). As the gum strands stick to the soil grains, they restrain the movement of soil from failure under controlled curing resulting in increased failure strain levels. In natural curing conditions, hardening of gum gel leads to formation of glassy like structure which imparts brittle nature to the soil therefore, reduction in failure strain is observed with increase in curing period.

Conclusions

The present study evaluates the secant modulus of xanthan gum (G1) and guar gum (G2) treated soil at various dosages under controlled and natural curing conditions. Failure strain values were also determined to understand the deformability behavior of gum treated soil under applied static loads. The conclusions drawn from the experimental study are mentioned below.

- With addition of biopolymer to soil, strain softening behavior is observed, and the mixture exhibits ductile characteristics.
- With an increase in the gum dosage, significant increase in secant modulus (E_S) values is observed under natural curing due to hardening of biopolymer gel; whereas, no such significant changes are observed under controlled curing.
- At any gum dosage with increase in curing period, no significant variation is observed in failure strain values under controlled curing due to non-hardening of gelatinous gel. With an increase in gum dosage, failure strain values increased for both curing conditions.

References

1. Lambe TW, Whitman RV (1969) Soil mechanics, edition Wiley, New York
2. Moghal AAB, Obaid AAK, Al-Refeai TO (2014) Effect of accelerated loading on the compressibility characteristics of lime-treated semiarid soils. J Mater Civ Eng, ASCE 26:1009–1016. https://doi.org/10.1061/(asce)mt.1943-5533.0000882
3. Briaud JL (2001) Introduction to Soil Moduli. Geotech. News., BiTech Publishers Ltd., Richmond, BC, Canada
4. Moghal AAB (2017) State-of-the-art review on the role of fly ashes in geotechnical and geoenvironmental applications. J Mater Civ Eng 29(8):04017072-1 04017072-14. https://doi.org/10.1061/(ASCE)MT.1943-5533.0001897
5. Mahedi M, Cetin B, White DJ (2018) Performance evaluation of cement and slag stabilized expansive soils. Transp Res Rec 2672(52). https://doi.org/10.1177/0361198118757439
6. Almajed A, Lateef MA, Moghal AAB, Lemboye KK (2021) State-of-the-art review of the applicability and challenges of microbial-induced calcite precipitation (MICP) and enzyme-induced calcite precipitation (EICP) techniques for geotechnical and geoenvironmental applications. Crystals 11(4), Article No. 370. https://doi.org/10.3390/cryst11040370
7. Moghal AAB, Vydehi KV (2021) State-of-the-art review on efficacy of xanthan gum and guar gum inclusion on the engineering behavior of soils. Innov Infrastruct Solut 6:108. https://doi.org/10.1007/s41062-021-00462-8
8. Vydehi KV, Moghal AAB (2021) Effect of biopolymeric stabilization on the strength and compressibility characteristics of cohesive soil. J Mater Civ Eng. https://doi.org/10.1061/(ASCE)MT.1943-5533.0004068
9. Chang I, Im J, Prasidhi AK, Cho G-C (2015) Effects of Xanthan gum biopolymer on soil strengthening. Constr Build Mater 74:65–72. https://doi.org/10.1016/2014.10.026
10. Singh SP, Das R (2019) Geo-engineering properties of expansive soil treated with xanthan gum biopolymer. Geomech Geoengin. https://doi.org/10.1080/17486025.2019.1632495
11. Soldo A, Miletić M, Auad ML (2020) Biopolymers as a sustainable solution for the enhancement of soil mechanical properties. Nat: Sci Rep 10, Article number: 267. https://doi.org/10.1038/s41598-019-57135-x
12. Biju MS, Arnepalli DN (2020) Effect of biopolymers on permeability of sand-bentonite mixtures. J Rock Mech Geotech Eng 12(5):1093–1102. https://doi.org/10.1016/j.jrmge.2020.02.004
13. ASTM D2166 (2016) Standard test method for UCS of cohesive soil. ASTM International, West Conshohocken, PA, USA
14. Muguda S, Booth SJ, Hughes PN, Augarde CE, Perlot C, Bruno AW, Gallipoli D (2017) Mechanical properties of biopolymer-stabilized soil-based construction materials. Geotech Lett 7:309–314. https://doi.org/10.1680/jgele.17.00081
15. Milas M, Rinaudo M, Tinland B (1985) The viscosity dependence on concentration, molecular weight and shear rate of xanthan solutions. Polym Bull 14:157–164. https://doi.org/10.1007/BF00708475
16. Casas J, Mohedano A, Garcıa-Ochoa F (2000) Viscosity of guar gum and xanthan/guar gum mixture solutions. J Sci Food Agric 12:1722–1727. https://doi.org/10.1002/1097-0010(20000915)80:123.0.CO;2-X

Chapter 24
Producing Biochar from Crop Residues for Safe and Environment-Friendly Waste Management and Using as an Innovative Material for Soil and Ground Improvement

Mahendra Pratap Choudhary, H. D. Charan, and Biswajit Acharya

Introduction

In India, a huge quantity of agricultural waste in the form of crop residues is generated annually. Crop residue management has become a big issue in the recent past. The Indian farmers of Punjab, Haryana, and other adjoining northern states generally practice open burning of the crop residues resulting in serious environmental health hazards including emission of harmful greenhouse gases and loss of soil fertility. The annual outburst of smog in and around New Delhi which is supposed to be one of the consequences of crop residue burning in nearby states has become a challenging task for the stakeholders to resolve. Harvesting of crops generates a large quantity of residues on-farm as well as off-farm. The Indian farmers have been making use of agricultural machinery in harvesting their crops for quite some time which has aggravated the issue of crop residue management. The direct and open crop residue burning not only causes a loss of nutrients like nitrogen, phosphorus, sulfur, and potassium in the soil and harms the beneficial soil organisms but also releases harmful greenhouse gases, aerosols, particulates, and hydrocarbons into the atmosphere.

In 2014, the Ministry of Agriculture, Govt. of India has implemented National Policy for Management of Crop Residues [1] highlighting the need of the hour to curb crop residue burning by way of technical interventions, promoting diversified use of crop residues, providing financial assistance and subsidy for purchase of improved machinery like happy seeders, etc., and capacity building and creation of awareness

M. P. Choudhary (✉) · B. Acharya
Rajasthan Technical University, Kota, Rajasthan, India
e-mail: choudhary_mp@yahoo.co.in; mpchoudhary@rtu.ac.in

H. D. Charan
Bikaner Technical University, Bikaner, Rajasthan, India

K. Muthukkumaran et al. (eds.), *Ground Improvement Techniques*, Lecture Notes in Civil Engineering 297, https://doi.org/10.1007/978-981-19-6727-6_24

among the farmers and the general public. Further, in 2015, the National Green Tribunal (NGT) strictly prohibited burning of crop residues in open fields, but Indian farmers are yet not complying with it in its strict manner.

There is a dire need to make the farmers aware in this regard and at the same time, they may be encouraged to produce biochar from the crop residues. The application of biochar in improving the crop yield and stabilization of expansive soils has a greater scope ahead. Therefore, transforming the agricultural waste into biochar in place of open field burning can be a more techno-economic and viable solution to air pollution caused by stubble burning. Long-term sequestration of carbon and allied beneficial soil effects can be ascertained by way of converting the residual biomass into biochar [2].

Biochar

The history of biochar dates back centuries ago to the Amazon Basin where charcoal was used to enhance the productiveness of agricultural soils. The black and fertile soil, known as *terra preta de indio,* was discovered by Wim Sombroek in the 1950's. The term "biochar" was coined by Peter Read in 2005 for the char obtained through the process of pyrolyzing the biomass in absence of oxygen. Recently, there has been more research interests toward biochar owing to its peculiar characteristics in terms of physical as well as chemical properties and has gained pertinence in sustainable agriculture and climate change mitigation [3]. Biochar is known to have high carbon-content which is produced by heating biomass either in absence or presence of small amounts of air at temperatures above 250 °C. The process of heating is called pyrolysis which is similar to the method adopted for preparing charcoal. Nevertheless, biochar is not the same product as the charcoal or any other product of carbon because it is envisioned for application as a soil amendment [4]. It has been recognized as a soil amendment because of its special characteristics of adsorption and stability which prove it exclusive among the other organic soil amendments [5]. The biochar production and its further usage as soil amendment has the potential to deal with other local as well as global environmental issues. But the only constraint has been the economic feasibility, and hence, biochar application in agricultural soils has not been much encouraging [6, 7]. The characteristics like pH, ash, large surface area, high water holding capacity, bulk density, and pore-volume are among the most peculiar properties of biochar [8]. High porosity is one of the distinct properties of biochar accountable for higher water retention capacity and enhanced retention of nutrients. Furthermore, the large surface area and pores in biochar create an advantageous situation for soil microorganisms by way of providing protection against drying and predation [9]. A lot of research has been carried out recently, and the literature highlights the major encompassing factors during production of biochar and its application as a soil amendment [2, 10–15]; but in case of India, we do not find such specific studies yet, as it is comparatively a new idea in terms of making use of crop residues to produce biochar and ensure its further application in soil improvement.

Materials and Methods

A Novel Method: BioCharan

During review of the available literature, it was found that there is no economic and feasible method available for producing biochar at the field in a short span of time. The available methods are either age-old conventional kilns used for making charcoal or the modern industrial setups based on pyrolysis requiring high capital investment and infrastructures. Looking at the large-scale incidences of crop residue burning in India and the vacuum of a suitable technological solution easily adoptable by the farmers was the true motivation behind this study [16].

A novel method named *BioCharan* for converting crop residues into biochar at the fields has been found which not only clears the crop fields post-harvesting, but also helps in extenuating the air pollution. Some other beneficial outcomes of the method include stabilization of the soil and fertility enhancement [17]. This method is an economic and feasible method as it does not require any technical skills to operate, and an individual farmer can himself apply this method for converting the residues into useful product biochar so as to resolve the issue of crop residue disposal and environmental air pollution through emission of greenhouse gases during crop residue burning. Another important contribution is the investigation of biochar application to soils to improve the geotechnical characteristics of soils. The studies carried out so far and available in the literature point out toward using biochar to enhance the agronomic parameters of soils related to improvement in crop yield and soil fertility. The present study is supposed to be one of the leading studies of its kind in India wherein application of biochar has been used to study the effect on geotechnical characteristics of the soils.

By understanding the concept of conventional methods of biochar production and then applying some indigenous wisdom, it was considered to produce biochar at the field level through a modified kiln made up of a cylindrical empty diesel drum. Looking to the absence of any cost-efficient alternative in Indian scenario, the farmers can produce biochar using this novel method of *BioCharan* and furthermore the biochar produced can be applied in the fields for increasing the soil fertility [18]. The various steps of the *BioCharan* method for producing biochar from cotton stalks in the field at village Kosana, district Jodhpur, Rajasthan are shown in Fig. 24.1a–e.

A cylindrical drum having both ends open with 3 ft height and 2 ft diameter has been utilized to make biochar from crop residues. The drum is supported on stones or bricks so that limited entry of air can be ensured from the bottom end. The drum is then filled up with the crop residues and once filled, it is ignited at the bottom.

Fig. 24.1 **a** Cylindrical drum used for making biochar; **b** filled with crop residues; **c** set to fire at the bottom; **d** The supporting bricks removed from the bottom and top is covered with a lid; biomass is allowed to heat inside for 10–15 min; and **e** raw biochar is crushed into pieces

After ignition, the residues are allowed to burn only for 2–3 min and the supporting stones or bricks are taken away, and the drum is covered at the top with a lid. The crop residue inside the drum is hence allowed to convert into biochar through heating process for a period of about 10–15 min. The drum is then removed from its place when it becomes cool and the raw biochar is collected and crushed into smaller pieces so as to mix with soil samples either for laboratory experimentation or for field trials.

A drum can handle approximately 15–20 kg of crop residue at a time and if 5–6 drums are used simultaneously, we can convert one quintal of stubble into biochar in one instance, i.e., 15 min and hence 1 ton of crop residues will usually take a time of about 3 h. The overall cost of conversion of crop residues into biochar comes out to be about Rs. 5 per kg considering the factors involved in cutting the residues up to the final production of biochar [19]. Therefore, we can say that the cost of biochar production is quite less since this method does not necessitate any expensive equipment or technical skills. The Indian farmers can produce it at their own fields.

Soil Sampling

The field samples of soil were obtained from the agrarian fields of village Kosana situated in district Jodhpur, Rajasthan. The standard procedure for collection of samples was adopted and after proper sealing in plastic bags, they were taken to the laboratory wherein further preparation for various tests as per the procedures laid down by the Bureau of Indian Standards in IS: 2720 (Part 1)–1983 [20] was carried out. Initially, the samples were put in thermostatically controlled oven for drying, then pulverized and finally sieved through a 4.75 mm size sieve. The grain size analysis, soil classification, and other soil characteristics were found out as per the relevant IS standards. The soil type was found as CI (Inorganic clay of intermediate plasticity) as per the Indian Standard Soil Classification System (ISSCS).

Experimental Investigation

All relevant experiments were carried out in the respective laboratories to work out the outcome of mixing biochar on the physical, chemical, and engineering properties of the soil. The samples of soil were mixed with varying proportions of biochar by weight (% w/w) starting with minimum 5% and then increasing in its multiples up to maximum 25%. The addition of biochar was in dry form to the soil, and it was ensured that the soil-biochar mix was mixed carefully [19]. The sampling of biochar as well as soil has been carried out with three replicates of each, and hence, the characteristic parameters of an individual sample are the mean of these three values. The experimental investigation of samples of biochar, soil, and their mixes for characterization and assessment of various parameters were carried out.

Results and Discussion

Biochar Characterization

Physico-chemical Characteristics and Grain Size Distribution

The characterization of biochar samples was carried out by assessing their physical and chemical characteristics as well as grain size distribution whose results are represented in Table 24.1.

Table 24.1 Biochar characterization

S. No.	Characteristics	Value
1	pH	9.80
2	Specific gravity	1.62
3	Moisture content (%)	8.10
4	Electrical conductivity (mS/cm)	5.12
5	Organic carbon (%)	1.56
6	Organic matter (%)	2.52
7	*Grain size distribution (%)*	
	(i) 2–4.75 mm (coarse)	23.6
	(ii) 0.425–2 mm (medium)	46.2
	(iii) 0.075–0.425 mm (fine)	30.2

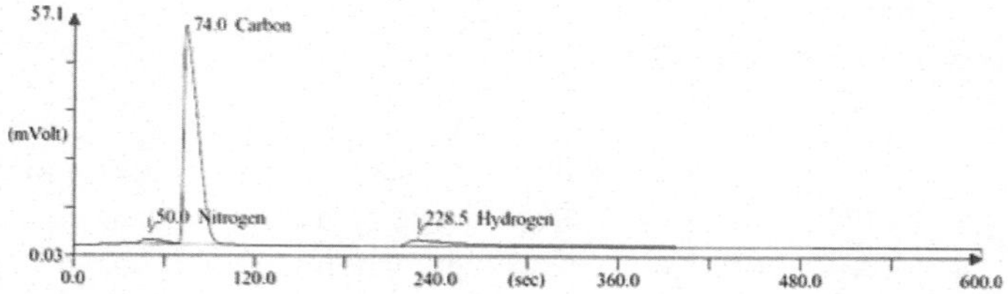

Fig. 24.2 CHN analysis for biochar sample

Elemental Analysis

The CHN analysis of biochar was carried out at SAIF, IIT Bombay as shown in Fig. 24.2. The major elements found in biochar are carbon (45.387%), hydrogen (1.88%), and nitrogen (1.162%), whereas the other minor elements like sulfur and oxygen were not found. Hence, the CHN analysis confirms that biochar samples are carbon-rich.

FTIR Spectroscopic Analysis

The Fourier Transform Infrared imaging of biochar was carried out at SAIF, IIT Bombay as shown in Fig. 24.3. The FTIR spectrum shows the functional groups present in a sample. The biochar sample has more than five peaks indicating that biochar is not a simple element. The spectrum shows C–H, O–H, and N–H as well as S=O, C=C, and O–Si–O stretching and bending vibrations in the range of 3727–457 cm^{-1} indicating the presence of various functional groups like phenolic, hydroxyl, methyl, methylene, alkane, alkyl, and silica.

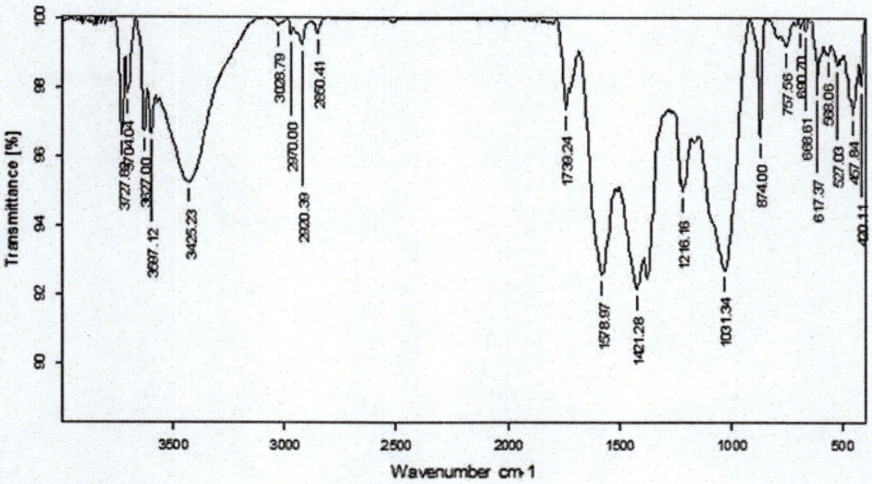

Fig. 24.3 FTIR Spectroscopic image for biochar sample

XRD Analysis

The X-ray diffraction analysis of biochar was performed at SAIF, IIT Bombay, as shown in Fig. 24.4. The XRD image helps to determine the crystalline phases in a mixture by comparing with position and intensity of reference patterns. The peaks in XRD pattern indicate the presence of graphite, SiO_2, CaO, and MgO. Hence, it can be concluded that biochar sample has a heterogeneous surface.

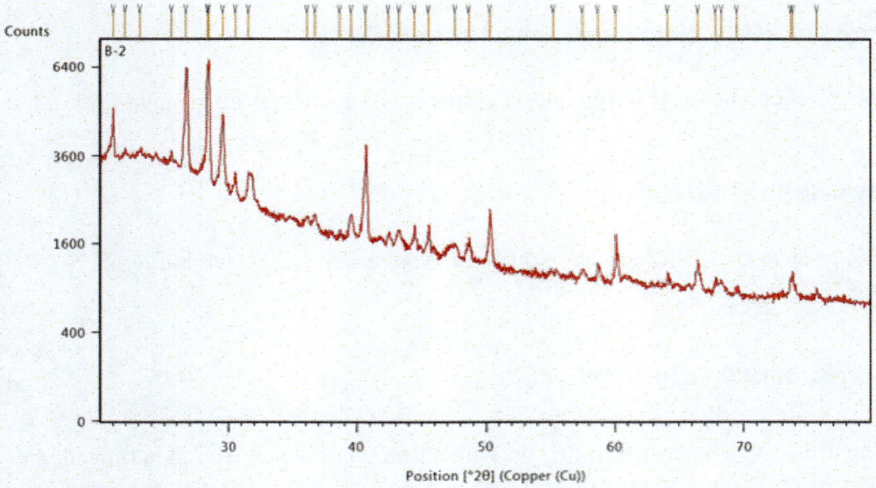

Fig. 24.4 XRD image for biochar sample

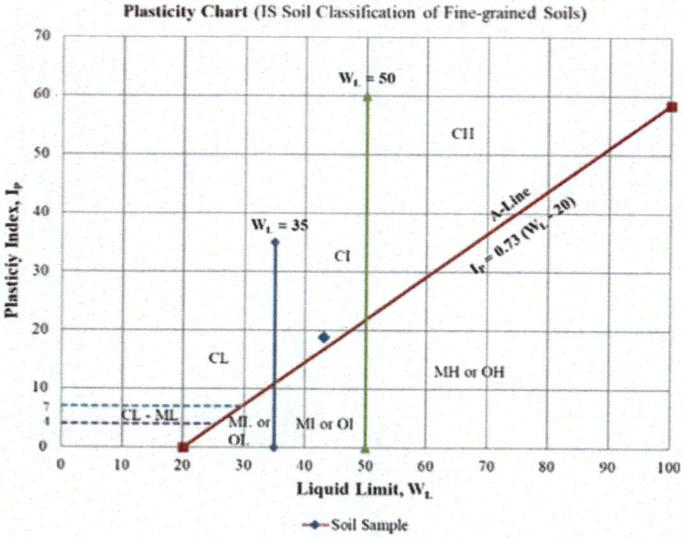

Fig. 24.5 Plasticity chart for soil sample

Soil Characteristics

Soil Classification and Plasticity Chart

The plasticity chart for the soil sample is represented in Fig. 24.5.

Physico-chemical and Engineering Characteristics

The physico-chemical and engineering characteristics of soil are given in Table 24.2.

Particle Size Distribution

The particle size distribution of soil and biochar sample is shown in Fig. 24.6.

Soil Amendment with Biochar

The initial parameters of virgin soil without mixing biochar were determined, and then different proportions of biochar were added in the soil samples, whose parameters are represented in Table 24.3. For the soil-biochar mix, the organic carbon increased by 26.79%, moisture content by 19.78%, water holding capacity by

Table 24.2 Physico-chemical and engineering characteristics of soil

S. No.	Name of characteristics	Value
1	pH	7.62
2	Specific gravity	2.56
3	Moisture content (%)	11.12
4	Electrical conductivity (mS/cm)	4.23
5	Organic carbon (%)	0.56
6	Organic matter (%)	0.98
7	Water holding capacity (%)	32
8	Optimum moisture content (%)	16.00
9	Maximum dry density (g/cc)	1.64
10	*Grain size distribution (%)*	
	(i) 2–4.75 mm	5.4
	(ii) 0.425–2 mm	17.0
	(iii) 0.075–0.425 mm	27.2
	(iv) 0.002–0.075 mm and <0.002 mm	50.4
11	*Consistency limits (%)*	
	(i) Liquid limit	43.12
	(ii) Plastic limit	24.25
	(iii) Shrinkage limit	17.63
	(iv) Plasticity index	18.87
12	California bearing ratio (%)	1.96
13	Unconfined compressive strength (kg/cm^2)	1.68
14	Shear strength (kg/cm^2)	0.84
15	Free swell index (%)	39

43.75%, unconfined compressive strength and shear strength by 10.71%, CBR value by 15.31%, whereas the free swell index decreased by 30.77% when biochar was mixed.

Discussion on Results

The results indicate that mixing of biochar to soil increased the organic carbon, organic matter, moisture content and water holding capacity, and at the same time, decreased the specific gravity and density of the soil. This resulted into an overall improvement in physical properties of the soil, which can be attributed toward the formation of macro-aggregates in due course of time resulting in enhanced inter-particle cohesion and improved resistance to slaking. For the soil-biochar mix, an increase in liquid limit has been observed up to a certain level due to distinct properties

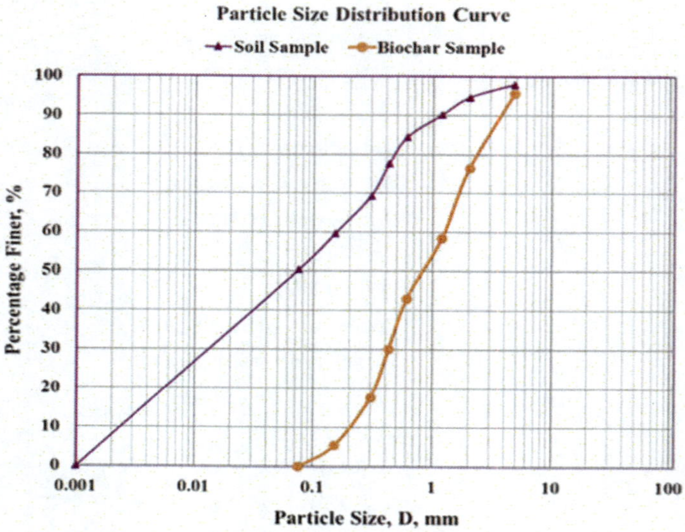

Fig. 24.6 Particle size distribution curve for soil and biochar sample

of biochar like large surface area, high porosity, and water absorbing capacity. The reason for increase in plastic limit may be the enhanced capacity of water absorbance of the soil-biochar composite. The value of OMC increased mainly due to moisture absorption by biochar, whereas the reduction in MDD may be owing to the low specific gravity of biochar as compared to that of the soil. Secondly, the soil particles covered by biochar resulted in large-sized particles having higher void ratio and lesser density. It also acted as one the reasons for increase in OMC. Other important parameters of expansive soil including unconfined compressive strength, shear strength, and swelling characteristics also improved upon addition of biochar. It has been observed that the free swell index of the soil decreased on addition to biochar which is due to the fact that with decrease in swelling potential of the compacted soil, it becomes more stable.

Table 24.3 Physico-chemical and engineering properties of soil amended with biochar

S. No.	Characteristics	Values for soil amended with biochar (BC)						% Variation
		Virgin Soil	Soil + 5% BC	Soil + 10% BC	Soil + 15% BC	Soil + 20% BC	Soil + 25% BC	
1	pH	7.62	7.65	7.68	7.72	7.76	7.79	*2.23*
2	Specific gravity	2.56	2.55	2.54	2.52	2.52	2.51	*−1.95*
3	Moisture content (%)	11.12	11.38	11.85	12.23	12.86	13.32	*19.78*
4	Electrical conductivity (mS/cm)	4.23	4.25	4.31	4.33	4.39	4.44	*4.96*
5	Organic carbon (%)	0.56	0.59	0.62	0.66	0.69	0.71	*26.79*
6	Organic matter (%)	0.98	0.99	1.02	1.03	1.05	1.05	*7.14*
7	Water holding capacity (%)	32	34	38	40	42	46	*43.75*
8	Liquid limit	43.12	44.57	45.62	47.24	48.63	49.26	*14.24*
9	Plastic limit	24.25	25.28	26.12	27.64	28.67	28.98	*19.49*
10	Shrinkage limit	17.63	18.86	19.62	20.47	21.67	22.45	*27.34*
11	Plasticity index	18.87	19.29	19.50	19.60	19.96	20.28	*7.49*
12	Optimum moisture content (%)	16.00	16.52	16.98	17.13	17.28	17.37	*8.56*
13	Maximum dry density (g/cc)	1.64	1.57	1.49	1.45	1.43	1.42	*−13.41*
14	Unconfined compressive strength (Kg/cm^2)	1.68	1.75	1.79	1.82	1.84	1.86	*10.71*
15	California bearing ratio (%)	1.96	2.03	2.08	2.16	2.21	2.26	*15.31*
16	Shear strength (Kg/cm^2)	0.84	0.88	0.90	0.91	0.92	0.93	*10.71*
17	Free swell index (%)	39	38	35	32	29	27	*−30.77*

Conclusion

The following conclusions can be drawn, based on the above findings and discussion:

(1) Biochar can be produced economically and efficiently at the fields by adopting the novel method of *BioCharan.*
(2) The production of biochar from crop residues and its further use in amending the soil properties have shown encouraging and enthusiastic results.
(3) Some of the significant engineering characteristics of expansive soils like Atterberg's limits, MDD, OMC, swelling characteristics, unconfined compressive strength, and shear strength of the soil have upgraded on addition of the biochar.
(4) The yearly outbreak of ambient air quality deterioration in northern India, especially in New Delhi due to stubble burning can be resolved through biochar looking to its promising prospective.
(5) Finally, *BioCharan* method can be called a multi-benefit method wherein the problem of crop residue burning is resolved by converting it to produce biochar, amendment in soil properties can be ensured by way of applying biochar to soils, and finally, environmental air pollution is minimized when there is no crop residue burning.

Acknowledgements The authors are highly thankful to Soil Testing Laboratory Jodhpur, Nakshtra Enviro Services Jaipur, Geotechnical Engineering Lab, RTU Kota and SAIF, IIT Bombay for the help and support provided for conducting experiments in their laboratories.

References

1. NPMCR (National Policy for Management of Crop Residues) (2014). Available online: http://agricoop.nic.in/sites/default/files/NPMCR_1.pdf. Last accessed 2021/07/01
2. Aditya P et al (2014) Biochar production from agro-food industry residues: a sustainable approach for soil and environmental management. Curr Sci 107(10):1673–1682
3. Lehmann J et al (2011) Biochar effects on soil biota—A review. Soil Biol Biochem 43(9):1812–1836
4. Lehmann J, Joseph S (2015) Biochar for environmental management: Sci. Tech. Implementation, 2nd edn. Routledge, London
5. Johannes L (2007) A handful of carbon. Nature 447:143–144
6. Lehmann J et al (2006) Bio-char sequestration in terrestrial ecosystems—a review. Mitig Adapt Strat Glob Change 11:403–427
7. Ranjan ST et al (2016) Use of biochar for greenhouse gas mitigation. Rashtriya Krishi 11(2):65–67
8. Shareef TME, Zhao BW (2017) The fundamentals of biochar as a soil amendment tool and management in agriculture scope: an overview for farmers and gardeners. J Agric Chem Env 6:38–61
9. Das SK et al (2016) Carbon-negative biochar from weed biomass for agricultural research in India. Curr Sci 110(11):2045–2046
10. Barus J (2016) Utilization of crops residues as compost and biochar for improving soil physical properties and upland rice productivity. J Degraded Min Lands Manage 3(4):631–637

11. Sarah C et al (2013) The Impact of biochar application on soil properties and plant growth of pot grown lettuce (Lactuca sativa) and cabbage (Brassica chinensis). Agronomy 3:404–418. https://doi.org/10.3390/agronomy3020404
12. Ding Y et al (2016) Biochar to improve soil fertility: a review. Agron Sustain Dev 36. https://doi.org/10.1007/s13593-016-0372-z
13. Gokila B, Baskar K (2015) Influence of biochar as a soil amendment on yield and quality of maize in Alfisols of Thoothukudi district of Tamil Nadu, India. Int J Plant, Anim Environ Sci 5:(1):152–155
14. Lydia F, Rianne V (2015) Biochar for soil improvement: evaluation of biochar from gasification and slow pyrolysis. Agriculture 5:1076–1115. https://doi.org/10.3390/agriculture5041076
15. Priit T et al (2016) Biochars in soils: towards the required level of scientific understanding. J Environ Eng Landsc Manag 25(2):192–207. https://doi.org/10.3846/16486897.2016.1239582
16. Choudhary MP et al (2021) A novel approach for disposing agriculture waste, minimizing air pollution and amending soil through biochar production and application. Nat Environ Poll Technol 20(1):421–427. https://doi.org/10.46488/nept.2021.v20i01.026
17. Choudhary MP et al (2021) Potential of biochar derived from crop residues in soil remediation and controlling air pollution due to stubble burning. Indian J Environ Prot 41(2):207–212
18. Choudhary MP et al (2021) Converting agricultural waste into biochar for improving physical properties of soil. In: Patel S, Solanki CH, Reddy KR, Shukla SK (eds) Proceedings of the Indian geotechnical conference 2019. Lecture Notes in Civil Engineering, vol 134. Springer, Singapore. https://doi.org/10.1007/978-981-33-6370-0_10
19. Choudhary MP et al (2021) Experimental investigation on biochar production and application: solution to air pollution due to stubble burning as well as amending soil consistency. J Inst Eng, Ser A, online published July 2. https://doi.org/10.1007/s40030-021-00548-7
20. IS 2720-Part 1 (1983) Methods of test for soils—Preparation of dry soil samples for various tests (Second Revision) (Reaffirmed May, 2020), Bureau of Indian Standards, New Delhi

Chapter 25
A Review on Comparative Study of Stabilization of Black Cotton Soil by Different Chemical Stabilizers

Narendra Sipani and Sukanya Sharma

Introduction

Stabilization process is the traditional process to improve the mechanical properties of expansive soils. In last few years, many raw materials and techniques are introduced to stabilize the soil as per the requirement. Black cotton soils are problematic soil for any civil engineering project and civil engineers. Large-scale damages are observed in many civil engineering projects due to the change in soil properties. Expansive soils are still a challenge to geotechnical engineers and society. Clay particles are initially accountable for the expansive nature of the soil. When particle comes in the contact with water, it shows the expansiveness. Different methods, namely stabilization, grouting, and reinforcement is employed to enhance the characteristics of these soils. This comprehensive review serves the practice of efficient stabilization using chemicals for pavements and other civil engineering structures.

In the process of chemical stabilization, the cementing agents create the bond with the soil particles to increase the strength of soil. The intermolecular forces develop between the particles to keep the bond. In this article, chemical additives, namely KCl, $CaCO_3$, $CaCl_2$, $AlCl_3$, $FeCl_3$, and NaCl are discussed for stabilizing the expansive soils.

N. Sipani
Department of Civil Engineering, Jaipur Engineering College and Research Centre, Jaipur, Rajasthan, India

S. Sharma (✉)
Department of Civil Engineering, Rajasthan Technical University, Kota, Rajasthan, India
e-mail: sskanya23@gmail.com

K. Muthukkumaran et al. (eds.), *Ground Improvement Techniques*, Lecture Notes in Civil Engineering 297, https://doi.org/10.1007/978-981-19-6727-6_25

Literature Study

Chemical improvisation is the intermixing of chemical additives with natural soils to eliminate moisture and improve soil resistance properties. Chemical processes play a vital role in every stabilization. Numerous researchers used different chemical stabilizers to improve the mechanical properties of soil. The published research work by various researchers is discussed below.

Sharma et al. [1] performed tests on black cotton soil using magnesium chloride for soil enhancement. They balanced out the black cotton soil with various levels 0, 2, 4, 6, and 10% $MgCl_2$ and performed CBR test for the soaked condition, proctor test, UCS test, index, and swelling parameters. From the conclusion, it is clear that the properties enhanced at 8% chemical. There is a reduction in liquid limit and swelling pressure. Likewise, an increase in the amount of the UCS values and dry density is achieved.

Rambabu and Bhavannarayana [2] investigated the work of stabilized soil with banana fiber and $MgCl_2$. This article studied the compaction parameters, cyclic plate load tests, CBR tests with 0.5, 1, and 1.5% magnesium chloride along with 0.25, 0.5, 0.75, and 1% Banana fiber. From the results, it is concluded that the free swell index of clay was decreased along with liquid limit as the percentage of $MgCl_2$ increased. On the other hand with the addition of $MgCl_2$ to 1% Banana fiber, it was noted that the plastic limit has been increased. The author further concluded that the CBR, shear strength, and ultimate load carrying capacity parameters of the soil are hence improved.

Ajay Raj et al. [3] investigated the improvement in performance of BCS along with NaCl and flyash. They played out Atterberg limit test, Proctor compaction test, unconfined compression test, California Bearing Ratio test-taking mix proportions 0, 5, 10, 15, and 20% fly ash with 0, 3, 6, 9, and 12% chemical stabilizer, i.e., NaCl. The tests outcomes summarized the liquid limit of BCS has been decreased from 48.32 to 27.15% while the plastic limit of a soil sample has been increased from 23.18 to 29.12%. The results of maximum dry density have increased and OMC has decreased. At 5 mm penetration, the maximum CBR value is 3.12 N/mm^2 and at 2.5 mm deflection, it is 3.98 N/mm^2, and hence, the BCS is thus stabilized with common salt and fly ash.

Zumrawi and Eltayeb [4] assessed the possessions of $CaCl_2$ as stabilizer for expansive soil. Proportions of $CaCl_2$ to be varied were 0, 2, 5, 10, and 15%. For expansive soil, the plasticity index was 36%, when added with 15% $CaCl_2$, the plasticity index was found to be 14%. For the same quantity of $CaCl_2$, the DFS is reduced to 49%. From 180% for expansive soil on addition of $CaCl_2$ at 5% UCS value increase almost 15% than that of an expansive soil, hence the optimum dose is 5% $CaCl_2$.

Kolaventi et al. [5] carried out the comparative analysis of improving the performance of BCS using salts. The proportion was chosen to be 0, 2, 4, 6, and 8%. It was found that plasticity index decreases as salt concentration increases and MDD raised. In this paper, we got the better results for calcium chloride than sodium chloride. The qualities of the black cotton soil are enhanced by the addition of NaCl and

$CaCl_2$ at 8%. Comparing the cost, NaCl is cheaper when compared to $CaCl_2$. So, from economic point of view, it is better to use NaCl for stabilizing large area.

Shukla et al. [6] investigated by using the potassium chloride in the soil the wetting and drying can be controlled. Mixing of KCl caused to substantial reduction in OMC, plasticity index and liquid limit of black cotton soil by various works covering different aspects of stabilization by chemicals. Shrinkage limit, MDD, and UCS are increased with addition of KCl. Further optimum quantity of KCl is approximately 7.0% as compared to earlier studies which is from 5 to 6%.

Dubey and Jain [7] determined the effect of NaCl on properties of BC soil. Proportion for mixing was chosen to be 0–8% at an interval of 2%. OMC has been decreased, and MDD is raised as the common salt is added further. MDD for expansive soil was 1.64 gm/cc which increased to a value of 1.79 gm/cc when added with 8% NaCl. For the same OMC decrease from 21.1 to 14.95% and DFS lowered from 41 to 19% which showed that the degree of expansiveness of sample is lowered from high to low. UCS increased by 62% when NaCl added at 8% proportion hence we can conclude that 8% NaCl is an optimum dose.

More et al. [8] studied the performance improvement of soil by potassium chloride. The proportions were chosen to be 0 N, 0.5 N, 1 N, and 1.5 N solution. It was found that gradual decrease in liquid limit, plastic limit, shrinkage limit, and optimum moisture content of soil with increasing potassium chloride proportions. Further gradual increase in MDD of soil with increasing potassium chloride proportions is observed.

Radhakrishnan et al. [9] studied the swelling parameters of soils mixed with chemicals ($MgCl_2$ & $AlCl_3$) were evaluated as a stabilizer of expansive soil with fly ash. Results indicating at various proportion of salt (0, 0.5, 1.0, 1.5, and 2%) that the soil is highly expansive. It was observed that the investigation of soil with $AlCl_3$ and fly ash at 1 and 10% is more effectual than the other. The proportion depletion in swell parameters is 63% for $MgCl_2$ and 68% for $AlCl_3$, whereas swell pressure is 69% and 73% for $MgCl_2$, $AlCl_3$, respectively.

Belabbaci et al. [10] focused on effect of KCL and $MgCl_2$ on physical and chemical property of expansive soil. Concentration of salts to be mixed in soil chosen to be 0.05, 0.1, and 0.2 mol/l. It was found that influence of salt solution has an effective on ES properties. Based on the results obtained, the clays which high plasticity seen after treatment with salts, consistence progress to show low plasticity index. This effect was found due to effective diffusion of solution.

Venkata Muthyalu et al. [11] enhanced the performance of expansive soil using different salts. The tests were conducted on soil samples accompanied by different salts like KCl, $FeCl_3$, and $CaCl_2$ with different percentages from 0 to 1.5%. The outcomes showed a remarkable change in consistency limits. For all the chemicals, it is noted that the decrease in the liquid limit is significant up to 1%. An increase in theoretical values of plastic limit and CBR is noted.

Ramesh et al. [12] checked the efficiency of sodium carbonate and calcium carbonate for BCS stabilizing property. Six samples of soil mixture were made with chemical content as 0, 0.5, 1, 1.5, 2, and 2.5%. Tests for Atterberg's limit, UCS, and CBR have been carried out. Results showed that the Na_2CO_3 and $CaCO_3$ are equally effective for considering 2% chemical for reducing the swelling parameters. But for the decrement in OMC and increment in the MDD, $CaCO_3$ is more effective than Na_2CO_3. Also, the CBR values of the mixture (soil + chemical) are higher than BCS. Hence, the author concluded that from a stability and strength point of view, $CaCO_3$ is better than Na_2CO_3.

Manoj Krishna et al. [13] checked the strength and FOS performance of BCS treated with calcium chloride. The overall laboratory testing was done with different $CaCl_2$ contents, i.e., 1, 2, 3, 4, 5, and 6%. For finding the optimum dosage of the chemical, compaction, and strength parameters w, the addition of 3% $CaCl_2$ to BCS decreases the MDD and increases the dry density. UCS at a 1.25 mm/min strain rate. In the study, the samples are tested with various curing periods in all the combinations of $CaCl_2$ and the optimum dosage of 3% is obtained on 30 days. Hence, the addition of $CaCl_2$ increases the safe bearing capacity of BCS.

Al-Omari et al. [14] used KCl at 0.0, 2.5, 5.0, and 7.5% to improve mechanical property of soil. A decrease in the Atterberg's limit like plasticity index and liquid limit was observed when KCl was mixed. KCl increased the MDD and reduces the OMC. The improvisation of free swell index and swelling pressure occur when KCl added was 5% and the values noted were 56% and 65%, respectively.

Srinivas and Prasada [15] investigated the outcomes of strong chemicals with soil on its swell parameters. They performed the laboratory test using three different chemicals KCl, $CaCl_2$, and $FeCl_3$. The results showed that the decrement in DFS values and swell potential has observed with the inclusion of the 1% chemical. In this study, comparatively, $FeCl_3$ gives the best results.

Abood et al. [16] examined of the effect of different chemical component on engineering parameters of soil containing silty clay is carried out. The proportions of salts are considered to be 2–8%. Varied tests like Atterberg's limit test, test for compaction and UCS was regulated. As the solution of NaCl is added, 4% gives the highest UCS, while the highest UCS value at 8% was given by adding calcium chloride. The work concluded that on comparing NaCl and calcium chloride, the results $CaCl_2$ is more efficacious.

Conclusion

From the study of various published article, the following conclusions are mapped:

(a) From the chemical stabilization, it has been observed that properties of expansive soil may be improved by adding the chemical stabilizer in the range of 1–16%.
(b) Addition of $CaCl_2$ has major contribution on compaction property of the stabilized soil. An increase in $CaCl_2$ content leads to an increase in dry density and decrease in optimum moisture content results in greater strength.
(c) KCl increases the MDD and decreases the OMC. This KCl addition decreases the moisture content by 18%.

(d) The UCS of the soil increases upon the inclusion of NaCl.
(e) UCS values are raised by 177% and 203% on the addition of $CaCl_2$ and $FeCl_3$, respectively at 1% treatment. It is further noted that the $FeCl_3$ is more effective thus selected electrolytes are effective in improving the properties of BCS.
(f) UCS of the mixed sample increases as the inclusion of NaCl increases. The stabilized samples of UCS increased from 73.54 to 119.64 KN/m^2.
(g) For 14 days curing period, the UCS values are raised by 133%, 171%, and 230% when the inclusion of 1% of KCL, $CaCl_2$, and $FeCl_3$, respectively.
(h) An increase in the values of UCS were noted down when soil mixed with chemicals increases.
(i) On the inclusion of common salt the soaked California Bearing Ratio values have increased. The California Bearing Ratio values raised from 1.43 to 3.10% on adding 8% NaCl into the soil.
(j) The index parameters like Liquid and Plastic Limit of BCS are decreased and increased, respectively, with the inclusion of NaCl.
(k) The materials NaCl, KCL, $CaCl_2$, and $FeCl_3$ are successfully implemented in the improvisation of BCS.
(l) As the chemical content $CaCl_2$, and $FeCl_3$ increases, a reduction in index parameter (liquid limit) is noted and an increase in plastic limit is notes, hence causing a net decrease in plasticity index.
(m) The possessions of BCS are enhanced by the inclusion of sodium chloride and $CaCl_2$ at 8%.
(n) Comparing the cost, NaCl is cheaper when compared to $CaCl_2$. So in economic point of view, it is better to use NaCl for stabilizing large area.
(o) Excessive amount not only reduced the cation exchange capacity but also increased the water absorption, and consequently, it reduced the efficiency of potassium chloride. Earlier studies found that optimum quantity of KCl were varying between 5 and 6%.

Finally, it has been concluded that the chemical can be used to improve the engineering properties of expansive soils. The chemical stabilizer develops the bond between the soil particles and due to this phenomenon, the swelling properties of soil decreases. It may also be concluded that these chemical stabilizers can be used in enhancement of pavement.

References

1. Sharma S, Kumar V, Bindlish A (2020) Enhancing the engineering properties of black cotton soil by using magnesium chloride. In: Proceedings of Indian geotechnical conference, Visakhapatnam Chapter 2020
2. Rambabu J, Bhavannarayana, Ch (2019) An experimental study of expansive soil stabilized with banana fiber and magnesium chloride. Int J Innov Eng Manag Res 8:162–168
3. Ajay Raj M, Ganapathy C, Vinay D, Suresh C (2018) Stabilization of black cotton soil by using sodium chloride and flyash. Int J Pure Appl Math 119(15), ISSN: 1314-3395

4. Zumrawi MME, Eltayeb KA (2016) Laboratory investigation of expansive soil stabilized with calcium chloride. Int J Environ Chem Ecol Geol Geophys Eng 10(2)
5. Kolaventi SS, Venigalla SG, Rakesh D (2016) Stabilization of black cotton soil using salts and their comparative analysis. Int J Geotech Environ 4(2), ISSN: 2321-9939
6. Shukla RP, Parihar NS Sood A (2015) Effect of potassium chloride on black cotton soil. History 40(184):241–245
7. Dubey P, Jain R (2015) Effect of common salt (NaCl) on engineering properties of black cotton soil. IJSTE-Int J Sci Technol Eng 2(01)
8. More, Manoj M, Landage AB, Konnur BA (2014) Laboratory studies on stabilization of an expansive soil by KCl. Int Inventive Multi J 2:1–5
9. Radhakrishnan G, Anjan Kumar M, Raju GVRP (2014) Swelling properties of expansive soils treated with chemicals and fly ash. Am J Eng Res 3(4):245–250
10. Belabbaci Z, Mamoune SMA, Bekkouche A (2013) Laboratory study of the influence of mineral salts on swelling (KCl, $MgCl_2$). Earth Sci Res 2(2):135
11. Venkata Muthyalu P, Ramu K, Prasada Raju GVR (2012) Study on performance of chemically stabilized expansive soil. Int J Adv Eng Technol 2(1):139–148
12. Ramesh P, Narasimha Rao AV, Krishna Murthy N (2012) Efficacy of sodium carbonate and calcium carbonate in stabilizing a black cotton soil. Int J Eng Technol Adv Eng 2:197–201
13. Manoj Krishna KV, Ramesh HN Strength and FOS performance of black cotton soil treated with calcium chloride. IOSR J Mech Civ Eng 2:21–25
14. Al-Omari R, Ibrahim S, Al-Bayati I (2010) Effect of potassium chloride on cyclic behavior of expansive clays. Int J Geotech Eng 4(2):231–239
15. Srinivas M, Prasada GVR (2010) Effect of Strong chemicals on the swell properties of an expansive clay. In: Indian geotechnical conference, IGS Mumbai Chapter & IIT Bombay, 2010, pp 613–616
16. Abood TT, Bin Kasa A, Bin Chik Z (2007) Stabilisation of silty clay soil using chloride compounds. J Eng Sci Technol 2(1):102–110

Chapter 26
Stabilization of Sub-grade Soil Using Shredded Waste Plastic Bags

U. Salini and A. Jegan Bharath Kumar

Introduction

Constructing pavement over soft soils is a great challenge for road engineers. In such cases, the general practice is to remove these weak soil and replace them with stronger borrow material. However, the replacement of existing soil with borrows material may not be feasible due to non-availability of good quality and cheap fill material. Therefore, it becomes necessary for the engineers to choose other options such as stabilization of weak in situ soil. Soil stabilization refers to the process by which different materials or additives are blended and mixed with soil so to improve its engineering properties. The mechanical stabilization of soil is carried out using different products such as polymers, geosynthetics, soil nailing, using industrial wastes like pond ash, fly ash, quarry dust, recycled wastes, etc. The research studies carried out on soil-fiber composites have shown that the use of fiber improved the California Bearing Ratio (CBR) of the sub-grade soil [1, 5, 8]. Another most commonly generated waste from household and commercial activities is the plastic covers and is being simply disposed along with the municipal solid waste. Utilization of plastic waste covers for soil stabilization may be a viable solution and needs examination. This will not only help in soil stabilization but also helps in the disposal of plastic waste. Only a few research works have been performed on using waste plastic in soil and most of them deal with the recycled product made from plastic waste. Jha et al. [5] examined the use of waste recycled product obtained from plastic waste fiber and brought out that the CBR value increased with an increase in the aspect ratio and amount of plastic strip. No study has been conducted on directly using the waste plastic covers in the improvement of CBR of the weak sub-grade soil. Therefore, the present research work examines the probable use of shredded

U. Salini (✉) · A. Jegan Bharath Kumar
NATPAC, Thiruvananthapuram, India
e-mail: saliniu@gmail.com

K. Muthukkumaran et al. (eds.), *Ground Improvement Techniques*, Lecture Notes in Civil Engineering 297, https://doi.org/10.1007/978-981-19-6727-6_26

waste plastic cover in weak soil to strengthen the soil. Modified Proctor compaction test, CBR tests, and unconfined compression (UCC) tests were performed on the soil remolded with shredded waste plastic covers.

Experimental Program

Materials

The soil was obtained from Thiruvananthapuram district, Kerala. The soil properties were obtained in accordance with the Indian Standards and are specified in Table 26.1.

The standard Proctor and modified Proctor compaction curves of the soil are presented in Fig. 26.1. X-ray diffraction analysis performed on soil showed that the clay fraction has kaolinite as the major mineral with traces of illite. The shredded waste plastics were obtained by shredding the waste plastic covers obtained from the local corporation disposal area in Thiruvananthapuram. The collected waste plastic covers were shredded and sieved into different sizes (600 μ, 1.18, 2.36, and 4.75 mm) using a shredder and 0.1% of each size of waste plastic was mixed with the soil separately to form composite specimens. Figure 26.2 shows pictures of various sizes of shredded waste plastics used in the investigation.

Table 26.1 Soil properties

Soil property	Value	
Grain size distribution	Gravel (%)	6
	Sand (%)	36
	Silt (%)	35
	Clay (%)	23
	Soil classification	MI
Atterberg limits	Liquid limit (%)	47
	Plastic limit (%)	34
	Plasticity index (%)	13
Compaction characteristics		
Standard Proctor compaction	OMC (%)	19
	MDD (g/cc)	1.65
Modified Proctor compaction	OMC (%)	13
	MDD (g/cc)	1.85
Specific gravity		2.55

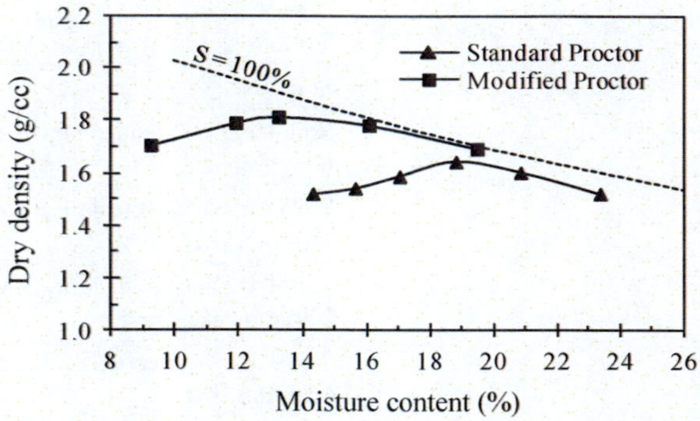

Fig. 26.1 Modified and standard compaction curves of MI soil

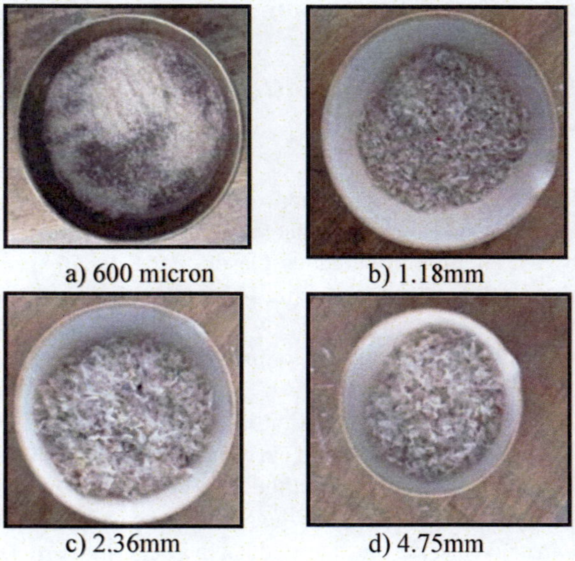

Fig. 26.2 Different sizes of shredded waste plastic used in the study

Test Method

Modified Proctor compaction tests were performed on soil blended with 0.1% of shredded plastic waste covers of various sizes in accordance with IS 2720 (Part 8) [2]. The details of the soil-plastic (S-P) composites are tabulated in Table 26.2. The CBR tests and UCC tests were conducted on these series of soil-plastic mixes at their OMC and MDD as per IS 2720 (Part 16) [3] and IS 2720 (Part 10) [4]. CBR tests were performed on both unsoaked specimens and specimens which were soaked in water for four days (96 h).

Table 26.2 Specimen details

Type	Specimen designation
Soil	S
Soil + 0.1% shredded plastic of 600 μ	S-P0.06
Soil + 0.1% shredded plastic of 1.18 mm size	S-P1.18
Soil + 0.1% shredded plastic of 2.36 mm size	S-P2.36
Soil + 0.1% shredded plastic of 4.75 mm size	S-P4.75

Results and Discussion

Effect of Shredded Waste Plastic on Compaction Characteristics

Figure 26.3 plots the compaction curves of soil and soil-plastic composites. It can be perceived from figure that the compaction characteristics of the soil were affected by the addition of shredded plastic. For S-P4.75 specimen, due to the addition of shredded plastic of larger size (4.75 mm), the compaction curve moved to the right, i.e., the maximum dry density (MDD) decreased, whereas the optimum moisture content (OMC) increased compared to S specimen. The increase in the surface area of the soil-plastic mix leads to increase in the amount of water required to reach to the OMC. With the reduction in the size of the shredded plastic in the soil-plastic composite, the compaction curve moved to left as can be seen for S-P2.36, S-P1.18, and S-P0.06 specimens. In the case of S-P1.18 and S-P0.06 specimens, the MDD obtained were higher than the S specimen showing that greater compaction can be achieved on using plastic shredded to a very fine particle. This may be due to the finer plastic filling the voids and forming a better interlocking between soil and plastic particles thereby leading to an increase in dry density.

Effect of Shredded Waste Plastic on the CBR Value

For different soil-plastic combinations, there is a difference in the soaked and unsoaked CBR value as can be seen in Fig. 26.4. It is visible that, for both unsoaked and the soaked conditions, the CBR of the soil increased on addition of 0.06 and 1.18 mm plastic and thereafter decreased with the increase in the size of the shredded waste plastic covers. The soaked CBR value increased from 13.2 to 18.4% showing an increase of 39.39% and the unsoaked CBR value increased from 22.4 to 37.8% showing an increase of 68.75% for soil remolded with 0.1% of shredded waste plastic of 1.18 mm in size.

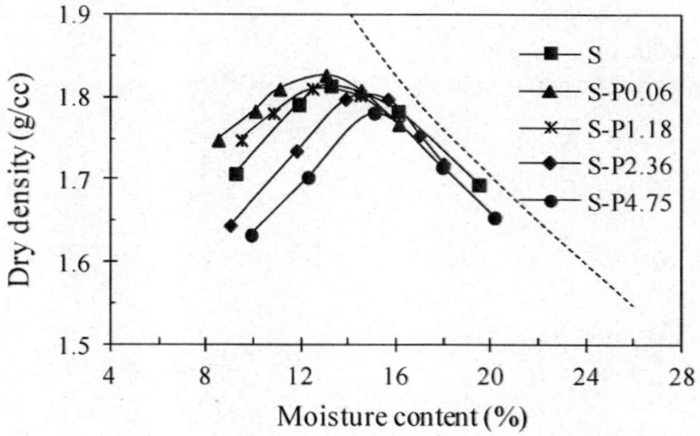

Fig. 26.3 Effect of shredded plastic on compaction behavior of different soil-plastic composites

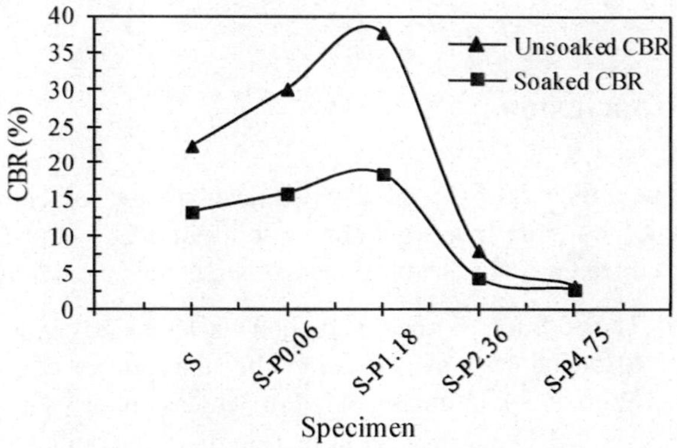

Fig. 26.4 Effect of shredded plastic on soaked and unsoaked CBR value of different soil-plastic composites

Effect of Shredded Waste Plastic on the Unconfined Compressive Strength (UCS)

Figure 26.5 shows the variation in UCS value for different soil-plastic composites. The UCS of the soil specimen increased on the addition of 0.1% of 0.06 mm size shredded plastic and thereafter with the increase in the size of shredded plastic, the UCS value gradually decreased. The UCS values increased from 222.29 to 249.42 kPa for S-P0.06 specimen. The increase in the UCS value is credited to the tensile strength of the plastic in the S-P composite. The reinforced soil can withstand more deformation, resulting in a higher stress at rupture, agreeing to the findings of Nataraj and McManis [6]. The tensile strength of soil sample is greatly affected by the randomly oriented fibers that further enhance the shear strength [7].

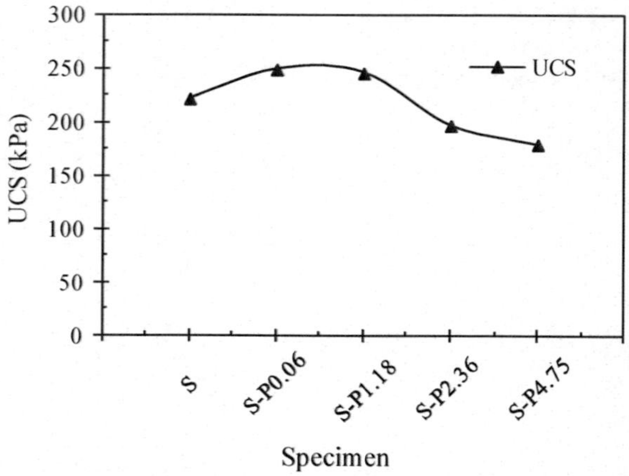

Fig. 26.5 Effect of shredded plastic on UCS of different soil-plastic composites

Conclusions

The influence of shredded waste plastic on the soil behavior was studied by carrying out laboratory investigations on soil remolded with 0.1% of shredded waste plastic of different sizes. The following conclusions were drawn as a result of the study.

- The addition of finely shredded waste plastic covers (0.06 and 1.18 mm size) led to the increase in dry density while the addition of coarsely shredded waste plastic (2.36 and 4.75 mm size) led to decrease in dry density.
- The CBR value also showed a similar trend but it was higher for soil with 0.1% of 1.18 mm size plastic compared to other mixes. An increase of 5% and 2.5% was visible for soaked CBR value of soil with 1.18 mm and 0.06 mm size shredded waste plastic, respectively.
- The effect of waste plastic fiber on the soil is influenced by factors like size of waste plastic fiber and plastic waste content.
- As observed in the case of dry density, the UCS value was higher for soil with 0.06 and 1.18 mm size shredded plastic (12% increase) compared to soil and thereafter with increase in shredded plastic size, UCS value decreased.

Therefore, the improvement of soil properties on the addition of finely shredded waste plastic covers suggest their potential application in soil stabilization and in eliminating the problem of waste disposal.

References

1. Chandra S, Viladkar MN, Nagrale PP (2008) Mechanistic approach for fiber-reinforced flexible pavements. J Transp Eng 134(1):15–23
2. IS 2720-Part 8 (1983) Methods of test for soils: determination of water content-dry density relation using heavy compaction, Bureau of Indian Standards, New Delhi, India

3. IS 2720-Part 16 (1987) Methods of test for soils: laboratory determination of CBR, Bureau of Indian Standards, New Delhi, India
4. IS 2720-Part 10 (1973) Methods of test for soils: laboratory determination of unconfined compressive strength. Bureau of Indian Standards, New Delhi, India
5. Jha JN, Choudhary AK, Gill KS, Shukla SK (2014) Behavior of plastic waste fiber-reinforced industrial wastes in pavement applications. Int J Geotech Eng 8(3):277–286
6. Nataraj MS, McManis KL (1997) Strength and deformation properties of soils reinforced with fibrillated fibers. Geosynthetics Int 4(1):65–79
7. Puppala AJ, Musenda C (2000) Effects of fiber reinforcement on strength and volume change in expansive soils. Transp Res Rec 1736(1):134–140
8. Yetimoglu T, Inanir M, Inanir O (2005) A study on bearing capacity of randomly distributed fibre reinforced sand fill overlying soft clay. Geotext Geomembr 23(2):174–183

Chapter 27
Studies on Consolidation Characteristics of Marine Clay Using Geodrain

Ashvini R. Mehta, S. P. Dave, and Shalini Singh

Introduction

Soft clay layers with a high-water content cover large region of the earth's surface. India's coastline, which stretches for over 6500 km, is often covered with soft marine clay deposits. The bearing capacity and compressibility of these soil deposits are both poor. The use of vertical drains with surcharge loading, on the other hand, is a well-established strategy for speeding up the initial consolidation of soft soils like marine clay in order to avoid costly post-construction settlements and gain necessary extra shear strength.

The aspect ratio (height: diameter) and the size of test specimens in geotechnical laboratory apparatus can have a significant impact on the compressibility and *k* values of peat and other soft soils. Side-wall friction, which forms along the inner wall surface of the confining cell and creates unequal compression along the specimen length, generally increases as the aspect ratio increases.

Purwondho and Djohan [1] conducted a case study in which the current container yard, which supported the Container Terminal of the Jakarta International Container Terminal, is located on terrain with a compressible clay layer from 5.00 to 22.00 m deep. The predicted consolidation settling without soil improvement is 1.418 m. In lieu of PVD, natural drain (80 mm, 8 mm) was shown to be beneficial in compression properties [1].

The behavior of a vertical drain constructed in soft clay was studied in the laboratory using a large-scale consolidation device. Fang and Yin [2] filled a stainless steel cylinder (300 mm in diameter and 450 mm in height) with soft remolded Hong Kong marine clay and placed a PVD (Colbonddrain-50 mm, 5 mm) using a 13 mm,

A. R. Mehta (✉) · S. P. Dave · S. Singh
L.D. Engineering College, Ahmedabad, Gujarat, India
e-mail: imashvini007@gmail.com

S. P. Dave
e-mail: principal_gecp@yahoo.co.in

K. Muthukkumaran et al. (eds.), *Ground Improvement Techniques*, Lecture Notes in Civil Engineering 297, https://doi.org/10.1007/978-981-19-6727-6_27

60 mm mandrel. Ngo et al. [3] also employed ultra-soft saturated soil from Mae Moh mine and PVD (Ali drain-50 mm, 5 mm) was placed using the 60 mm, 10 mm mandrel. The smear impact of PVD is investigated in these publications using the PVD dimension.

In this publication, researchers investigate the consolidation properties of marine saturated soil in a large-scale model consolidometer with an aspect ratio of 1:0.564(H:D) and a vertical drain with a (re/rw) value of 10. PVD is utilized as a vertical drain, and soil is taken from two distinct Saurastra regions: Una and Bhavnagar.

Experimental Investigation

Testing Program

Total four test performed

(1) Standard Oedometer Test on Una region soil
(2) Standard Oedometer Test on Bhavnagar region soil
(3) Large size model consolidation test on Una region soil with PVD
(4) Large size model consolidation test on Bhavnagar region soil with PVD.

Testing Material

(1) Soil-The soil sample is made of Una and Bhavnagar region soil with mixed deaired distilled water, twice liquid limit to form slurry the properties of soil is in Tables 27.1 and 27.2
(2) PVD-The properties of PVD are in Table 27.3.

Table 27.1 Geotechnical properties of marine clay procured from Una

S. No.	Description	Indian Standard	Symbol	Determination
1	Specific gravity	IS:2720 part-3	G	2.67
2	Liquid limit	IS:2720 part-5	LL	51.16
3	Plastic limit	IS:2720 part-5	PL	24.42
4	Shrinkage limit	IS:2720 part-6	SL	14.75
5	Plasticity index	IS:2720 part-5	PI	26.74
6	Hydrometer	IS:2720 part-4	Clay%	45%
			Silt%	51%
7	Soil type	IS:1498-1970	–	CH

Table 27.2 Geotechnical properties of marine clay procured from BHAVNAGAR

S. No.	Description	Indian Standard	Symbol	Determination
1	Specific gravity	IS:2720 part-3	G	2.71
2	Liquid limit	IS:2720 part-5	LL	48.67
3	Plastic limit	IS:2720 part-5	PL	26.05
4	Shrinkage limit	IS:2720 part-6	SL	13.58
5	Plasticity index	IS:2720 part-5	PI	22.62
6	Hydrometer	IS:2720 part-4	Clay%	43%
			Silt%	57%
7	Soil type	IS:1498-1970	–	CI

Table 27.3 Index and engineering properties of PVD

Property	Unit	Test Standard	Determination
Grab tensile strength	m/s	ASTM 4632	$\geq$200
Trapezoidal tear strength	μm	ASTM 4633	$\geq$100
Permeability test	μm	ASTM 4491	$\geq 15 \times 10^{-3}$
Pore size	μm	ASTM 4751	85 ± 25

Conventional Oedometer Test

The classic oedometer test is explained by the Terzaghi one-dimensional hypothesis. The test is performed on a saturated soil cylindrical specimen with a diameter of 75 mm and a thickness of 20 mm. This sample is packed with porous stone and filter paper in the leaver arm loading mechanism. The weight is maintained for 24 h, during which time the soil consolidates due to the porous stone drainage. At each level, the applied load is gradually increased by doubling the applied pressure. After each load is applied, the deformation is measured at 6, 15, 30 s, 1, 2, 4, 8, 16, 30 min, and 1, 2, 4, 8, and 24 h, respectively. Two dial gauges with a sensitivity of 0.001 mm were utilized to measure soil settlement.

Large-Scale Consolidation Test Setup

A 10 mm thick circular tank was created for the compressive stress test apparatus, which also included a mechanical jack and proving ring. During the consolidation process, the tank has an internal diameter of 254 mm, a height of 450 mm, and four drain holes of dimension 25 mm for bottom drainage. Two dial gauges with a sensitivity of 0.001 mm were utilized to measure soil settlement (Fig. 27.1).

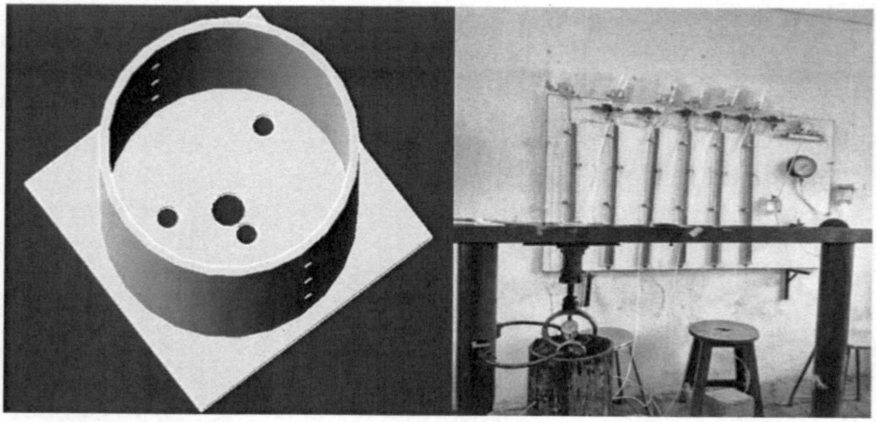

Fig. 27.1 Consolidation test setup

Preparation of Test Sample and Test Procedure

A thick slurry of coastal soil was poured to generate a 400-mm-high soil sample. The water content and bulk density of the soil were measured before each test sample. In large consolidation test samples, these are the only variables that can be controlled and compared. The initial void ratio was calculated based on the soil's initial water content and bulk density. The loading plate was placed on top of the sample, causing a vertical pressure of 1 kPa to be evenly distributed throughout the entire surface, and the sample was allowed to stabilize for two days. The load was applied to the loading plate using a mechanical jack. The loading was done in phases to avoid pumping up the sludge. A 10 kPa vertical drain was installed during preconsolidation if testing was required. Vertical loads of 10, 20, 40, 160, and 320 kPa were gradually increased to produce 10, 20, 40, 160, and 320 kPa stresses. Under varying applied pressures, marine clay was allowed to consolidate for three days at a time. Settlement measurements at regular time intervals were made with help of two dial gauge of sensitivity of 0.001 mm.

Theoretical Background

The most detailed solution to the problem of radial consolidation utilizing drain wells was Barron's [4]. Throughout the consolidation process, horizontal sections must stay horizontal for equal stain conditions. The governing equation for radial drainage consolidation [4] is provided by, which considers flow into and out of an infinitesimal cylindrical element. The average degree of consolidation, U_h, in the soil body is given by,

$$U_{\mathrm{h}} = 1 - \exp\left(\frac{-8T_{\mathrm{h}}}{F(n)}\right). \quad (27.1)$$

The drain spacing factor, $F(n)$, is calculated as follows:

$$F(n) = \frac{n^2}{n^2 - 1}\ln(n) - \frac{3n^2 - 1}{4n^2}. \quad (27.2)$$

The time factor T_{h} is defined as follows,

$$T_{\mathrm{h}} = \frac{c_{\mathrm{h}} \cdot t}{4r_e^2}. \quad (27.3)$$

The radial drainage consolidation coefficient, C_{h}, is expressed by,

$$C_{\mathrm{h}} = \frac{k_{\mathrm{h}}(1 + e)}{a_v \gamma_w}. \quad (27.4)$$

In the case of vertical drain and constant surcharge, a hyperbolic approach for predicting total primary settlement is used. Plot the hyperbolic form of the Test settlement data as *t*/ versus *t*, where *t* represents time and *t* represents settlement from the start of constant load application. Determine the slope of the first linear segment immediately after the initial concave downward segment of the curve using the hyperbolic plot. Between U50 and U90%, hyperbolic graphs are linear. The slope of the lines radiating from the origin to the U50% point is (1/0.5 = 2.0), and the slope of the lines radiating from the origin to the U90% point is (1/0.9 = 1.11). As a result, for any settlement, the ratio of the slopes of these radiating lines to the slope of the linear component of the hyperbolic plots determines the U50 and U90%. Drains and surcharges were used to keep track of the data. The importance of the hyperbolic technique is greater since the quantity of secondary compression in field consolidation data is unclear.

Experimental Result and Discussion

Figure 27.2 depicts Degree of consolidation verses Logarithmic time recorded for large-scale consolidation for Una and Bhavnagar marine soil with central PVD $n = 10$. Theoretical curve given by Barron for equal strain condition is almost similar to experimental plot.

For large size consolidation, the time necessary for 50% degree of consolidation was determined using the hyperbolic approach in the absence of pore water pressure. Figures 27.3 and 27.4 depict the Degree of Consolidation versus the Logarithm of Time Factor for large and conventional size specimens, respectively. The marine soil is highly compressible, and the response reported for both soils were in the shape

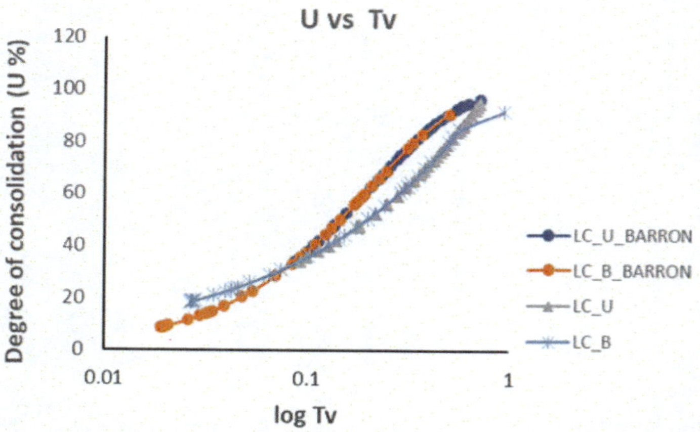

Fig. 27.2 Degree of consolidation versus logarithmic time

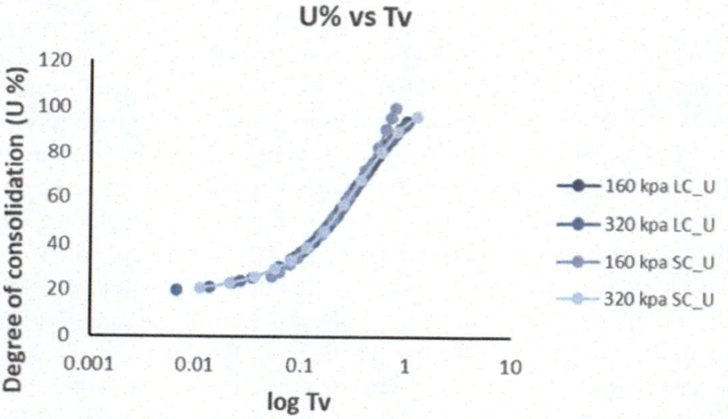

Fig. 27.3 Degree of consolidation versus logarithmic time for Una soil

of a classic s shape, with initial consolidation finished and secondary compression underway.

Effect of Aspect Ratio on Compressibility of Large Specimen

Despite the vastly different aspect ratio, the strains measured in large-scale model consolidation tests for Una soil (lc_u) and traditional oedometer apparatus (sc_u) were identical (1:0.5 and 1:3, respectively). Under 320 kPa, the big and small specimens exhibited almost perfect strain of around 10.77 and 19.64% for Una area soil, and 8.11 and 17.05% for Bhavnagar region soil.

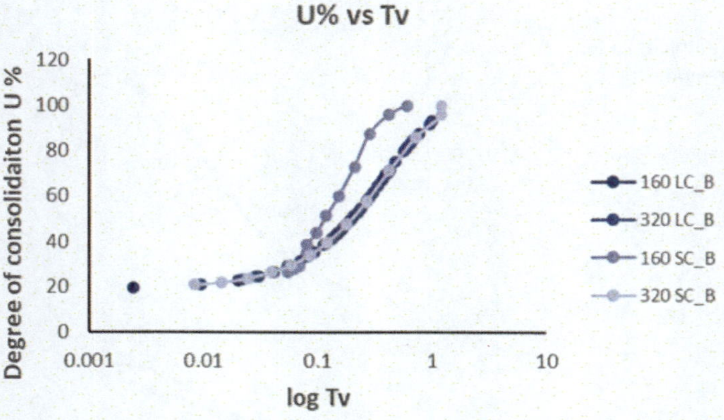

Fig. 27.4 Degree of consolidation versus logarithmic time for Bhavnagar soil

Large-scale consolidation tests encountered lesser strain under the same applied stress due to higher side wall friction and specimen boundary effects. The initial specimen compression reported at the beginning of each load stage was negligible, showing that the marine soil specimen had been fully saturated for both Una and Bhavnagar area soil for practical purposes (Figs. 27.5 and 27.6).

Fig. 27.5 Cumulative strain versus logarithmic time for Una region soil

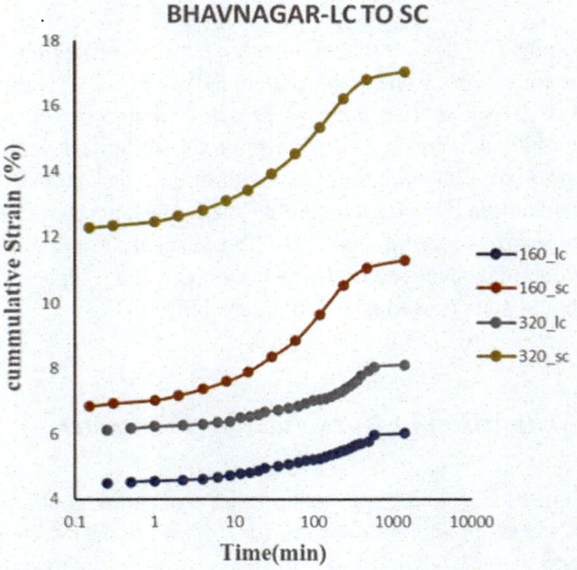

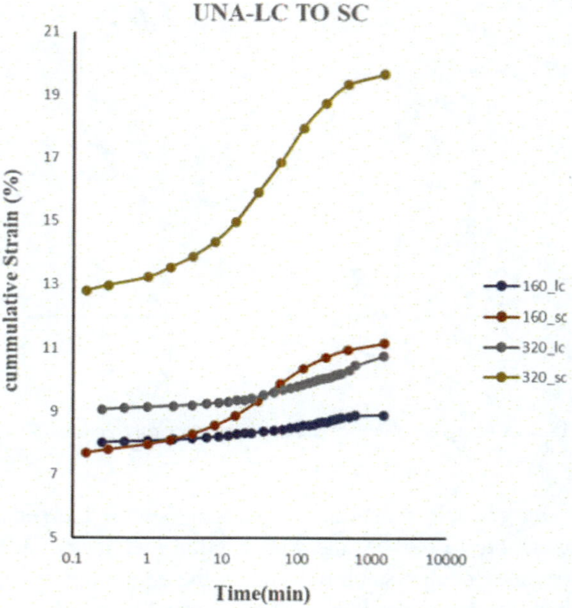

Fig. 27.6 Cumulative strain versus logarithmic time for Bhavnagar region soil

Comparison of Large Specimen Settlement

Figure 27.7 depicts the relationship between settlement and time for marine soil from various regions and drain material PVD for a "*n*" value of 10. The structure of the plot reveals that as the time of consolidation grows, settlement progresses, which is obvious. When comparing the total settlement values and rate of settlement for large size consolidation tests with drain (for Una and Bhavnagar region soil), it is evident that Una region soil has the highest amount of total settlement compared to large size consolidation tests with vertical drain and large size consolidation tests with Bhavnagar region soil. However, when a higher load was applied, Bhavnagar region soil showed some compressibility.

Comparison of Consolidation Parameter

Figures 27.8 and 27.9 illustrate the plot of the coefficient of vertical consolidation (Cv) versus applied pressure for the traditional oedometer test and the large-sized model consolidation test. For the purposes of analysis, the time necessary for 50% consolidation is taken for all applied pressures. For any applied pressure, the value of the coefficient of consolidation (Cv) in a typical (small size) oedometer test with an aspect ratio of 1:3 remains constant. However, in the instance of a large-scale

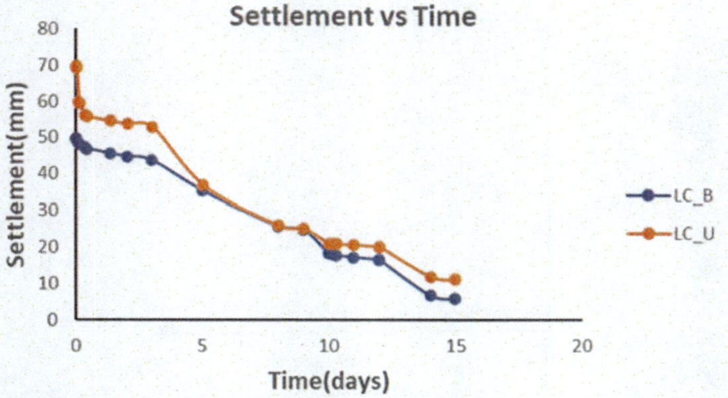

Fig. 27.7 Settlement versus time in large-scale consolidation

model consolidation test with an aspect ratio of 1:0.564, the value of the coefficient of consolidation (Cv) rises as the applied pressure rises.

Figures 27.10 and 27.11 depict a classic curve of regularly consolidating virgin soil. Una marine clay has a compression index (Cc) of 1.54 for traditional oedometer apparatus and 0.132 for big size consolidation apparatus. For conventional oedometer equipment, the compression index (Cc) of Bhavnagar marine clay is 0.267, while for large size consolidation apparatus, it is 0.143.

The plot of Vertical permeability coefficient versus Figs. 27.12 and 27.13 show the applied pressure for a large-scale consolidation test as well as a traditional (small-scale) oedometer test. By using a theoretical calculation, the vertical permeability value for big size specimens is 150–300 times greater than for small size specimens.

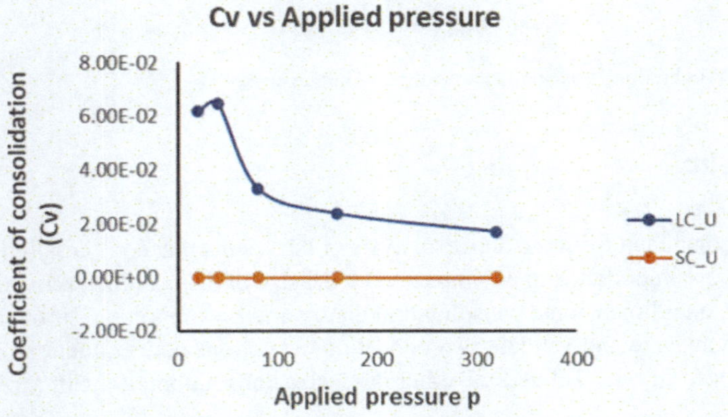

Fig. 27.8 Coefficient of consolidation versus applied pressure for Una region soil

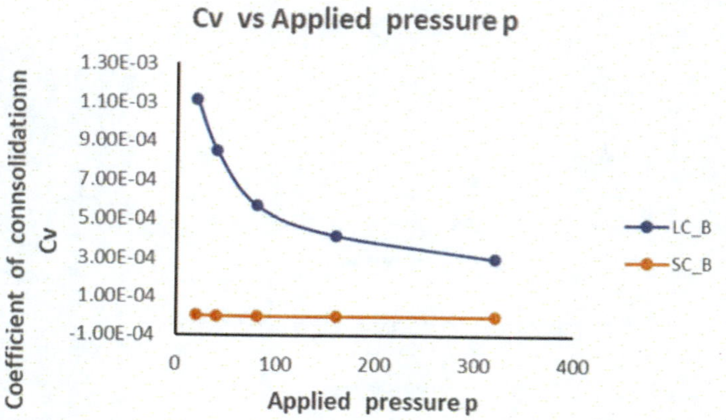

Fig. 27.9 Coefficient of consolidation versus applied pressure for Bhavnagar region soil

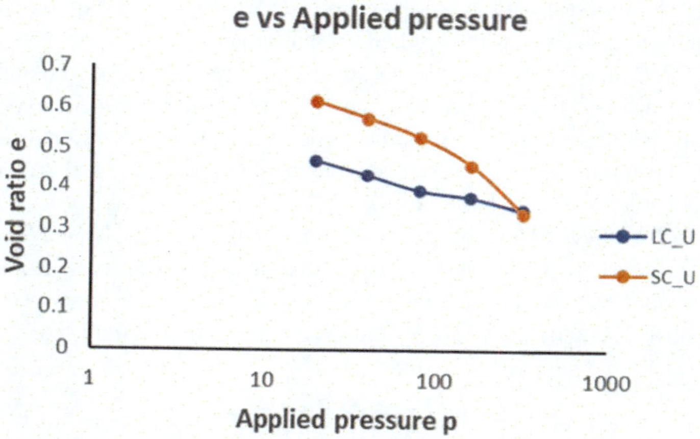

Fig. 27.10 Void ratio versus applied pressure for Una region soil

Conclusion

It can be concluded from the above analysis of the results that compressibility characteristics are important and are impacted by the size of the oedometer. In this work, Terzaghi's one-dimensional consolidation theory and Barron's equal strain theory is utilized to analyze and validate two consolidometers, one conventional size (small size) with aspect ratio 1:3 and the other large size consolidometer with aspect ratio 1:0.564.

- When compared to a conventional oedometer apparatus under the same applied stress, the large-scale consolidation model suffered less strain due to larger impacts of side wall friction and specimen boundaries. In comparison to small

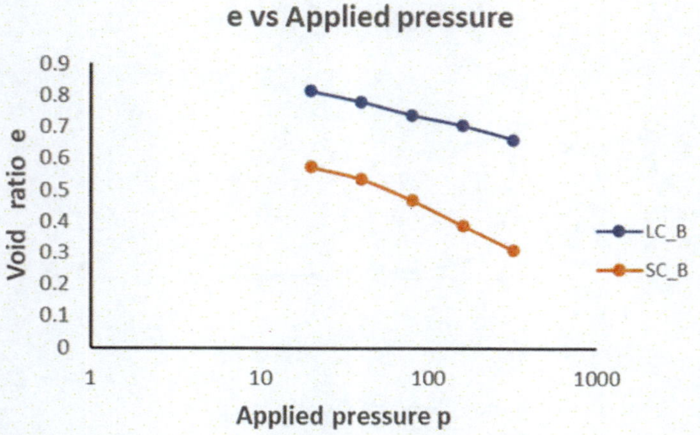

Fig. 27.11 Void ratio versus applied pressure for Bhavnagar region soil

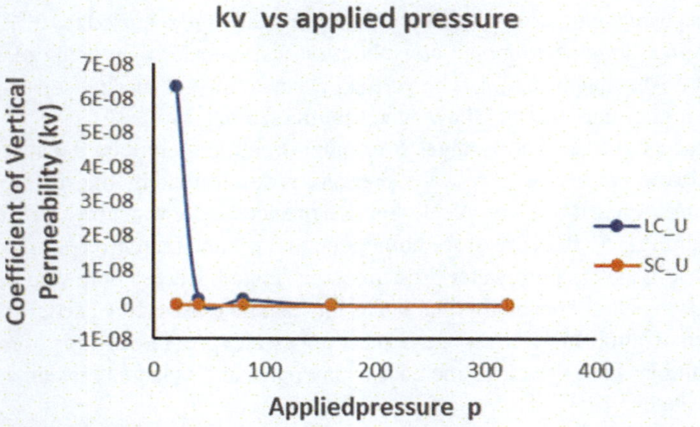

Fig. 27.12 Coefficient of vertical permeability versus applied pressure for Una region soil

size consolidometers, large size consolidometers show a significant influence on aspect ratio.

- In comparison to small size consolidometers, large size consolidometers show a significant influence on aspect ratio. When the quantity of vertical compressibility due to pore water expulsion in the inward radial direction is combined across the whole drain length, the compressibility with hydrodynamic lag is larger. When pore water travels vertically, it takes longer than when it travels horizontally.
- Experimental plots of Degree of Consolidation versus Time Factor for "*n*" value of 10 and all pressures are found to be equivalent to theoretical plots obtained by Barron's equal strain hypothesis. So that Barron's equal strain theory may be utilized to determine various consolidation parameters from large-scale model consolidation test results.

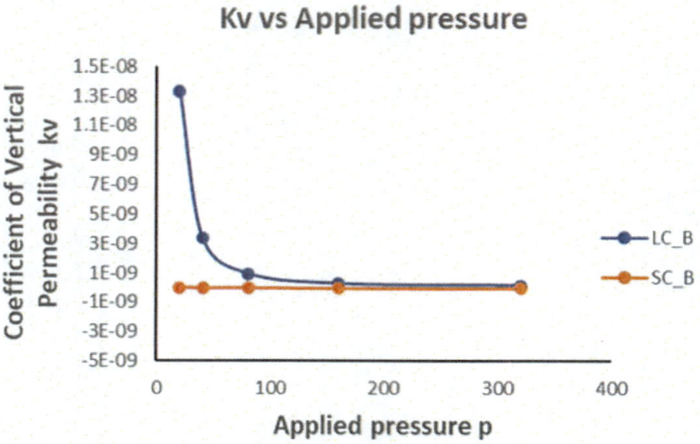

Fig. 27.13 Coefficient of vertical permeability versus applied pressure for Bhavnagar region soil

- Because using marine soil from the Una region, the time required for 50% consolidation in a large-scale model consolidation test is faster than using marine soil from the Bhavnagar region. The vertical permeability of marine soil in the Una area is higher than that of Bhavnagar, ranging from 1.9 to 2.3.
- PVD has a number of advantages over other draining materials. Because the rate of dissipation of excess pore water pressure is determined by the surface contact between clay and drain, and late dissipation of excess pore water pressure increases the clogging rate. Because of the homogeneous size and increased permeability of coastal soil, excess pore water pressure is dissipated. Excess pore water pressure dissipation is higher at beginning, but as the load increases, clay particle blockage between neighboring pvd has been excess pore water pressure dissipation is higher at beginning, but as the load increases, clay particle blockage between adjoining pvd has been observed.
- The results indicate that in order to completely appreciate the impact of aspect ratio, considerable mathematical modeling as well as comprehensive measurements of pore water pressure are necessary. Under steady-state flow circumstances, the potential head created by pore water may be bigger, but pore water can only be dissipated at a quicker rate if the same amount of water is penetrated at a faster rate.

Acknowledgements Authors are thankful to Prof. Dr. C. R. Sanghavi Head of Applied Mechanics Department and Prof. Dr. R. K. Gajjar Principal of L.D. College Of Engineering for providing all required research facilities for this project.

References

1. Purwondho R, Djohan R (2018) Research on the acceleration of settlement by installing vertical drain and preloading with sand. In: The 2nd International conference on eco engineering development 2018
2. Fang Z, Yin JH (2016) Physical modelling of consolidation of Hong Kong marine clay with prefabricated vertical drains. Can Geotech J 43
3. Ngo DH, Horpibulsuk S, Suddeepong A, Hoy M, Udomchai A, Doncommul P, Rachan R, Arulraja A (2020) Consolidation behavior of dredged ultra-soft soil improved with prefabricated vertical drain at the Mae Moh mine, Thailand. Geotext Geomembr
4. Barron RA (1948) Consolidation of fine-grained soils by drain wells. Trans ASCE 113:718–742
5. IS: 2720-4-1975: Grain size analysis
6. IS: 2720-5-1970: Determination of liquid and plastic limits
7. IS: 2720-6-1972: Determination of shrinkage factors
8. IS: 2720-15-1986: Determination of consolidation properties
9. IS: 2720-40-1977: Determination of free swell index of soil
10. IS 15284 (Part 2) (2004) Design and construction for ground improvement-guidelines, pre-consolidation using vertical drain

Chapter 28
Mechanical Behavior of Silty Soil Reinforced with Carbon Fibers

Nadeem Gul, Bashir Ahmed Mir, and K. M. N. Saquib Wani

Introduction

To address the issue of land scarcity and mitigation of marginal soil deposits, researchers and engineers have proposed several methods which have been put to practice. Chemical stabilization by means of additives like cement, lime, and fly ash have rendered easy and affordable soil stabilization solutions [1, 2]. However, sustainability development demands ecofriendly methods of ground improvement [3]. Moreover, the paucity of conventional materials and their exorbitantly expensive nature has motivated engineers to find sustainable alternatives. The use of discrete natural or synthetic fiber reinforcement offers an effective and prudent solution of ground modification. Thus, fiber-reinforced soils have attracted a good deal of research in the form of both field and laboratory studies [4–9]. Carbon fibers (sometimes also referred as graphite fibers) are lightweight high-performance fibers with exceptional chemical resistance. They are lighter and stiffer as compared to glass fibers but significantly more expensive. Generally, they are manufactured from polyacrylonitrile (PAN), a carbon polymer which is hard and has high melting point. Given their high strength, high modulus, low density, high temperature, and corrosion resistance, carbon fibers have been well received in industry with wide range of applications in aircraft, automotive, and aerospace industry [10]. In civil construction, CFs are known to affect and improve the behavior of concrete beams, plates, and columns in tension, compression, and shear. Researchers have shown that CFs produce better reinforcement effects at relatively low concentrations and are therefore cost-effective [11].

Given the diverse applications of CF, researchers have attempted to explore its potential in stabilization of marginal soils. The efficacy of CF in augmenting the mechanical properties of soft soils is well researched [12]. Gao et al. [13] studied the

N. Gul (✉) · B. A. Mir · K. M. N. Saquib Wani
National Institute of Technology Srinagar, Jammu and Kashmir, Srinagar 190006, India
e-mail: nadeemgul@nitsri.ac.in

K. Muthukkumaran et al. (eds.), *Ground Improvement Techniques*, Lecture Notes in Civil Engineering 297, https://doi.org/10.1007/978-981-19-6727-6_28

effect of carbon fiber on unconfined compression strength of clay and also discussed the underlying mechanisms. They have concluded that there is a substantial improvement in the strength of carbon fiber-reinforced clay. Zaimoglu and Yetimoglu [14] investigated the effect of polypropylene fibers on fine-grained soils through UCS, DST, and CBR tests and reported significant improvement. Bao et al. [15] recently conducted a laboratory study on mechanical performance of CF reinforced clay. They have reported a maximum benefit of reinforcement of 6 mm fiber length at 3%. From the overview of the literature, it was noted that there are inconsistencies with regards to the optimum fiber dosage in laboratory studies, which needs special attention for field applications of CF reinforced soil. The carbon fiber has been recently used for ground improvement which is also suggested by the literature perusal as there are only a few reported studies. This study is, therefore, pursued with the objective to further substantiate the literature that would open up the ways for field application of carbon fiber in strengthening of soft soils.

Materials and Methods

Soil

For the purpose of this study, the soil samples were collected from a site located at Sempora, Pampore in Jammu and Kashmir, the place famous for its saffron fields. Soil samples were procured at a depth of 0.5 m from the ground level and were placed in the plastic bags to preserve the moisture content. After performing in situ tests, the disturbed samples were air dried, sieved through IS 2 mm sieve. The soil sample was then characterized and the physical properties were determined conforming to the relevant IS standards [16–21]. The soil properties so obtained are summarized in Table 28.1.

Carbon Fibers

Commercially available microcarbon fibers of 6 mm length and 8 μm diameters were used in this study as shown in Fig. 28.1. The properties of the CFs used are listed in Table 28.2. The selection of short fibers was done considering the fact that they are relatively easier to mix and disperse uniformly with the soil matrix.

Table 28.1 Index and engineering properties of soil

S. No.	Properties		Values
1	Soil color		Light brown
2	Specific gravity (G_s)		2.61
3	Sand (%)		1.5
4	Silt (%)		81.0
5	Clay (%)		17.5
6	Liquid limit (%)		36
7	Plastic limit (%)		25
8	Shrinkage limit (%)		13
9	Plasticity index (%)		11
10	P.I, A-line		12
11	P.I, U-line		25
12	Classification		MI
13	In situ DST	Cohesion, c (kN/m^2)	26
		Angle of internal friction, ϕ (°)	15.32
14	In situ UCS, q_u (kN/m^2)		49
15	Optimum moisture content (%)		19.15
16	Maximum dry unit weight (kN/m^3)		16.68

Fig. 28.1 Soil and the carbon fibers used in the present study

Table 28.2 Properties of carbon fiber used in the study

Fiber length (mm)	Filament dia (μm)	Elongation (%)	Tensile strength (MPa)	Tensile modulus (GPa)	Carbon content (%)
6	8	1.75	3500	230	95

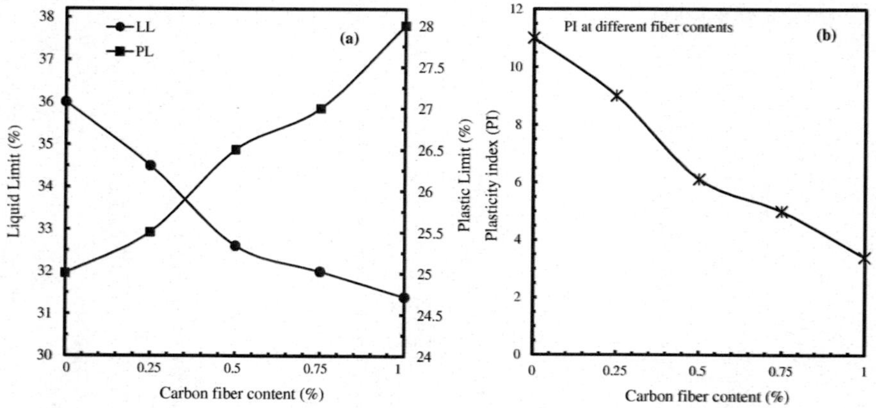

Fig. 28.2 **a, b** Variation of Atterberg's limits with CF content

Testing Methodology

Soil fiber composite samples and unreinforced samples were prepared corresponding to same initial conditions of dry unit weight and moisture content in order to make the comparison rational. Fibers were initially mixed in dry condition, and desired water quantity was subsequently added. The samples, thus prepared, were reasonably homogenous samples with good fiber dispersion in soil. The prepared samples were stored in desiccator to avoid moisture loss prior testing.

Atterberg Limits Liquid limit and plastic limit tests were performed on oven dried soil samples passing 425 μm sieve according to the Casagrande's method. Thread rolling method was used to determine the plastic limit. These limits represent the critical moisture contents corresponding to which soil changes its states from liquid to semi-solid. The determination of these parameters is essential for speculating the engineering behavior of fine-grained soils.

Direct Shear Test Direct shear test is an expedient and convenient procedure for determining the shear strength parameters (c and ϕ) which are necessary for calculation of bearing capacity and stability analysis of slopes, foundations, and retaining walls. In multiple layers, the soil–fiber mix was compacted in a shear box of size 60 mm × 60 mm × 25 mm to MDU, and tests were performed under a sequence of normal stresses 50, 100, and 150 kPa. Performing DST test under different vertical stresses allows the determination of both c and ϕ using Mohr–Coulomb failure criterion.

Unconfined Compression Test To evaluate the stress–strain response under uniaxial compression, UCS tests were performed on samples of dimension 50 mm × 100 mm samples for both reinforced and unreinforced cases. Static compaction method was employed to prepare UCS samples in a cylindrical mold. Three samples corresponding to each fiber content were tested to establish the reliability of test results.

Results and Discussion

An extensive experimental program was run, and tests were performed on soil with and without carbon fiber reinforcement. The inclusion of carbon fiber was studied in terms of its effect on Atterberg limits, direct shear test, and UCS test results. The findings are discussed here.

Atterberg Limits

Atterberg limits are basically indicative of soil consistency, any change in these limits would change the soil consistency and in turn the engineering behavior of soil. From the test results, it was seen that fiber inclusion led to a continuous drop in liquid limit from 36 to 31.4% with increase in fiber content from 0 to 1%. On the contrary, plastic limit increased with fiber percentages, although the increase is marginal (25–28%). Kinjal et al. [22] in their study on effect of polyester fibers on Atterberg limits of expansive soil have reported similar findings. The cumulative effect of the changes in liquid and plastic limit resulted in reduction of plasticity index which in itself is a good sign of improvement.

Direct Shear Test

The shear strength parameters (c and ϕ) of both unreinforced and reinforced samples at varying proportions of fiber content are summarized in Table 28.3. As is evident, there is a significant improvement in shear strength with increase in percentage of fiber. The cohesion intercept showed a continuous improvement from 36 to 56.33 kPa with increase in fiber content from 0 to 1%. Friction angle, on the other hand,

Table 28.3 Summary of test results of direct shear test

S. No.	Mix proportion (by dry weight of soil)	Cohesion (kPa)	Friction angle (°)	Equation of failure envelope	*R*-square (COD)
1	Soil + 0% CF	36.00	18.26	$y = 0.33x + 36$	0.9758
2	Soil + 0.25% CF	41.33	19.29	$y = 0.35x + 41.333$	0.9868
3	Soil + 0.5% CF	47.00	21.30	$y = 0.39x + 47$	0.9826
4	Soil + 0.75% CF	54.67	23.26	$y = 0.43x + 54.667$	0.9389
5	Soil + 1% CF	56.33	20.81	$y = 0.38x + 56.333$	0.9854

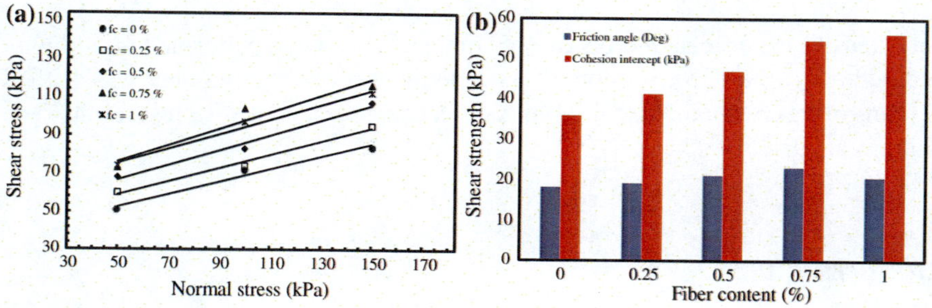

Fig. 28.3 **(a)** Mohr failure envelopes and **(b)** variation of cand ϕ with CF content

increased in a small range from 18.26° to 23.26° up to $f_c = 0.75\%$ beyond which it decreased. This trend is in agreement with the findings of Wang et al. [23]. The possible explanation to this behavior could be that carbon microfibers being small in length occupy the voids in the soil samples and are also long enough to form interlocking network between the soil grains which restrain the slippage of soil particles when loaded. At higher fiber percentages, there was a visible agglomeration of fibers resulting in the formation of low-density pockets. The Mohr failure envelopes of various mixes and variation of shear strength parameters is shown in Fig. 28.3a, b.

Unconfined Compression Strength Test

Before inclusion of fibers in the soil, baseline data of stress–strain relationship and UCS values of parent soil was taken for comparison with reinforced soil samples. The stress–strain and UCS trend plots for different mix proportions of samples are shown in Fig. 28.4. From the graphs, it was seen that there is a substantial improvement in terms of both strength and deformation behavior of reinforced samples than that of unreinforced soil samples. There is a continuous increase in UCS values up to the fiber content of 0.75 % and beyond which there is marginal decrease in UCS values. However, it is seen that reinforced samples for all percentages show a continuous increase in failure strain up to 1% indicating better deformation characteristics. This improved mechanical performance of reinforced samples under compression could be attributed to the soil–fiber interaction which comes into place due to the surface roughness of the fibers [24]. Moreover, the carbon microfibers occupy the voids within the specimens and thus impart integrity to the soil samples.

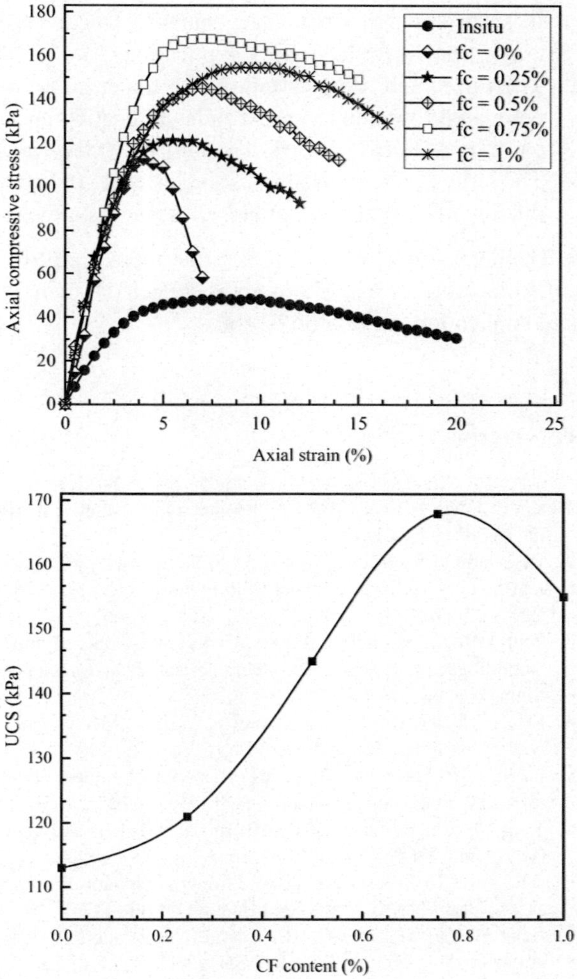

Fig. 28.4 Stress–strain plots and UCS behavior of reinforced and unreinforced samples

Conclusions

On the basis of the results obtained, the following conclusions were drawn:

1. The fiber incorporation improved the consistency behavior of the soil due to reduction in the plasticity index. The plasticity index was reduced from 11 to 3.6%.
2. The shear strength parameters (c and ϕ) were substantially increased with effect being more pronounced on cohesion intercept. There was a marginal decrease in shear strength when fiber content exceeded 0.75% indicating an optimum concentration for laboratory application.
3. The reinforced samples failed at a larger strain under compression testing as compared to unreinforced samples indicating an increase in energy absorption capacity of failed samples. The initial stiffness, however, remained unchanged.
4. The failure pattern of samples suggested a gradual transformation from brittle to ductile mode as the percentage of fiber was increased.

The above conclusions suggest that there are promising benefits of reinforcing soil with carbon fibers. The study is expected to help in developing better insight in field applications of reinforced soils.

References

1. Jan OQ, Mir BA (2018) Strength behaviour of cement stabilised dredged soil. Int J Geosynth Ground Eng 4(2):1–14
2. Belawadikar S, Patil DS, Juneja A (2021) Use of fly ash as weak cementing agent to strengthen marine clay. In: Proceedings of the Indian geotechnical conference 2019: IGC-2019 volume II, vol 134. Springer, p 85
3. Biju MS, Arnepalli DN (2019) Biopolymer-modified soil: prospects of a promising green technology. In: Geotechnical characterisation and geoenvironmental engineering. Springer, Singapore, pp 163–169
4. Estabragh AR, Bordbar AT, Javadi AA (2011) Mechanical behavior of a clay soil reinforced with nylon fibers. Geotech Geol Eng 29(5):899–908
5. Kaniraj SR, Havanagi VG (2001) Behavior of cement-stabilized fiber-reinforced fly ash-soil mixtures. J Geotech Geoenviron Eng 127(7):574–584
6. Tang CS, Shi B, Zhao LZ (2010) Interfacial shear strength of fiber reinforced soil. Geotext Geomembr 28(1):54–62
7. Butt WA, Mir BA, Jha JN (2016) Strength behavior of clayey soil reinforced with human hair as a natural fibre. Geotech Geol Eng 34(1):411–417
8. Jiang H, Cai Y, Liu J (2010) Engineering properties of soils reinforced by short discrete polypropylene fiber. J Mater Civ Eng 22(12):1315–1322
9. Tang CS, Wang DY, Cui YJ, Shi B, Li J (2016) Tensile strength of fiber-reinforced soil. J Mater Civ Eng 28(7):04016031
10. Chung DD, Chung D (2012) Carbon fiber composites. Elsevier
11. Cui H, Jin Z, Bao X, Tang W, Dong B (2018) Effect of carbon fiber and nanosilica on shear properties of silty soil and the mechanisms. Constr Build Mater 189:286–295
12. Rout SS, Sahoo RR (2021) Enhancing strength properties of soft soil using carbon fiber. In: Proceedings of the Indian geotechnical conference 2019. Springer, Singapore, pp 333–344

13. Gao L, Zhou Q, Yu X, Wu K, Mahfouz AH (2017) Experimental study on the unconfined compressive strength of carbon fiber reinforced clay soil. Mar Georesour Geotechnol 35(1):143–148
14. Zaimoglu AS, Yetimoglu T (2012) Strength behavior of fine-grained soil reinforced with randomly distributed polypropylene fibers. Geotech Geol Eng 30(1):197–203
15. Bao X, Huang Y, Jin Z, Xiao X, Tang W, Cui H, Chen X (2021) Experimental investigation on mechanical properties of clay soil reinforced with carbon fiber. Constr Build Mater 280:122517
16. Bureau of Indian Standards (1980a) IS 2720-Part 3(1): determination of specific gravity of fine-grained soils. Bureau of Indian Standards, New Delhi
17. Bureau of Indian Standards (1985a) IS 2720-Part 4: determination of grain size distribution. Bureau of Indian Standards, New Delhi
18. Bureau of Indian Standards (1985b) IS 2720-Part 5: determination of Atterberg limits. Bureau of Indian Standards, New Delhi
19. Bureau of Indian Standards (1980b) IS 2720-Part 7: determination of water content-dry density relation using light compaction. Bureau of Indian Standards, New Delhi
20. Bureau of Indian Standards (1986a) IS 2720-Part 10: laboratory tests procedure for unconfined compressive strength test. Bureau of Indian Standards, New Delhi
21. Bureau of Indian Standards (1986b) IS 2720-Part 13: direct shear test. Bureau of Indian Standards, New Delhi
22. Kinjal S, Desai AK, Solanki CH (2012) Experimental study on the Atterberg limits of expansive soil reinforced with polyester triangular fibers. Int J Eng Res Appl 2(4):636–639
23. Wang Y, Zhang X, Zhao X, Chen Y (2016) Compaction and shear strength tests of clay soils reinforced by carbon fibers. Electron J Geotech Eng 21(20):6689–6698
24. Gul N, Mir BA (2022) Parametric study of glass fiber reinforced fine-grained soil with emphasis on microstructural analysis. Int J Geotech Eng 16(6):716–728. https://doi.org/10.1080/19386362.2022.2049524

Chapter 29
Evaluation on the Shear Strength Characteristics of Soil Reinforced with Randomly Distributed Areca Fibers

Femy M. Makkar, Shilpa Babu, Riya Maria George, and I. Shifana

Introduction

The demand for civil engineering structures is increasing due to rise in urbanism and industries. Because of excessive settlement and weak shear strength, it is highly risky to do construction on soft soils. Soil reinforcement and soil stabilization are some ground improvement techniques to improve the in stiffness and strength of soft soil. Introduction of a tension resisting material in the form of natural fibers to improve the strength of soil is found to be environmental friendly and very cost-effective ground improvement technique. The fibers can be used in random manner on soft, or problematic soils are one of the techniques to improve the strength. In addition, this technique reduces the weak planes in the direction of oriented reinforcement and strength isotropy. Another advantage of this technique is that the interaction or bonding between the soil and fiber is significantly improved the stiffness and strength and enhances axial strain at failure and crack reduction capacity of the soil composite.

Many researchers have studied the shear strength behavior of fiber-reinforced soil by considering fiber-soil mixture as a composite material [1–3, 5–10]. Most of these studies are done on synthetic fibers. Synthetic fires such as metal fibers, metal strips, and polymeric fires are preferred as soil reinforcement due to their reproducibility and uniform material properties. Nowadays, due to cost effectiveness, strength, bulk availability, and environmental friendliness, natural fibers are getting more attention in the field of soil reinforcement. However, among the plenty of available natural fibers, most of the current researches are concentrating only on coir, bamboo, and jute fibers.

F. M. Makkar · S. Babu (✉) · R. M. George · I. Shifana
Department of Civil Engineering, Amal Jyothi College of Engineering, Kanjirappally, Kottayam, Kerala, India
e-mail: shilpababu015@gmail.com

F. M. Makkar
e-mail: femymmakkar@amaljyothi.ac.in

K. Muthukkumaran et al. (eds.), *Ground Improvement Techniques*, Lecture Notes in Civil Engineering 297, https://doi.org/10.1007/978-981-19-6727-6_29

Table 29.1 Basic properties of soil

Basic properties	Value
Specific gravity	2.63
Effective particle size, D_{10}	0.29
Average particle size, D_{50}	0.74
Uniformity coefficient, C_u	2.8
Coefficient of curvature, C_c	0.96
Maximum dry unit weight (kN/m^2)	18.04
Minimum dry unit weight (kN/m^3)	14.26
Unit weight for 50% relative density (kN/m^3)	16.25

In the present investigation, natural areca fiber was chosen as the reinforcing material. Areca fiber shows higher tensile properties with lowest density compared to other natural fibers like coconut and palm fibers. The higher crystallinity index of areca fibers compared to that of other natural fibers indicates that they have low tendency for water absorption. It ensures good interfacial bonding with other materials, thus increases composites strength [4].

To assess the suitability of areca fiber as a soil reinforcing element, a series of experiments were carried out on cohesionless soil with different percentages of fiber inclusion.

Materials Used for Study

Sand

The natural river sand is used for the study. According to Indian Standard specifications, the basic properties of the sand were determined. The basic properties of soil sample used in the study are given in Table 29.1. Figure 29.1 shows the particle size distribution curve for the sand. According to unified soil classification system (USCS) system, the soil is classified as poorly graded sand (SP).

Areca Fiber

Areca fiber was used as the reinforcing element. The collected areca nuts were soaked in water for 24 h, and the fiber was extracted manually and air dried. The fiber type is single, and the length of the areca fiber used was randomly measured between 2 and 3 cm. Figure 29.2 shows the areca fiber used in the study.

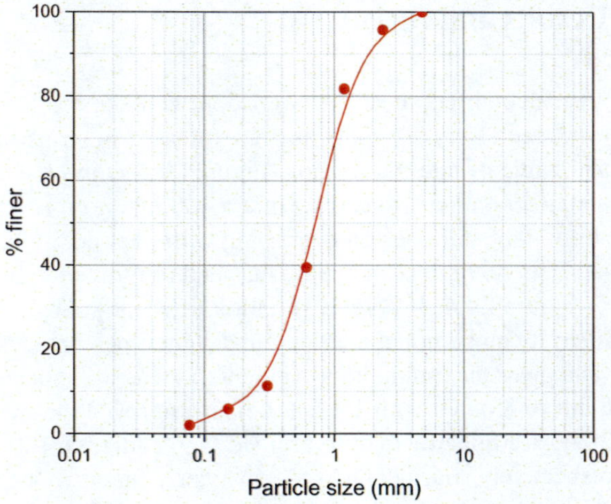

Fig. 29.1 Particle size distribution curve for sand

Fig. 29.2 Areca fiber

Experimental Program

The shear strength characteristics of unreinforced sand and areca fiber-reinforced sand (AFRS) were studied by conducting direct shear tests. The size of the direct shear box used is in the study was 6 cm × 6 cm × 2.7 cm. All the tests were conducted at a relative density of 50%. In the case of reinforced sand, the fiber content was varied as 0.5, 1, and 1.5%. The required amount fiber was calculated, and it was randomly mixed with sand. At normal stresses of 50, 100, and 150 kPa, the tests were conducted. The shearing of the test specimen was done at a shear displacement rate of 1.25 mm per min. The shear force was measured by a proving ring connected to the upper half of the box, and horizontal displacement was measured using a dial gage which connects the front face of the shear box.

Results and Discussion

A series of direct shear tests were done to understand the shear behavior of areca fiber-reinforced sand. The results are expressed in terms of peak shear stress, cohesion, and angle of internal friction. Figure 29.3 shows the shear reaction of unreinforced and areca fire-reinforced sand at a 100 kPa normal stress. It can be seen that shear strength of soil goes on increasing with fiber addition. The maximum improvement in shear strength is shown by the sand reinforced with 1% of areca fiber.

Figure 29.4 shows the variation of shear strength with different percentages of fiber. It can be noted that among different percentages of fiber, 1% fiber-reinforced sand gives maximum improvement in shear strength compared to unreinforced sand. Sand reinforced with 1% areca fiber improved the shear strength 34.72, 30.36, and 30.63% compared to unreinforced sand under 50, 100, and 150 kPa normal stresses, respectively. Further increase in fiber content reduces the improvement in shear strength. It may be due to the segregation of fibers in sand which leads to improper mixing of fiber with sand.

Figure 29.5 depicts the failure envelope of unreinforced sand and reinforced sand with different percentage of fiber. Failure envelop was plotted using Mohr Coulomb criterion. From the plot, it can be seen that the shear strength of unreinforced sand is only due to the friction between sand particles, and it does not exhibit any cohesion. Whereas, the inclusion of fiber in sand exhibits an apparent cohesion for all percentages of fiber content. It can be noted that the angle of internal friction also improved with the addition of areca fibers. Table 29.2 shows the shear parameters obtained for unreinforced and areca fiber-reinforced sand. At an optimum fiber content of 1%, the cohesion was increased to 23.26 kPa, and friction angle was increased to 42. 39°.

Direct shear test is one of the most common methods used to find out the shear strength of the soil. Shear strength of a soil is its maximum resistance to shearing

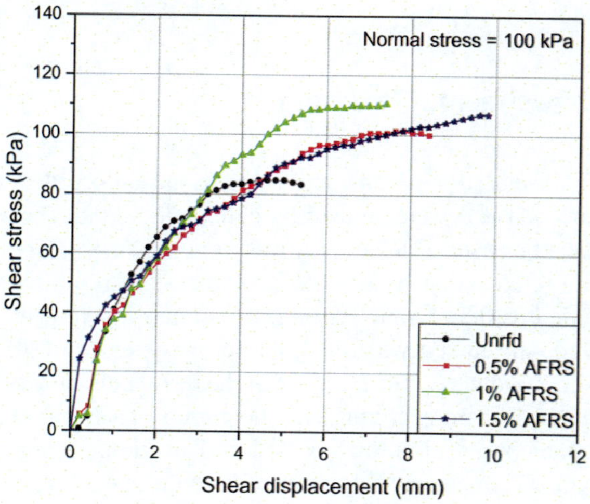

Fig. 29.3 Shear stress-displacement curve for a normal stress of 100 kPa

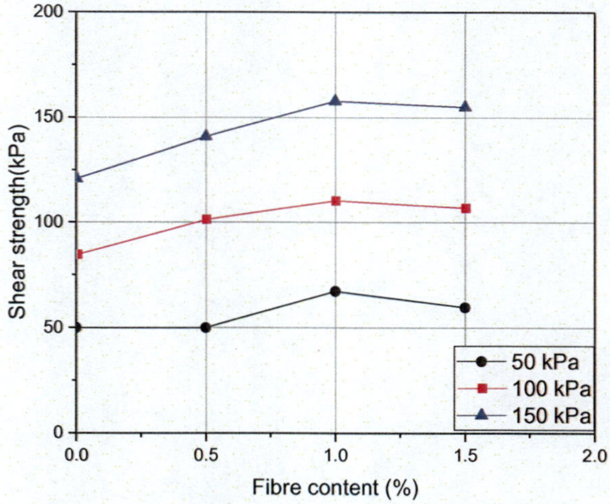

Fig. 29.4 Variation of shear strength with % fiber

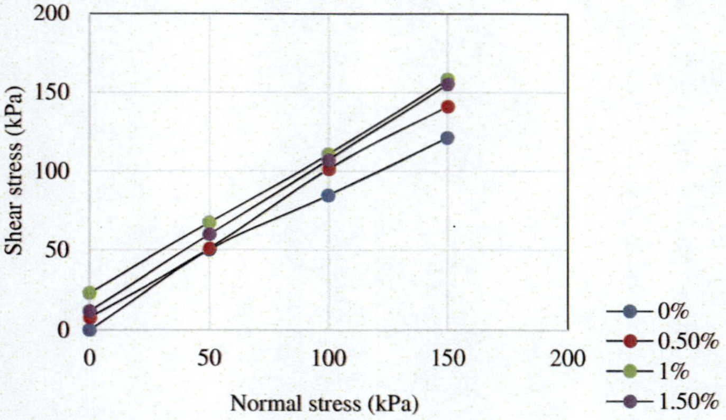

Fig. 29.5 Failure envelop for sand reinforced with different % fiber

stresses. In a direct shear test, the soil is sheared along a predetermined horizontal plane by moving one part of the soil container relative to the other.

To find out the shear strength behavior and shear strength parameters of the soil, direct shear test was conducted at three different normal forces 50, 100, and 150 kPa on both unreinforced and reinforced soil. 0.5–1.5% areca fiber is added as reinforcement with an increment of 0.5%. Fiber is mixed in random manner. Table 29.2 shows the obtained values of τ (kPa), c (kPa), and ϕ (°).

Table 29.2 Shear parameters for unreinforced and reinforced sand

Normal stress, σ (kPa)	Unreinforced			0.5%AFRS			1% AFRS			1.5% AFRS		
	τ (kPa)	c (kPa)	ϕ (°)	τ (kPa)	c (kPa)	ϕ (°)	τ (kPa)	c (kPa)	ϕ (°)	τ (kPa)	c (kPa)	ϕ (°)
50	50	0	39.72	50.69	7.86	41.82	67.36	23.26	42.39	59.72	11.94	42.67
100	84.72			101.39			110.42			106.94		
150	120.83			140.97			157.85			155		

Conclusions

Based on the present investigation, the following conclusions have been drawn.

- The addition areca fiber in random manner improved the shear strength of soil significantly.
- The optimum amount of areca fiber content was found to be 1% of fiber by weight of soil.
- Cohesion of the soil improved by 7.865, 23.36, and 11.94 kPa for 0.5, 1, and 1.5% of fiber addition.
- Angle of internal friction of unreinforced soil was observed to be 36.72. Its value got improved by 41.83, 42.39, and 42.67 for 0.5, 1, and 1.5% of fiber content
- At the optimum fiber content, the shear strength of the sand improved by 34.72, 30.36, and 30.63% compared to unreinforced sand under normal stresses of 50, 100, and 150 kPa, respectively.
- The fiber-reinforced soil exhibits an apparent cohesion, and the angle of internal friction also gets improved.
- The arecanut fiber soil stabilization will be more economical since it is naturally available as an agricultural waste.

References

1. Anagnostopoulos CA, Tzetzis D, Berketis K (2014) Shear strength behaviour of polypropylene fibre reinforced cohesive soils. Geomech Geoeng 9(3):241–251
2. Cai Y, Shi B, Ng CWW, Tang C (2006) Effect of polypropylene fiber and lime admixture on engineering properties of clayey soil. Eng Geol 87:230–240
3. Gray DH, Ohashi H (1983) Mechanics of fiber-reinforcement in sand. J Geotech Eng 1093:335–353
4. Loganathan TM, Hameed Sultan MT, Jawaid M et al (2020) Physical, thermal and mechanical properties of areaca fibre reinforced polymer composites-an overview. J Bionic Eng 17:185–205
5. Manohar DR, Anbazhagan P (2021) Shear strength characteristics of geosynthetic rubber sand mixtures. Geotext Geomembr 49:910–920
6. Patel SK, Singh B (2017) Shear strength response of glass fibre-reinforced sand with varying compacted relative density. Int J Geotech Eng 1–13
7. Prabakar J, Sridhar RS (2002) Effect of random inclusion of sisal fibre on strength behaviour of soil. Constr Build Mater 16(2):123–131
8. Shewbridge SE, Sitar N (1989) Deformation characteristics of reinforced sand in direct shear. J Geotech Eng 1158:1134–1147
9. Tingle JS, Santoni RL, Webster SL (2002) Full-scale field tests of discrete fiber-reinforced sand. J Transp Eng 1281:9–16
10. Yetimoglu T, Salbas O (2003) A study on shear strength of sands reinforced with randomly distributed discrete fibers. Geotext Geomembr 21:103–110

Chapter 30
Design of Foundation on Erratic Landfill with Ground Improvement Techniques—A Case Study

R. J. Satchithananda Satheesh and S. Selvakumar

Introduction

There are several challenges are faced due to reclaiming a landfill site for the purposes of construction activities which includes liability considerations and practical difficulties such as settlement and low bearing capacity [1, 2]. In recent times, there is an increase in previously utilized commercial locations or landfill areas are restored in case of construction purposes [3–5]. The main considerations such as analyzing the locations for the different structures and the original process of design are assessed prior to construction [6]. Earlier, if any construction has built on closed landfills that may experience foundation problems related to excessive settlement. Extensive foundation problems can result unless the landfill has been stabilized before structures are placed on it [7, 8].

Erratic landfills are known as filled with different types soils in one area; each soil properties varies because of that so many problems encountered such as soils with low bearing capacity, poorly compacted soil, change in moisture content, and soil consolidation [9–11]. Generally, the soils with low bearing capacity are considered as weak soil or newly filled up soil having highly compressible nature, and buildings constructed on such type of soils require special footings or any ground improvement techniques prior to construction. In recent times, the global communities are shifted toward sustainable practices which results in use of waste materials for the purposes of ground improvement, soil stabilization, soil reclamation, etc.

In this study, the pond ash is selected and used as the stabilizing materials for the reclamation of landfill sites. Pond ash is the sediment waste particles found from

R. J. Satchithananda Satheesh (✉) · S. Selvakumar
Department of Civil Engineering, Vel Tech Rangarajan Dr. Sagunthala R&D Institute of Science and Technology, Avadi, Chennai, Tamil Nadu 600062, India
e-mail: satchuerd@yahoo.com

S. Selvakumar
e-mail: selvakumars@veltech.edu.in; selva.geotech@gmail.com

K. Muthukkumaran et al. (eds.), *Ground Improvement Techniques*, Lecture Notes in Civil Engineering 297, https://doi.org/10.1007/978-981-19-6727-6_30

the combustion of coal products. After burning the coal, there is occurrence of two types of combustible products such as bottom ash and fly ash [12, 13]. According to the origin and the constituents in clinker, the evaporative residue consists of silica (Si), alumina (Al), and calcium (Ca) [14]. The size of circular-shaped pond residues varies between 0.5 and 300 μ. For the determination of pond ash compaction characteristics, generally at practice, standard Proctor compaction and modified Proctor compaction test methods have been followed. Some of the researchers evaluated the waste materials compaction characteristics and indicated that the test procedure could influence compaction parameters [15, 16]. Also, observed the compacting duration during standard Proctor method, difference in pond ash compaction characteristics was observed when samples are in saturated condition [17, 18].

The main objective of this research is to get an ultimate solution for design of foundation in erratic land fill by adopting pond ash. The utilization of waste lands which are erratically filled by adopting latest techniques, proper planning and execution. To get economic benefit by utilizing the filling lands that by reduce the cost of the project, this ultimately leads to the project economically viable.

Materials and Methods

General Geology of Chennai District

The geological alterations in the Chennai district range from ancient Archaean to contemporary alluvium. The district's geological foundations are divided into three categories: (1) Archaean crystalline rocks, (2) cemented Gondwana and tertiary sediments, and (3) recent alluvium. The pre-Cambrian rocks of the Chennai district cover chief pond of charnockites, schists, in addition to the related mafic and ultra-mafic intrusive. They are crystalline in nature. They are disintegrated and connected. The level and magnitude of disintegration differs at various locations, and the breadth of disintegrated mantle also differs till 12 m. The Gondwana shale which has jointed/fractured nature is dark gray or black in color. The major portion of the district is filled with alluvium deposits. The alluvium deposits consist of loam or parts of soil, sand, and clay. An optimum depth of about 28 m is encountered in northern part of Chennai located not far from Perambur circle has been extended in the thickness of alluvium deposits. A 24-m-thick alluvium deposits have been found in areas of Kilpauk water works. Kodungaiyur comes under Perambur Taluk of Chennai city, Tamil Nadu, India. Kodungaiyur which is situated in Perambur Taluk, Chennai, Tamil Nadu. India has been used as a dumping yard for municipal solid wastes for the last four decades. In this study, Kodungaiyur municipal yard has taken as case study; the latitude and longitude are 13° 08′ 02″ N and 80° 16′ 09″ E, respectively. The advancement of the study area has about 7 m above from the mean sea level. Figure 30.1 shows the photographic view of Kodungaiyur erratic landfill area prior to construction.

Fig. 30.1 Kodungaiyur erratic landfill area

Soil Exploration

The soil exploration was carried out in the erratic land filling area at Kodungaiyur to find out the sub-soil strata. Three exploratory boreholes of 150 mm diameter were drilled up to a depth varying from 27.0 m for BH1, 22.0 m for BH2, and 28.50 m for BH3 using rotary drill adopting mud circulation technique at the location adjacent to existing sewage treatment plant. Standard penetration tests were conducted at 1.0–1.5 m intervals or at change of stratum as per Indian Standard IS2131-1981 using split spoon sampler as per Indian Standard IS:9640-1980, and standard penetration resistance (*N* values) was noted. Refusal was considered if the blow count exceeds 50 for the penetration of 15 cm. Using split spoon sampler, the soil samples in disturbed state were collected at regular intervals. Also, it is noted that ground water level exists at a depth of 5.80 m, 5.50 m, and 5.0 m at the time of soil exploration. Ground water level fluctuates may occur due to seasonal variations.

Standard Penetration Tests (SPT)

Sub-soil investigation was conducted at erratic landfill area to identify the sub-soil profile and properties of soil. The sub-soil stratification based on field tests, laboratory tests, visual identification, and engineering characteristics of soils was derived from SPT-value correlation. Three exploratory boreholes of 150 mm diameter was drilled up to a depth varying from 27.0 m for BH1, 22.0 m for BH2, and 28.50 m for BH3 using rotary drill adopting mud circulation technique at the location. From the exploratory borehole, soil specimens were taken at 1 and 1.5 m interval using standard penetration test method (SPT), and also, at the depth of 2 m, four test pits at different location were excavated, and sample was collected. The collected samples were tested in the laboratory.

Table 30.1 Physical composition of pond ash

Parameters	Range
Color	Light gray
Optimum moisture content, OMC (%)	33–46%
Maximum dry density, MDD (gm/cc)	0.856–1.247
Specific gravity, G	2.19
Plasticity index, I_P	Non-plastic
Coefficient of curvature (C_c)	1.27
Uniformity coefficient (C_u)	5.69

Pond Ash

India is the second global coal producer and third greatest consumer of electricity in the world. Out of 3,82,730 MW power produced, the thermal power generation capacity using coal alone stands as 2,02,675 MW as on 30.04.2021 (as per data from Ministry of Power, GOI). The ash content of coal produced in the whole nation normally ranges between 25 and 45% whereas the imported coal has an average ash content that varies in the range of 10% to 20%. The coal consumption in Thermal Power Plants in India is at about 9,66,288,693 tons which leads to enormous amount of ash produced in India. Particles of pond ash are very finely powdered, lightweight (density in the range of 1.97–2.89 g/cc) and sphere shaped (having bulk volume varying from 4000 to 10,000 cm^2/g; radius 0.5 to 75 μ), refractory and pozzolanic nature. Pond ash is generally differing from gray to blackish gray in color, and it relies on type of color combustion process. The physical characteristics of pond residues are calculated. The values are given in Table 30.1. Figure 30.2 explains the stages of filling pond residue on the site.

Results and Discussions

Plate Bearing Tests

The plate bearing test is commonly used to determine the allowable and safe bearing capacities of the soil and probable settlements. This study involves rate of application of static load on a rigid apparatus at in situ for the probable proposed founding level. It is a physical experiment where the allowable bearing capacity of soil and respective settlements under the applied load were determined. The loading steel plate is placed at the level of foundation, and the respective settlements for every applied load in an increasing order were noted from the dial gauge reading. The load applied through steel plate is raised slowly until the steel plate begins to drop at a faster rate. Dividing the total plate load at that stage by the steel plate area, the allowable bearing capacity

Fig. 30.2 Stages of filling pond ash at the site

of soil is obtained. The value of safe bearing capacity of soil is attained by dividing the obtained value by factor of safety [2, 3].

A hydraulic jack provided with a mounted pressure gauge was inserted between the platform and the steel plate; the remaining gap was made up by using a compression column pipe with appropriate length and stiffness. The direction of the vertical load is maintained throughout the test with the help of a ball and socket assembly. Before initiating the test, the minimum seating pressure of about 70 g/cm^2 was applied and then removed. The vertical load was gradually increased in a cumulative order. The settlement was measured after the gradual increase of each load with respect to two

Table 30.2 Load versus settlement data for the plate load test-1

Incremental stage no.	Applied load in tones	Applied pressure in T/m^2	Settlement in mm
1	0	0	0
2	0.384	2.419	0.065
3	0.769	4.383	0.327
4	1.154	7.257	0.59
5	1.54	9.677	0.765
6	1.924	12.09	0.955
7	2.309	14.515	1.15
8	2.694	16.933	1.38

dial gauges of 0.01 mm accuracy which are located diametrically opposite to the ends of the steel plates. A linear scale curve was obtained from the graph of load plotted against settlement was subjected to zero correction which is provided by the intersection of the early straight lines or the nearly straight-line part of the curves.

On conducting a plate load test at site, the allowable bearing pressure of the soil can be determined. For proper performance of tests, the plates of circular or square plates whose size ranges between 3 and 6 m with 25 mm breadth were used. With the help of hydraulic jack, loads were applied gradually on the steel plate. The applied load reaction of a hydraulic jack was distributed at two ends with respect to the suitable beam or steel truss imbedded. By using a set of four dial gauges with a sensitivity of 0.01 mm, the settlement of the plate can be measured. During the test, the undisturbed independent supports are fixed with the dial gauges. Tables 30.2 and 30.3 summarize the load versus settlement values of the plate load tests.

A linear scale curve was obtained from the graph of load plotted against settlement is subjected to zero correction that is provided by the intersection of the early straight lines or the nearly straight-line part of the curves. Figures 30.3 and 30.4 show the

Table 30.3 Load versus settlement data for the plate load test-2

Incremental stage no.	Applied load in tones	Applied pressure in T/m^2	Settlement in mm
1	0	0	0.00
2	0.384	2.419	0.02
3	0.769	4.383	0.42
4	1.154	7.257	0.80
5	1.54	9.677	1.06
6	1.924	12.09	1.36
7	2.309	14.508	1.64
8	2.694	16.926	1.90

curve attained from graph of load plotted against settlement of the plate load tests 1 and 2.

The safe bearing pressure for cohesion less soil can be found, with respect to the plate settlement (S_p), that be obtained by calculating the result from Eq. 30.1:

$$S_f = \left(B\left(B_p + 0.3\right)\right) / \left(B_p(B + 0.3)\right)^2 \tag{30.1}$$

where B = Size of footing in m, B_p = Size of test plate in m, S_p = Settlement of test plate in m, and S_f = Settlement of footing in m.

From this formula, total settlement of footing (S_f) is calculated taking S_p as observed total settlement of plate. S_f = Settlement of footing is taken as 5 mm and equivalent size of square footing = 2.0 m × 2.0 m.

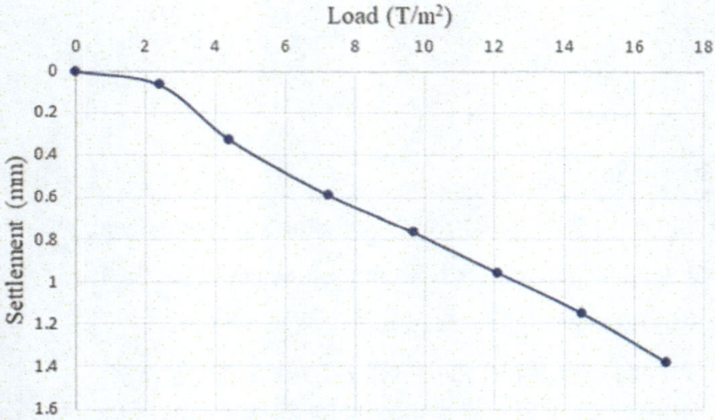

Fig. 30.3 Load versus settlement curve for plate load test-1

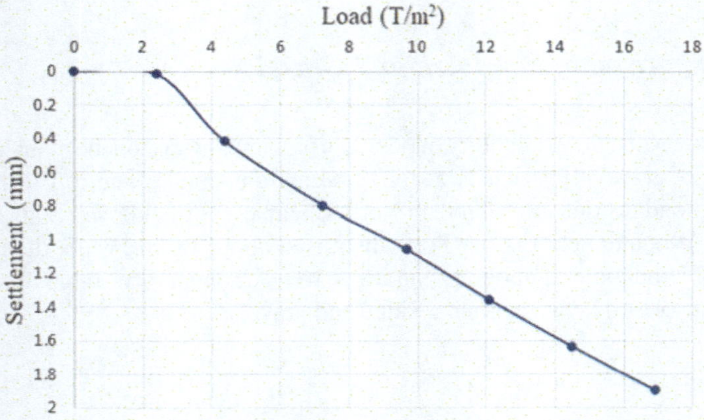

Fig. 30.4 Load versus settlement curve for plate load test-2

Estimation of Allowable Bearing Capacity from Plate Load Tests

Due to the unpredictable landfill and noticeable modifications in the compressibility characteristics of assisting layers, that resulted in occurrence of minor differential settlement. Depending upon the ability of the supporting soil strata, it could not able to absorb the maximum permissible settlement which resultant in differential settlements to the superstructure. For the suggested foundation, the estimations and analyses have been made. Based on the computation (Eq. 30.2), the plate load test results, there was no failure observed till the applied load is maximum. It is evident from Figs. 30.3 and 30.4, which demonstrates the graph of load plotted against settlement for the plate load tests.

$$S_f = S_p(2.0(0.45 + 0.3)/0.45(2 + 0.3))^2 \tag{30.2}$$

where

$S_f = S_p \times (1.5/1.035)^2 = S_p \times 2.100.$

$S_p = 5/2.100 = 2.381$ mm.

$S_p = 2.381$ mm is the total settlement of plate for 5 mm settlement.

Load applied in plate load test for 1.38-mm settlement is 16.934 T/m^2.

So, the safe bearing capacity is 1.38 mm.

Settlement of foundation = (16.934/1.38) × 2.381 = 29.22 T/m^2.

For the applied load of 16.934 T/m^2, the total settlement of the plate calculated was 1.38 mm. Limiting the settlement of footing by 5 mm, the safe bearing capacity is taken as 29.22 T/m^2. Hence, the suggested safe bearing capacity was 25 T/m^2.

Cost Comparison of Pond Ash and Gravel

It is well known that the availability of pond ash is abundant, and on the other hand, the availability of gravel is restricted due to socio-environmental concern. The abundant quantity of pond residue is available in North Chennai Thermal Power Station which is situated Athipattu near Ennoor Port which is at about 18 km, while the gravel is available in Uthukottai at about 55 km away from Chennai city. Table 30.4 shows the cost comparison between pond residue and gravel.

Table 30.4 Cost comparison of pond residue and gravel

Particulars	Pond residue	Gravel
Cost	Rs. 40/m^3	Rs. 900/m^3
Availability	Abundant quantity	Minimal quantity
Environmental impact	By using pond ash, we reduce pollution, and the environment is saved	Cutting of gravel increases socio-environmental problem

Conclusion and Recommendations

Conclusion

Gleaned from the outcomes presented on above sections, the conclusions can be summarized. These conclusions are limited to the materials used and test conditions under which the tests are conducted.

1. Replacement of weak top layer in the erratic land fill with pond ash and by proper confinement to prevent leaching in case of water percolate in to the filling and using latest compacting roller to get sustainable layer compaction and thus the required SBC is achieved.
2. Hence, by utilizing the abundantly available pond ash which are produced in power plants, which once upon labeled as waste material, by utilizing them in effective way and also reduce the charges of cutting of the good earth for filling and reduce the project cost.
3. Thus, utilizing the erratically filled land effectively to achieve social and economically viable projects in the present stage of faster urbanization in developing countries.

Recommendations

The following points may be taken into consideration for further research:

1. It is recommended to use pond ash in all construction activities, and necessary and adequate research is to be done considering typical site requirements to reduce the overall cost of the projects.
2. It is recommended to use pond ash wherever possible, thus utilizing the large quantities of pond ash available in all thermal power plants.
3. It is recommended to improve the material with draining case to study the leaching effects.
4. It is recommended to study the consequences of soaking duration on collapse potential of soil particles improved with bentonite with different percent.

5. It is recommended to make extensive research to mix pulverized fuel ash, bottom ash, and other bonding agents toward cost economic projects and environmental protection.

References

1. Gandhi SR, Ambily AP (2004) Experimental and theoretical evaluation of stone column in soft clay. In: ICGGE-2004, International conference on geosynthetics and geoenvironmental engineering, IIT, Bombay, pp 201–206
2. Raju VR (2011) QC methods for deep vibro techniques. In: Proceedings of the 14th Asian regional conference on soil mechanics and geotechnical engineering, Hong Kong, pp 1–31
3. Odud C (2000) Current state of the practice of construction on closed landfill sites. In: 16th International conference on solid waste technology and management, PA, USA, pp 1–8
4. Jedele LP, Buschmeier B (2013) Ground improvement for redevelopment of former landfill. In: 7th international conference on case histories in geotechnical engineering. Chicago, Illinois, USA, pp 1–10
5. Ronga F, Zhaoguib G, Tugenb F (2011) Analysis of stability and control in landfill sites expansion. Procedia Eng 24:667–671
6. Ashford SA, Visvanathan NH (2000) Design and construction of engineered municipal solid waste landfills in Thailand. Waste Manage Res 18(5):462–470
7. Yee YW, Chua CG Yandamuri Keller HK (2009) Foundation works for a sewage treatment plant using ground improvement methods in Malaysia. In: Ground improvement technologies and case histories, Singapore, pp 677–684
8. Mishra B (2016) A study on ground improvement techniques and its applications. Int J Innov Res Sci Eng Technol 5(1):72–86
9. Ayele S (2017) A case study on ground improvement techniques and its applications. Int J Sci Eng Res 8(9):1288–1305
10. Kerry Rowe R, Caers CJ, Reynolds G, Chan C (2000) Design and construction of barrier system for the Halton landfill. Can Geotech J 1:662–675
11. Sharma HD, De A (2007) Municipal solid waste landfill settlement. J Geotech Geoenviron Eng 133(6):362–370
12. Bera AK, Ghosh A, Ghosh A (2007) Compaction characteristics of pond ash. J Mater Civ Eng 19(4):349–357
13. Gray DH, Lin YK (1972) Engineering properties of compacted fly ash. J Soil Mech Found Div 98(4):361–380
14. Gurtug Y, Shridharan A (2004) Compaction behavior and prediction of its characteristics of fine grained soils with particular reference to compaction energy. Soils Found 44(5):27–36
15. Omar M, Abdallah S, Basma A, Barakat S (2003) Compaction characteristics of granular soil in the United Arab Emirates. Geotech Geol Eng 21(3):283–295
16. Osman S, Togrol E, Kayadelen C (2008) Estimating compaction behaviorof fine-grained soils based on compaction energy. Can Geotech J 4(6):877–887
17. Patra CR, Sivakugan N, Das BM, Rout SK (2010) Relative density and median grain-size correlation from laboratory compaction tests on granular soil. Int J Geotech Eng 4(1):55–62
18. Singh SP, Sharan A (2013) Strength characteristics of compacted pond ash. Geomech Geoeng: Int J 9(1):1–9

Chapter 31
Improvement of Soil Subgrade with Shredded Rubber Tire Waste

Shuvankar Chowdhury and Saroj Kundu

Introduction

The rapid development of metropolitan areas, and thus the increase in construction activity, has resulted in a shortage of land with ideal soil conditions, necessitating the use of locally accessible weak soils for construction via stabilization techniques. Due to large-scale construction of roads in India, the requirement of fill material is huge, and the availability of high strength and low compressibility material near the construction sites is very rare. Such soil required the addition of some strengthening materials to increase the existing strength and reduce compressibility. Low strength and bearing capacity, poor drainage, high compressibility, and high liquefaction potential are all major geotechnical issues associated with these sorts of soils. In the context of these circumstances, a suitable and effective ground improvement technology is expected. The techniques of soil stabilization are often categorized in different ways like vertical drains, vibrations, surcharge load, consolidation, admixture grouting, and reinforcements, etc. Geotechnical engineers all over the world are looking for the latest alternative materials, which are needed for both cost-effective ground improvement and the preservation of limited natural resources.

The number of vehicles is increasing day by day as the Indian economy becomes more globalized and more focused is placed on infrastructural development. Apart from the noise and air pollution, increased expansion has begun to generate pollution in the form of waste tire stockpiles. Waste rubber tires are already prohibited from being disposed of in sanitary landfills in many countries. The use of scrap rubber tires as fuel is now outlawed by the Indian government due to environmental concerns.

S. Chowdhury
Senior Geotechnical Engineer, Pioneer Geoscience, Kolkata 700156, India

S. Kundu (✉)
Brainware University, Kolkata 700125, India
e-mail: kundu.saroj44@gmail.com

K. Muthukkumaran et al. (eds.), *Ground Improvement Techniques*, Lecture Notes in Civil Engineering 297, https://doi.org/10.1007/978-981-19-6727-6_31

In order to eliminate the bad effect of these materials, depositions and in terms of sustainable development, there is great interest in recycling those non-hazards solid wastes. In recent years, the use of shredded tires in geotechnical engineering to improve soil behavior has gotten a lot of attention. If scrap tires are used as a construction material instead of being burned, it could be beneficial to civil and environmental engineering. Scrap tires have been used for a variety of uses in recent years, including lightweight fill, insulation beneath roads for possible highway building, lightweight backfill for retaining walls, and drainage improvement.

So, a study has been undertaken to evaluate the beneficial use of rubber when applied to weak subgrade fine-grained material for using it in a pavement in a proper way.

Literature Review

Rubber tire recycling is a potential option for rubber tire waste management. Many studies have recently been conducted on the effective utilization of rubber tire waste in civil engineering structures.

Prasad et al. [1] carried out test to asses CBR tests and direct shear tests for finding the optimum percentages of waste plastics and waste rubber tire in gravel sub-base material. With maximum percentage of waste plastics of 10% and waste rubber of 5% by weight of soil in gravel sub-base. It was founded that 22.85% of CBR increased when the gravel sub-base was reinforced compared to unreinforced sub-base. Soni and Sagane [2] studied the use of shredded rubber tires in embank construction. Tire shredded can be used to construct embankment on weak, compressible foundation soils. The authors opined that tire shreds were viable in those application due to their lightweight. The weight of the tire shreds reduces horizontal pressure and allows for construction of thinner, less expansive walls. Humphrey [3] made an investigation exploring the effectiveness guidelines on use of tire-derived aggregate as lightweight embankment fill for design guidelines. Three projects build in 1995 were considered that experienced internal heating reaction, and it was planned to properly design and construct TSD fill as per the author. Sabat and Pattnaik [4] carried out tests to assess improvement of expansive soil by use of lime with shredded tire. Three materials were used for that experimental program, expansive soil, shredded waste tire (SWT), and lime. SWT was added to the expansive soil up to 20% by weight of soil at an increment of 5% by replacement of expansive soil with SWT. Standard proctor test, UCS, and soaked CBR test were conducted on expansive soil–SWT mixes. The optimum percentages of lime for stabilization with optimum percentage (10%) SWT stabilized expansive soil was found to be 5%. Bavrej and Kumar [5] carried out test to assess the stabilization effect on clay soil with the use of shredded rubber with marble dust. With the addition of rubber in the soil, the maximum dry density decreased for all the samples and the OMC did not show much changes. But with addition of marble dust in the soil, the MDD started to increase with a decrease in OMC which might be due to specific gravity of marble dust and low plasticity. Waste shredded rubber mixture showed an improvement in UCS value after the addition of marble dust in to the soil by 19.32% a maximum. Reddy et al. [6] (2016) conducted test to assess improvement of a fine-grained back cotton soil characteristics by use of shredded rubber along with cement in different percentages. The scrap tire shreds were added to the soil cement of 2% and 4% by weight rapidly included and was tested to determine the improvement of soil behavior by conducting UCS test and CBR test both in soaked and in unsoaked conditions. Both the values were found increased with the increase in cement content at an optimum fiber content of 5%.

Scope of the Study

The objective of the present study is to investigate the effect of shredded rubber tire and plastic sacks randomly mixed with a clayey soil for estimating its improvement effect to on geotechnical properties finding its utilities in different geotechnical applications. The following scope of study is considered for fulfillment of the objective-

- Performance of waste shredded rubber tires of different size and different percentages by weight of a clayey soil is to be studied in order to understand the extent of improvement of the soil characteristics.
- Series of tests like standard proctor test, California bearing ratio test, unconfined compressive strength test of fined-grained soil mixed with different proportions of shredded rubber tire, are to be conducted to evaluate the optimum use of those waste materials for maximum improvement of soil behavior.

Experimental Studies

Material Used

The present investigation was carried out on soil samples collected from local area of New Town, Kolkata, West Bengal, at a depth of 6.0 m below the ground surface. The silt was transported to the laboratory using scaled plastic bags. Standard test methodologies outlined in the appropriate IS Code of Practice [7] were used to establish various geotechnical parameters of silty soil. The grain size distribution curves for silty soil used in the study are shown in Fig. 31.1, and Table 31.1 highlights the fundamental features of silty soil.

Shredded tire material was obtained from the waste generated from tire rethreading industries J.P. Tire shop, Barabazar, Kolkata. Rubber tires were cut into specific sizes of 1cmX1cm, 1cmX2cm, 1cmX3cm, and 1 cm × 4 cm. The shredded tire has a thickness ranging from 2 to 3 m, and they do not contain any steel wire or nylon fibers. The specific gravity of shredded tire obtained with a pycnometer test

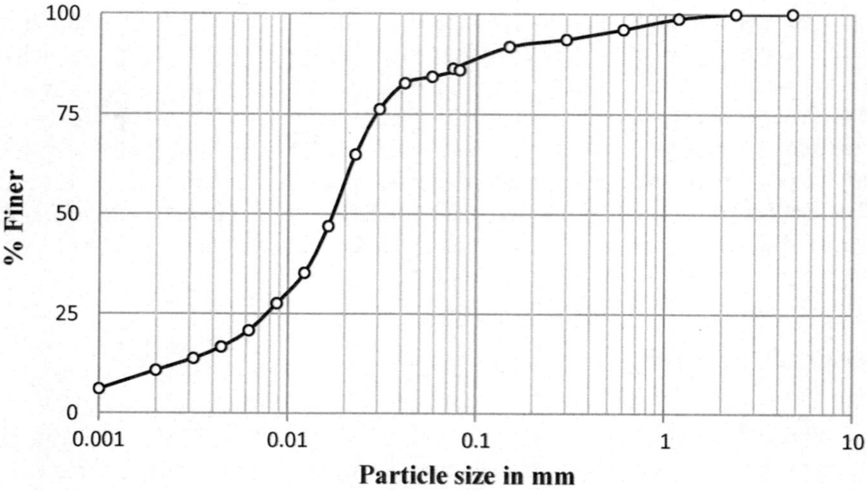

Fig. 31.1 GSD curve for soil used in study

Table 31.1 Properties of soil used in the study

Property	Value	IS code
Specific gravity (G)	2.61	IS 2720-Part 3
Liquid limit (W_l)	46.90%	IS 2720-Part 5
Plastic limit (W_l)	32.25%	IS 2720-Part 3
Plasticity index (Ip)	14.4%	IS 2720-Part 5
IS classification of soil	MI	IS 2720-Part 4
MDD	1.74 gm/cc	IS 2720-Part 7
OMC	15%	IS 2720-Part 7
CBR (unsoaked)	6.826	IS 2720-Part 16
CBR (soaked)	3.641	IS 2720-Part 16
UCS	9.5 kg/cm^2	IS 2720-Part 10

ranges from 0.80 to 0.13. The view of the shredded tire utilized in the study is shown in Fig. 31.2.

Methodology

A systematic experimental program has been undertaken for different soil-shredded tire composite to determine the optimum percentage of tire to be added with soil for obtaining enhancement of different geotechnical properties if any.

(a) (b)

Fig. 31.2 (**a**) Shredded rubber tires chips; (**b**) soil and shredded rubber tire mixture

Rubber tires were cut into specific sizes of 1cmX1cm, 1cmX2cm, 1cmX3cm, and 1cmX4cm. These shredded tires were mixed randomly by hand with 5 kg of soil at different percentages of 1, 3, 5, 7% to the respective weight of soil taken. Then, the following tests were conducted with the tire-mixed soil samples.

- Standard proctor compaction test
- California bearing ratio test–unsoaked and soaked condition
- Unconfined compressive strength test

After determining the optimum moisture content for different size and percentages of added rubber tires, the California bearing ratio test unconfined compressive test with shredded rubber tires were conducted by adding water corresponding to respective OMC, so that the effect of addition of shredded rubber tire on improvement of soil properties can be fairly adjudged.

Result and Discussion

Standard proctor compaction test, California bearing ratio tests, unconfined compressive strength tests are conducted for the soil by mixing with shredded tires of different sizes, i.e., 1cmX1cm, 1cmX2cm, 1cmX3cm, and 1cmX4cm with the varying rubber percentages, i.e., 1, 3, 5, and 7% and the results were noted and compared as shown in the Table 31.2.

Table 31.2 Geotechnical properties of tested soil and soil-shredded rubber tire mix of varying percentages

Materials	% of shredded tire chips	OMC %	MDD (gm/cc)	CBR (unsoaked)	CBR (soaked)	UCS (kg/cm^2)
Soil	0	15	1.74	6.826	4.369	9.50
Soil + tire of size 1 cm × 1 cm	1	15.3	1.72	7.08	4.57	9.95
	3	15.8	1.72	7.33	5.32	9.06
	5	16.3	1.63	6.07	3.64	8.29
	7	18.2	1.52	5.27	3.56	6.43
Soil + tire of size 1 cm × 2 cm	1	17.1	1.7	7.58	4.67	10.75
	3	17.4	1.69	8.01	5.77	10.37
	5	19	1.67	5.07	2.96	7.34
	7	21	1.64	4.67	2.76	5.38
Soil + tire of size 1 cm × 3 cm	1	18.5	1.68	4.97	3.26	8.12
	3	18.9	1.6	5.37	3.06	7.20
	5	21.3	1.46	4.46	2.56	6.24
	7	22.8	1.34	3.81	2.31	5.01
Soil + tire of size 1 cm × 4 cm	1	19.7	1.52	4.41	2.96	7.43
	3	21.3	1.36	3.71	2.81	6.74
	5	21.8	1.27	3.41	2.26	4.49
	7	23	1.121	3.11	2.00	3.65

Compaction Characteristics of Soil Tire Mix

Figure 31.3 shows the variation of optimum moisture content (OMC) with different percentages of shredded rubber tire chips for each sample. The value of OMC increases with addition of different percentages of shredded rubber tires of different sizes. Maximum value of increase of OMC is 53.33% for addition of 7% of shredded rubber tires of size 1 cm × 4 cm. Shredded rubber tires are occupying more space in soil and the pores get filled with water due to high absorptive capacity of shredded rubber tires for water.

Figure 31.4 shows the variation of maximum dry density (MDD) with different percentages of shredded rubber tire chips for each sample. MDD is found to decrease with increase of percentages and sizes of shredded rubber tire chips. Maximum value of decreases of MDD is found 35.57% for the soil for addition of 7% of shredded rubber tires of size 1cmX4cm. The shredded rubber tire chips are lightweight in nature and they have low specific gravity compared to soil. Resilience of shredded tire also reduces compaction efficiency which may lead to reduction of maximum dry density.

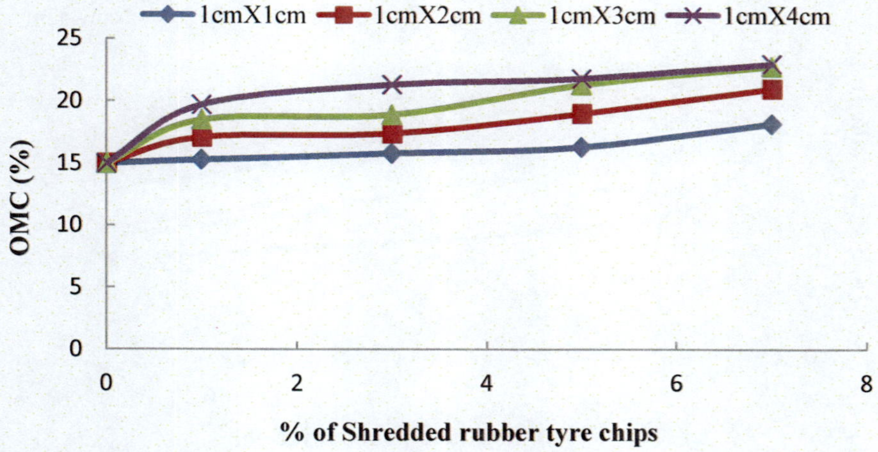

Fig. 31.3 OMC value with different percentages shredded rubber tire

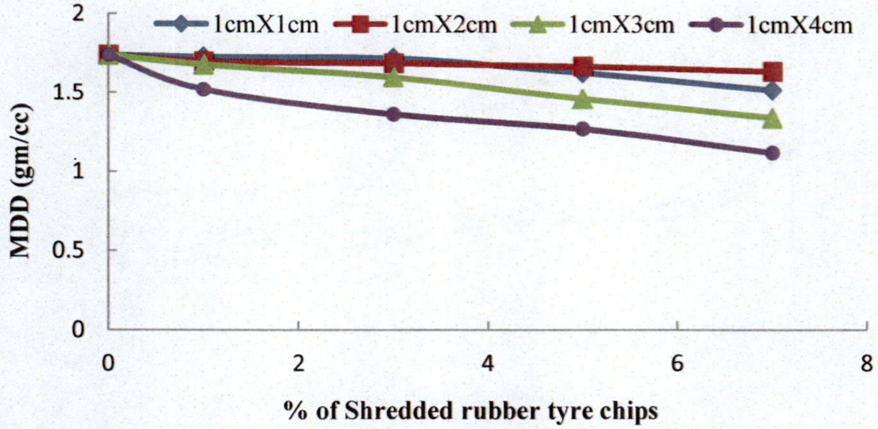

Fig. 31.4 MDD value with different percentages shredded rubber tire

California Bearing Ratio of Soil Tire Mix

Figures 31.5 and 31.6 show the variation of California bearing ratio (CBR) values for unsoaked and soaked, respectively, with different percentages of shredded rubber tire chips for each sample. Value of California bearing ratio (unsoaked and soaked) increases gradually for a particular percentage of added shredded rubber tire for a particular size, and then starts to decrease with increase of percentages of shredded rubber tire chips. Maximum value of increase of CBR is 16.96 and 32.18% for unsoaked and soaked condition, respectively, for addition of 3% of shredded rubber tires of size 1 cm × 2 cm.

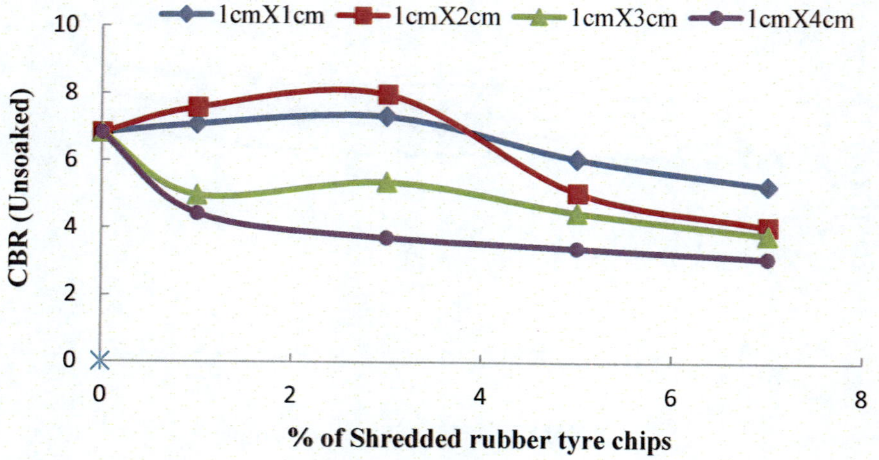

Fig. 31.5 CBR (unsoaked) value with different percentages shredded rubber tire

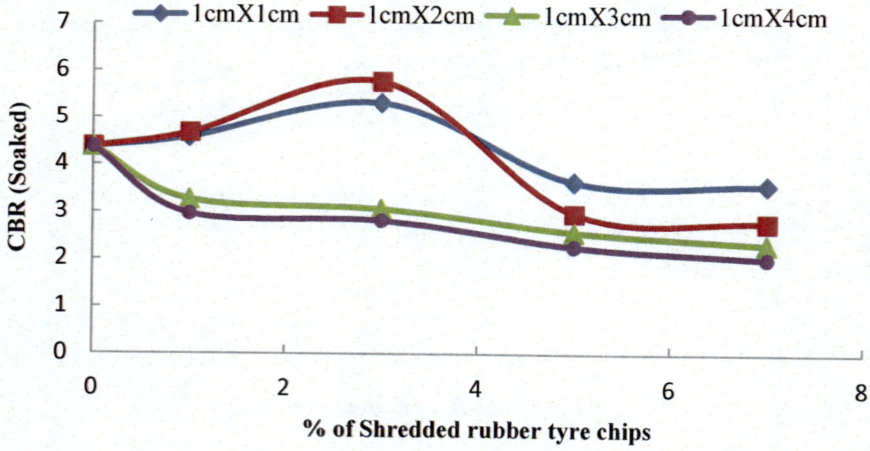

Fig. 31.6 CBR (soaked) value with different percentages shredded rubber tire

Unconfined Compressive Strength of Soil Tire Mix

Figure 31.7 shows the variation of unconfined compressive strength (UCS) value for with different percentages of shredded rubber tire chips for each sample. UCS values increase when a certain percentage of shredded rubber tires of particular size is added and then start to decreases with increase of percentages of shredded rubber tire chips. Maximum increase of value of UCS is 13.15% for addition of 1% of shredded rubber tires of size 1cmX2cm.

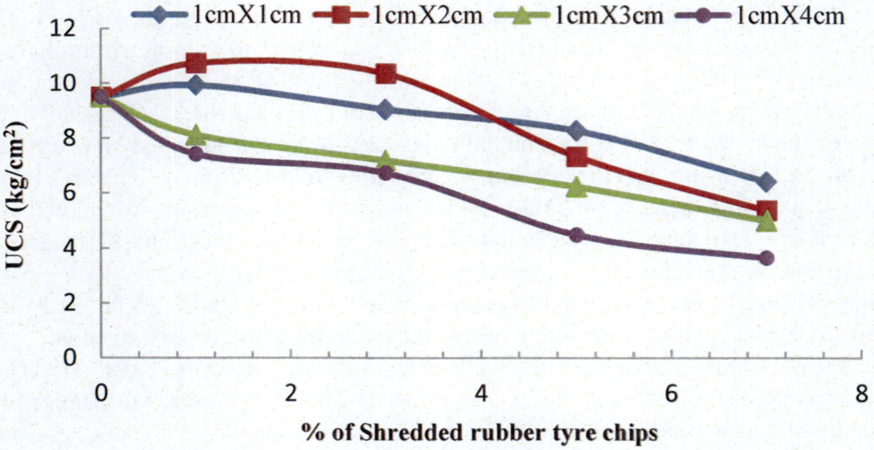

Fig. 31.7 UCS value with different percentages shredded rubber tire

Conclusions

Based on the experiments carried out on soil and tire mixtures, the subsequent observations conclusions are drawn:

1. The influence of shredded rubber tire waste on soil characteristics is mostly determined by strip size and tire content. It is more desirable to add shredded rubber in the shape of strips to improve the strength qualities.
2. With the addition of shredded rubber tire in the soil, the optimum moisture content increases for all the samples and the maximum dry density decreases for all the samples. This may be due to the lightweight nature of the shredded rubber tire.
3. Maximum value of increase of optimum moisture content is 53.33% for addition of 7% of shredded rubber tires of size 1 cm × 4 cm.
4. Maximum decrease of maximum dry density is found 35.57% for the soil for addition of 7% of shredded rubber tires of size 1 cm × 4 cm.
5. In both the unsoaked and soaked conditions for each size of sample, tire waste material mixed with soil showed an increase in CBR value with the addition of tire content up to 3%, and then dropped with additional increases in the content. As a result, in both unsoaked and soaked conditions, the optimal value of waste tire content is 3%.
6. Maximum value of increase of California bearing ratio is 16.96 and 32.18% for unsoaked and soaked condition, respectively, for addition of 3% of shredded rubber tires of size 1 cm × 2 cm. Increasing the CBR of stabilized soil can significantly reduce the overall thickness of the roadway, thereby reducing the overall cost of road construction.
7. Maximum increase of value of UCS is 13.15% for addition of 1% of shredded rubber tires of size 1 cm × 2 cm.

On the other hand, IRC 36 [8] has specified that MDD value in the range from 1.55 to 2.12 gm/cm^3 and OMC from 10 to 20% for silty soil with low plasticity, similarly, IRC SP 20 [9] has suggested that MDD of not less than 1.68 gm/cm^{3}of a material to be selected for subgrade layer. And as per MoRTH [10] guidelines, MDD shall not be less than 1.75 gm/cm^3 of a material to be selected for the subgrade layer. From Table 31.2, it is seen that the dry unit weight varies from 1.12 to 1.74 gm/cm^3. It is apparent that the values of MDD are a bit lower than the minimum density required as per MoRTH [10] guidelines, but conforms to IRC SP 20 [9] and IRC 36 [8] standers. Based on the compaction characteristics, it can be concluded that shredded rubber tires of some sizes with some percentages mixed with soil qualify to be used for subgrade construction, even without resorting to pre-treatment or modification.

Table 31.1 shows that low strength does not meet the standards of IRC 37 [11], which calls for 1.5–3.0 N/mm^2 for untreated soils and 0.75–1.5 N/mm^2 for chemically modified soils in the construction of flexible pavement subgrades. This suggests that soil mixed with shredded rubber tire must be stabilized or solidified.

A reasonable high CBR is another crucial criteria for material used in the subgrade layer of a pavement or a road, in addition to strength. According to IRC SP 20 [9], the required values are 15% under wet conditions and 8% according to IRC 37 [11].

The recorded CBR values in soaked condition, for soil mixed with 3% of shredded rubber tires of size 1 cm × 2 cm, only meet the specifications. So this size of shredded rubber tire soil mixture is most effective for subgrade constructions.

The findings demonstrate the utility of tire waste recycling in the geotechnical element of waste management. As a result, ineffective soil material costs are reduced, and tire waste is recycled. The findings significantly support the recycling of tire scrap as reinforcing materials in the form of shredded rubber tire combined with soil to address environmental challenges. However, the selected shape of tire chips would be suitable for backfill case also, to use in subgrade, the tires have to be shredded like fibers. And for the long-term sustainability and durability of subgrade stabilized tire chip, several groundwater conditions are to be investigated, before using them in practice.

References

1. Prasad DSV, Prasada Raju GVR, Ramana Murthy V (2008) Use of waste plastic and tyre in pavement systems. J Inst Eng India Civil Eng Div 89(AOU):31–35
2. Soni KK, Sagane SC (2008) Use of shredded tyres in embankment construction- an overview. Int J Emerg Eng Res Technol 2(4):264–272. ISSN 2349-4395
3. Humphrey DN (2005) Effectiveness of design guidelines for use of tire derived aggregate as lightweight embankment fill. In: Recycled materials in geotechnics, pp 61–74
4. Sabat AK, Pattnaik U (2015) Geo-engineering properties of lime stabilized expansive soil-shredded waste tyre. Int J Appl Eng Res 10(18):38943–38947
5. Kumar S, Parkash V, Kumar V (2016) Stabilization of clayey soil using steel slag. Int J Res Technol Stud 3(11):2348–1439
6. Reddy VR, Reddy IS, Prasad DSV (2016) Improvement of soil characteristics using shredded rubber. IOSR J Mech Civ Eng 13:44–48

7. (i) IS: 2720- Part 3-1980, Bureau of Indian Standards New Delhi, Feb (1981)). Determinations of specific gravity of soil soilds. (ii) IS: 2720- Part 5-1985, Bureau of Indian Standards New Delhi, August (1985). Laboratory method for determination of LL and PL of soil. (iii) IS: 2720-Part 4-1985, Bureau of Indian Standards New Delhi, January (1986). Laboratory method for grain size analysis. (iv) IS: 2720- Part 7-1980, Bureau of Indian Standards New Delhi, December (1980). Laboratory determination of CBR value. (v) IS: 2720-Part 10-1991, Bureau of Indian Standards New Delhi, May (1992). Determination of Unconfined Compressive Strength
8. IRC 36 (1970) Recommended practice for the construction of earth embankments for road work. In: Indian Road Congress, New Delhi, India
9. IRC SP 20 (2002) Rural roads manual. In: Indian Road Congress, New Delhi, India
10. MoRTH (2000) Specifications for road and bridge works, 3rd revision. In: Indian Road Congress, New Delhi, India
11. IRC: 37 (2012) Guidelines for the design of flexible pavements. In: Indian Road Congress, New Delhi, India

Chapter 32
Use of Nylon Fibers in Improving the Strength of Weak Laterite Soil Blended with Metakaolin and POFA

L. N. V. N. Varaprasad and R. DayakarBabu

Introduction

Lateritic soils are soil types rich in aluminum and iron, existing in many parts of the world. The soil is formed by the weathering of basalt, granite, gneiss, breccias, and conglomerate. Due to laterite soils weathered under varied climate conditions like high temperature and humidity in alternate wet and dry conditions results in poor engineering properties as a result, due to the soils encountered seasonal volume changes, the use of these types of soils is possessed difficulty in handling. The high plasticity of the laterite soil induces cracks in the foundations causing damages to the structures founded on them. Hence, the need for soil improvement enough by any of soil modification techniques may be required to obtain the desired results of the properties.

There is a growing interest in stabilization of soils in recent times due to low cost and convenience, particularly in the soil related projects that require a high volume of soil improvement. Soil stabilization by adding various admixtures is one of the best alternatives to overcome to the problems associated with different types of soils.

Based on many investigations, metakaolin and palm oil fuel ash have been regarded as the effective materials to improve the soil properties and their stabilization mechanisms and can be used as a soil stabilizing material [1]. The properties of lateritic soils could be revamped with the addition of metakaolin acts as a barricade for the containment of municipal solid waste [2].

L. N. V. N. Varaprasad (✉)
Department of Civil Engineering, University College of Engineering Kakinada, JNTUK, Kakinada 533003, India
e-mail: prasadcivil109@gmail.com

R. DayakarBabu
Kakinada 533003, Andhra Pradesh, India
e-mail: rdbgkk25@gmail.com

K. Muthukkumaran et al. (eds.), *Ground Improvement Techniques*, Lecture Notes in Civil Engineering 297, https://doi.org/10.1007/978-981-19-6727-6_32

Table 32.1 Properties of laterite soil

S. No	Property	Value
1	Specific gravity	2.46
2	Differential free swell index (%)	75
3	Liquid limit (%)	32.6
4	Plastic limit (%)	16.7
5	Plasticity index (%)	15.9
6	Sand size particles (%)	28
7	Silt and clay size particles (%)	40
8	Gravel size particles (%)	32
9	Max. dry density (g/cc)	1.65
10	Optimum moisture content (%)	17.8
11	CBR—soaked (%)	4.1
12	CBR—unsoaked (%)	9.9
13	Unconfined compressive strength (kN/m^2)	102

Methodology

Laterite Soil

Laterite soil collected from Orissa region at 1.0 m depth is used for the study. After collecting, the soil is dried for 2 weeks. The properties of the laterite soil assessed based on relevant I.S. Code provisions are given in Table 32.1.

Metakaolin

Metakaolin is a high quality pozzolonic material produced by process of calcination from pure kaolinite clay temperatures between 650 and 850 °C. Once the burning process gets completed, it is properly grinded to desired fineness to improve the properties of soil. It has high reactivity compared to other pozzolana due to presence of aluminosilicate pozzolana that is rich in silica and alumina (Table 32.2).

Table 32.2 Properties of metakaolin

Property	Value
Moisture content	0.18%
Specific gravity	2.65
Bulk density (kg/m^3)	710
pH	7.0

Table 32.3 Properties of palm oil fuel ash

Property	Value
Specific gravity	1.96
Sand size particles	71
Silt and clay size particles (%)	29
Max. dry density (g/cc)	1.3
Optimum moisture content (%)	14.1

Palm Oil Fuel Ash (POFA)

POFA is also an industrial waste results due to burning processes of the palm waste used in the palm oil industry [3]. It has a good content of the pozzolanic materials represented with the silicates and aluminates which can be promising for the pozzolanic reaction due to the glassy phase of its silicates [4, 5]. The palm oil fuel ash is collected from Ruchi industries, Samalkot, A.P (Table 32.3).

Nylon Fibers

It is the fiber that is used in several applications and has always been in great demand. Nylon is the first ever fiber that is completely synthetic. Nylon fibers used in this study are in threaded form. For the present study, nylon fibers were used as reinforcement in laterite soil. The nylon fibers used in this study was purchased from the local factory.

Results and Discussion

Various laboratory experiments were conducted by combination of various percentages of metakaolin and palm oil fuel ash in the lateritic soil. Compaction, California bearing ratio, and unconfined compressive strength tests were conducted with a view to determine the optimum combination of metakaolin and POFA and further with nylon fiber as reinforcement. CBR and UCS are conducted for curing studies.

Effect of Metakaolin and POFA in Improving the Compaction Properties of Laterite Soil

As per IS: 2720 (Part VII), 1980 [6] Standard Proctor Compaction test was conducted on soil. The test is conducted on soil alone and soil with combination of POFA and metakaolin. The compaction test is done immediately after treating it with the stabilizer. The test results of compaction test with soil alone and soil with combination of palm oil fuel ash (POFA) and metakaolin dosages were shown in the graphs in Figs. 32.1 and 32.2. However, the maximum dry density values increased with increase in metakaolin and POFA content, because the soil with lower specific gravity was replaced by higher specific gravity of metakaolin material. The increase in maximum dry density also be attributed to the flocculation and agglomeration of the clay particles due to cation exchange.

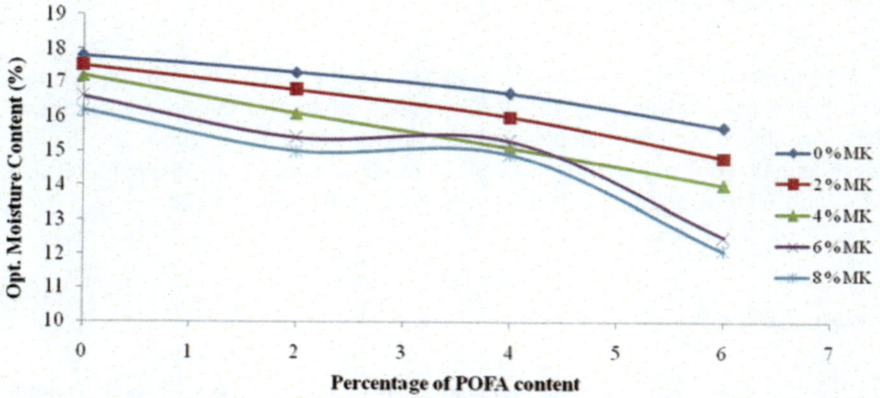

Fig. 32.1 The variation in optimum moisture content with percentage of POFA content

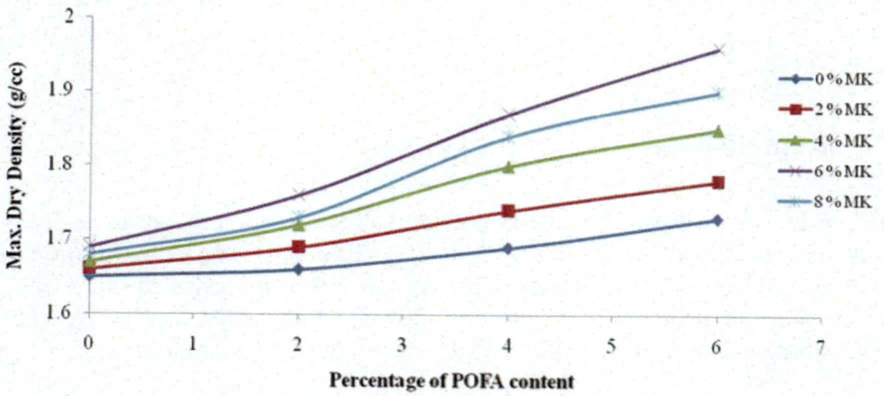

Fig. 32.2 The variation in maximum dry density with percentage of POFA content

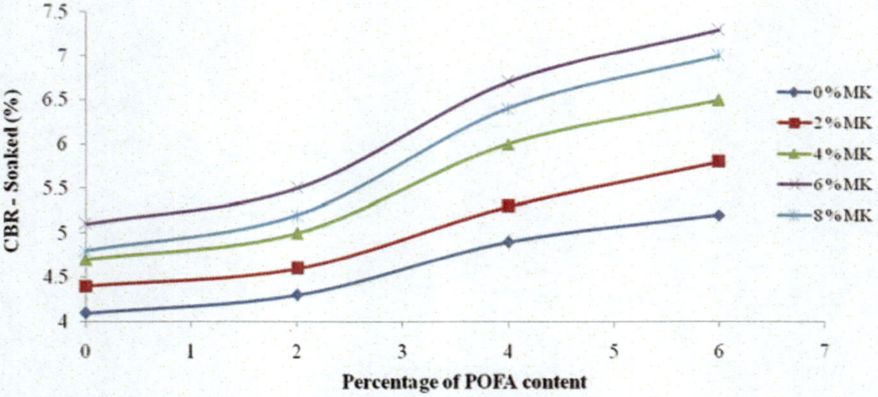

Fig. 32.3 The variation in CBR values with percentage of POFA content

Effect of Metakaolin and POFA on the California Bearing Ratio (Soaked) Values

The CBR curves for different mix proportions of soil mixed with metakaolin and POFA for soaked CBR graphs for all mixes and corresponding values are shown in Fig. 32.3. From the Fig. 32.3, addition of metakaolin and POFA has shown increment in CBR values. From above figures, it can conclude that laterite soil when combined with 6% POFA and 6% metakaolin had shown more pronounced improvement. This development in strength characteristics may be attributed to the secondary cementitious substances as a result of reaction between soil, POFA, and metakaolin. This reaction also contributed to interparticle bonding.

Effect of Metakaolin and POFA on the Unconfined Compressive Strength of Laterite Soil

Figure 32.4 shows the change of UCS values with different percentage of combination of POFA and metakaolin. From the Fig. 32.4, addition of metakaolin and POFA has shown increment in UCS values. From above figures, it can conclude that laterite soil when combined with 6% POFA and 6% metakaolin had shown more pronounced improvement. The observed increase in all the UCS values was primarily due to the reaction between the calcium and the pozzolana in POFA and metakaolin to form a compound known as (CASH) with time the compound is gradually forms in to a well-crystalline phase of CSH and CAH, these two compound hardens with time to form external compound that blind the soil particle. The gradual decrement on further increment of 6% metakaolin and 6% POFA content was as a result of increased

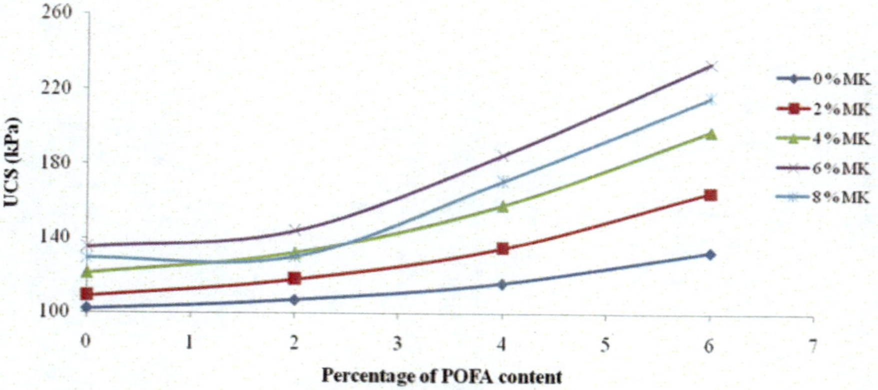

Fig. 32.4 Variation in UCS values with percentage of POFA content

surface area caused by excess metakaolin and POFA content, thereby making the mixture to require more water for hydration process.

From the above results, the combined optimum content of POFA + MK for improving the weak laterite soil (WLS) is 6% POFA + 6% MK. Since the CBR requirement as per IRC 37-2012 which is soaked CBR of 8% is not achieved for the optimum combination, use of nylon fibers is chosen and added to the optimum combination samples as discrete reinforcing elements to enhance the strength and penetration characteristics.

Effect of Nylon Fiber Added to the Optimum Combination on Compaction Properties of Laterite Soil

From the Figs. 32.5 and 32.6, the addition of nylon fiber has shown increment in MDD values from 1.96, 1.97, 1.98, 2.01, 2.00, 1.99 g/cc at nylon fiber percentage of 0%, 0.3%, 0.6%, 0.8%, 1.2%, and 1.5%, respectively, and OMC values decreased from 12.5%, 12.4%, 12.3%, 12.2%, 12.1%, and 12% for the same percentage of nylon fiber. This may be due to the addition of fiber strengthens the capacity of the laterite soil against compression forces. Further, beyond about 0.8% nylon fiber content, which there is a drops in values.

Effect of Nylon Fiber Added to the Optimum Combination on CBR and UCS Values of Laterite Soil

From the Figs. 32.7 and 32.8, the addition of nylon fiber has shown increment in CBR values from 7.3%, 7.5%, 7.8%, 8.2%, 7.9%, 7.6% at nylon fiber percentage of

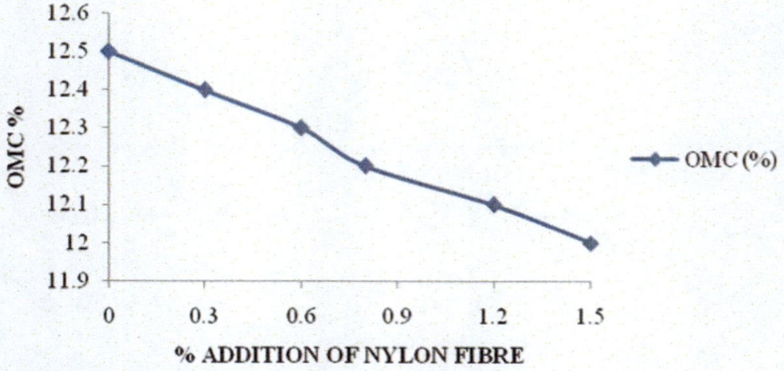

Fig. 32.5 Variation of OMC for laterite treated with different percentages of nylon fiber

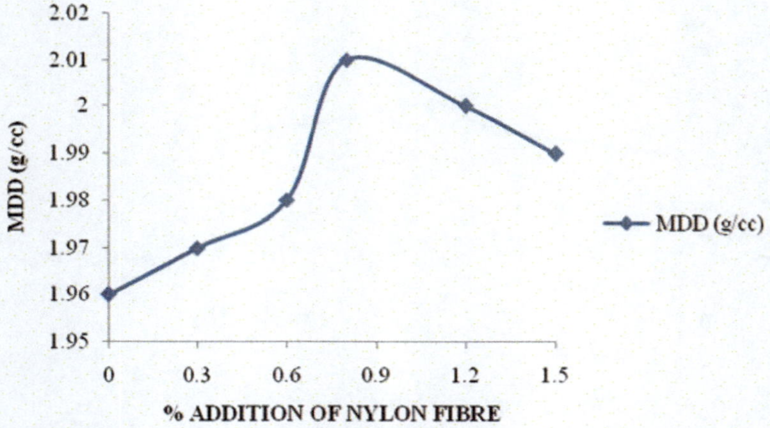

Fig. 32.6 Variation of maximum dry density for laterite soil treated with different percentages of nylon fiber

0%, 0.3%, 0.6%, 0.8%, 1.2%, and 1.5%, respectively, and UCS values also increased from 234, 247, 263, 289, 270, 258 kPa for the same percentage of nylon fiber. From the results, it can be show that the UCS is directly proportional to the nylon fiber content up to the maximum stress, further which the unconfined compressive strength reduction in value with increase in nylon fiber content.

From the above results, the combined optimum content of POFA + MK + NF for improving the weak laterite soil (WLS) is 6% POFA + 6% MK + 0.8% NF. Since the CBR requirement as per IRC 37-2012 which is 8% is achieved.

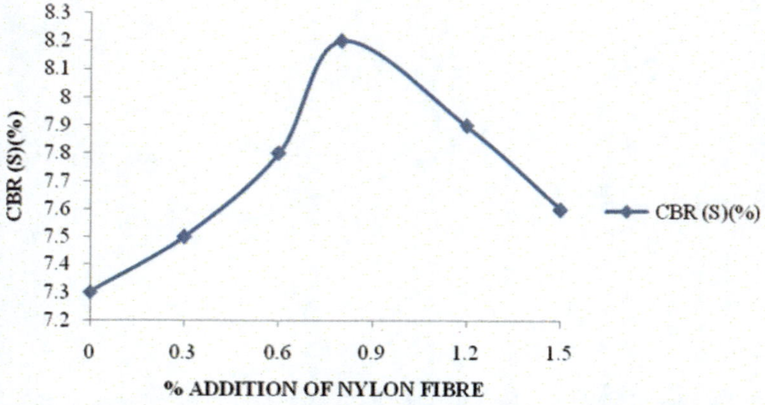

Fig. 32.7 Variation of soaked CBR values for laterite soil with different percentages of nylon fiber

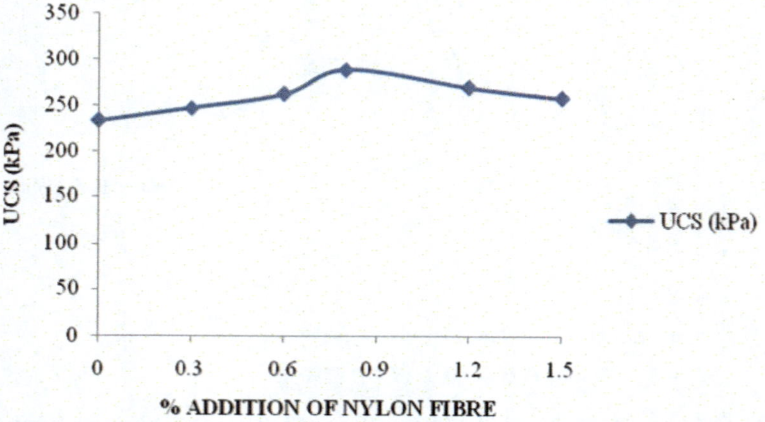

Fig. 32.8 Variation of soaked UCS values for laterite soil with different percentages of nylon fiber

Effect of Curing on Samples Prepared with Combined Optimum Content of POFA + Metakaolin + Nylon Fiber

Figures 32.9 and 32.10 show the variation of penetration and shear characteristics for different curing periods. From above figures, it can conclude that laterite soil when cured with 28 days had shown more pronounced improvement.

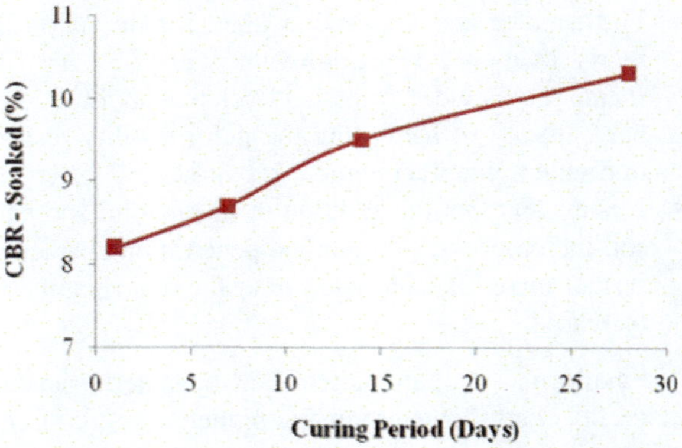

Fig. 32.9 The variation in CBR values with different curing periods

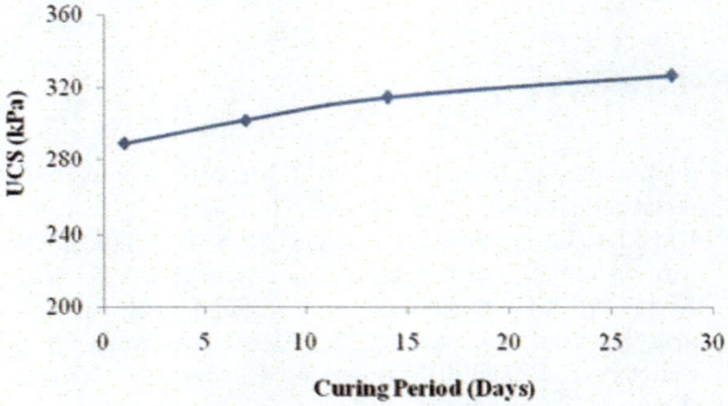

Fig. 32.10 The variation in UCS values with different curing periods

Conclusions

Metakaolin and palm oil fuel ash treated soil were examined intensively and drawn the following conclusions.

- It was observed that laterite soil treated with metakaolin and POFA has moderately improved the laterite soil.
- The compaction characteristics of the soil showed increase in MDD and decrease in OMC with subsequent addition of POFA and metakaolin content.
- It is observed that for the addition of 6% POFA + 6% MK, there is a gradual increase in maximum dry density about 18.79%.
- There is an improvement in strength characteristics with POFA (%) + metakaolin (MK) (%) as added to the weak laterite soil (WLS). There is an improvement of 78.05% in CBR and 129.05% in UCS values. A strength gain of the soil-metakaolin POFA mixtures shows that admixture has a long duration strength improving ability.
- The compaction, CBR, and UCS tests indicated that at 6% metakaolin and at 6% palm oil fuel ash content, the values are more pronounced.

- Further blending with nylon fiber, for the nylon fiber content of 0.8%, MDD increases about 2.55% improvement of 12.3% in CBR and 23.50% in UCS values. It can be concluded from this work that addition of nylon fiber to the lateritic soil has capacity of increasing the deformability resistance ability of the soil. The optimum nylon fiber content is 0.8%.
- It is evident that the addition of metakaolin and palm oil fuel ash to the laterite soil increases the strength and penetration characteristics to some extent and on further mixed it with nylon fiber the deformability resistance ability of the soil increased.

Finally, it can be summarized that the materials metakaolin, palm oil fuel ash, and nylon fiber had shown promising influence on the strength characteristics of laterite soil.

References

1. Konstantinos G, Kolovos et al (2013) Mechanical properties of soilcrete mixtures modified with metakaolin. Constr Build Mater 47:1026–1036
2. Umar SY, Elinwa AU, Matawal DS (2015) Hydraulic conductivity of compacted lateritic soil partially replaced with metakaolin. J Environ Earth Sci 5(4)
3. Aprianti E (2017) A huge number of artificial waste material can be supplementary cementitious material (SCM) for concrete production—a review part II. J Cleaner Prod 142:4178–4194
4. Calligaris GA et al (2015) Assessing the pozzolanic activity of cements with added sugar cane straw ash by synchrotron X-ray diffraction and rietveld analysis. Constr Build Mater 98:44–50
5. Jafer H, Atherton W, Sadique M, Ruddock F, Loffill E (2018) Stabilisation of soft soil using binary blending of high calcium fly ash and palm oil fuelash. Appl Clay Sci 152:323–332
6. IS 2720-Part VII (1980) Methods of test for soils—determination of water content—dry density relation using light compaction, Bureau of Indian Standards, New Delhi

Chapter 33
Amelioration of Strength Characteristics of Expansive Soil Treated with Calcium Chloride and Terrasil

K. Ramu, R. DayakarBabu, and K. Abhiram

Introduction

In India, swelling soils are also called as "Expansive soils or Black cotton soils." Any soils have swelling and shrinkage nature, it depends upon its presence of clay minerals. There are four types of clay minerals mainly montmorillonite, illite, kaolinite, and halloysite. The soil is showing expansive or shrinkage nature due to presence of basic mineral montmorillonite.

The structures that are built on these types of soils are severely damaged due to differential settlement in a structure. The damage of structure caused by expansive soil has been reported from different locations in the clay plain. So, there is a need to study and find the problems due to expansive soils and prevent damage to the structures. Soil stabilization with various additives is one of the techniques which reports lot of success for the improvement in engineering properties of soil. The stabilization with pozzolana and especially with silica-rich admixtures can be explained with the formation of gelation process when mixed with the cementing agents like lime, etc. The present day stabilization is focused on the use of nanosilica like terrasil. In the present work, laboratory experimentation has been made to check the effect of different percentages of calcium chloride and terrasil on strength properties of expansive soil.

K. Ramu
Department of Civil Engineering, UCEK, JNTUK, Kakinada 533003, India

R. DayakarBabu (✉)
Kakinada 533003, India
e-mail: rdbgkk25@gmail.com

K. Abhiram
Department of Civil Engineering, UCEK, JNTUK, Kakinada 533003, India

K. Muthukkumaran et al. (eds.), *Ground Improvement Techniques*, Lecture Notes in Civil Engineering 297, https://doi.org/10.1007/978-981-19-6727-6_33

Review of Literature

Expansive soils are causing severe damages to the structures such as buildings and pavements built over them due to their high degree of swelling and shrinkage. This phenomenon will result in a rutting mode of failure requiring immediate strengthening. Thus, for safe design, such soils need to be improved before construction. The study discusses the potential use of calcium chloride and terrasil in the field of geotechnical engineering.

Nandan et al. [1] investigated on the usage of terrasil for soil stabilization. It was found that with the addition of terrasil, liquid limit and plastic limit decrease. Permeability and free swell index value reduce considerably. This study has shown that soil treated with (0.041%) of terrasil decreases thickness up to 2%. Cost of construction also reduces. Johnson [2] done an investigation on the use of terrasil for stabilization of soil and concluded that plasticity index decreases with increase in dosage of terrasil and also there is a decrement in permeability with the increment in the dosage of terrasil into the expansive soil.

Rajshekhar [3] carried out a research on the ways to improve sub-grade properties of pavement by the chemical terrasil. The study concludes that there is an increase in CBR value, reduction in sub-grade thickness, plasticity index reduced, and dry density increases.

Methodology

Expansive Soil

Expansive soil was taken from Turupulanka near Amalapuram at a depth of 2 m below the ground level. The soil is dark gray to black in color. The taken soil was air dried, pulverized, and passing through 4.75 mm IS sieve was used. The properties of the expansive soil are given in Table 33.1.

Calcium Chloride

Calcium chloride was collected from M/s. Sri Saibalaji enterprises Ltd. (chemical dealers), Vijayawada. The properties of calcium chloride are furnished in Table 33.2.

Table 33.1 Properties of expansive soil

S. No	Property	Value
1	Specific gravity	2.61
2	Differential free swell index (%)	120
3	Liquid limit (%)	72.57
4	Plastic limit (%)	30.43
5	Plasticity index (%)	42.14
6	Sand size particles (%)	8.0
7	Silt and clay size particles (%)	27 and 65
8	IS soil classification	CH
9	Max. dry density (g/cc)	1.42
10	Optimum moisture content (%)	28.24
11	CBR—soaked (%)	1.8
12	CBR—unsoaked (%)	3.1
13	Unconfined compressive strength (kN/m^2)	38.8

Table 33.2 Properties of calcium chloride

Calcium chloride	$Cacl_2$
Molecular weight	110.98gm/mole
Density	2.15 gm/cm3
Boiling point	1935^{0}C
Melting point	772^{0}C

Terrasil

Road surfacing cracking caused by expansive soil can be reduced the use of terrasil, and the patented molecule present in the terrasil forms a bond with expansive soil and also forms a water repellent nanolayer and makes the roads or soil impermeable to water (Table 33.3).

Table 33.3 Properties of terrasil

Chemical name	Terrasil
Physical appearance	Pale yellow
Specific gravity	1.01
Solid content	68 + 2%
Solubility forms	White clean solution
Flash point	Flammable at 12 °C

Results and Discussions

The collected soil was air dried, pulverized manually, and soil passing through 4.75 mm IS sieve was used. Studies were carried out for salient geotechnical characteristics in the laboratory. Compaction, California bearing ratio, and UCS tests were conducted by using various percentages of terrasil and calcium chloride mixed with expansive soil materials with a view to find the optimum percentage.

Effect of Calcium Chloride and Terrassil on the DFS Values

For finding the differential free swell value for the sample used in the investigation IS Code of practice [IS:2720 (Part-40): 1977 was followed. DFS values decreased continuously upon the addition of calcium chloride and terrasil as shown in Figs. 33.1 and 33.2.

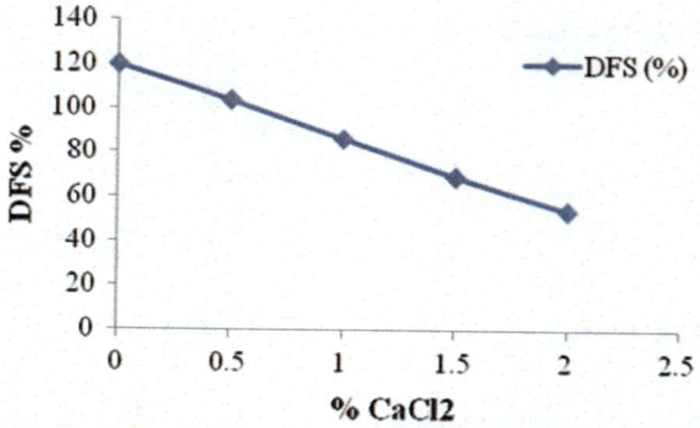

Fig. 33.1 Variation DFS with different percentage of calcium chloride

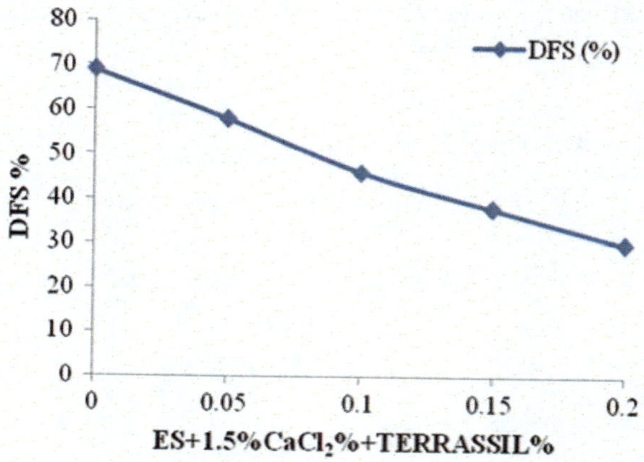

Fig. 33.2 Variation of DFS of expansive soil blended with optimum percentage

From Figs. 33.1 and 33.2, addition of calcium chloride and terrassil has shown decrement in the value of DFS from 69% to 30. % in addition of calcium chloride in proportions of 0%, 0.5%, 1%, 1.5%, 2% and terrassil in proportions of 0%, 0.05%, 0.10%, 0.15%, 0.20%.

Calcium chloride and different percentage of terrassil.

Effect Calcium Chloride and Terrassil on the Atterberg's Limit Values

To know the index properties of the samples, IS Code of practice [IS: 2720 (Part-5)—1985 [4]; IS: 2720 (Part-6)—1972 [5]] was followed. Liquid limit values were continuously decreased and plastic limit values were increased with increasing percentage of calcium chloride, which resulting in decreased PI as shown in the Figs. 33.3 and 33.4.

From the Fig. 33.3, addition of calcium chloride has shown decrement in liquid limit value from 72.57%, 70.14%, 57.45%, 66.82%, and 62.60% at with 0%, 0.5%, 1.0%, 1.5%, and 2.0% calcium chloride, respectively, and plastic limit values increased from 30.43%, 32.75%, 33.51%, 38.07%, and 39.80% for the same percentages of calcium chloride; eventually decreasing the plasticity index values of expansive soil.

From the Fig. 33.4, the addition of 1.5% calcium chloride with terrassil in increments of 0%, 0.05%, 0.10%, 0.15%, and 0.20% has shown decrement in liquid limit value from 72.57%, 60.70%, 57.45%, 54.82%, and 54.35%, respectively, and the plastic limit values increased from 39.13%, 40.14%, 40.81%, and 41.16% for the

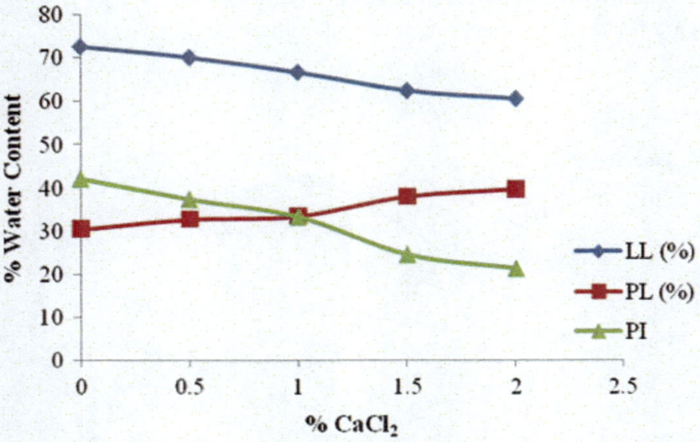

Fig. 33.3 Variation of LL, PL, and PI of expansive soil blended with different percentage of calcium chloride

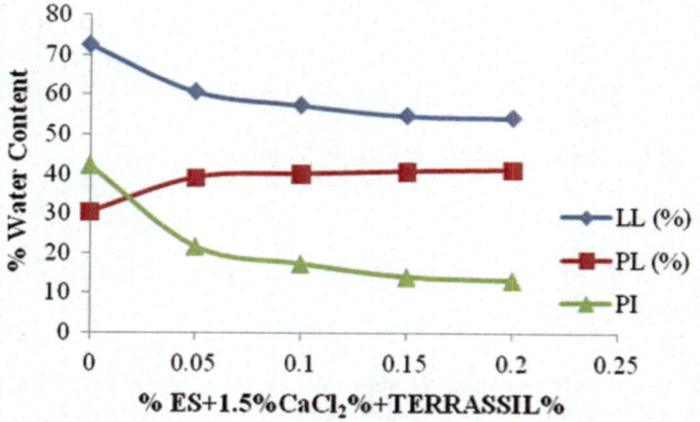

Fig. 33.4 Variation of LL, PL, and PI of expansive soil blended with optimum percentage calcium chloride and different percentage of terrassil

same percentages of terrassil eventually decreasing the plasticity index values of expansive soil.

Effect of Calcium Chloride and Terrasil on Compaction Characteristics

From the Fig. 33.5, the addition of calcium chloride has shown increment in MDD values from 14.18%, 14.92%, 15.74%, 16.32%, 15.97 at $CaCl_2$ of 0%, 0.5%, 1%, 1.5%, 2%, respectively, and OMC values decreased from 28.24, 27.11, 25.60, 23.90, and 23.55 g/cc for the same percentage of calcium chloride.

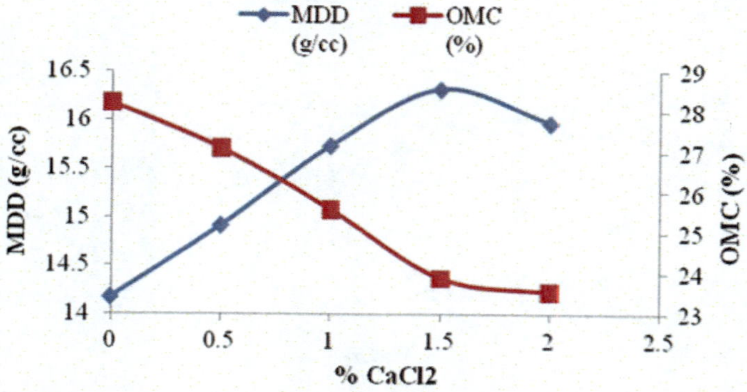

Fig. 33.5 Variation of OMC and MDD for expansive soil treated with different percentage of calcium chloride

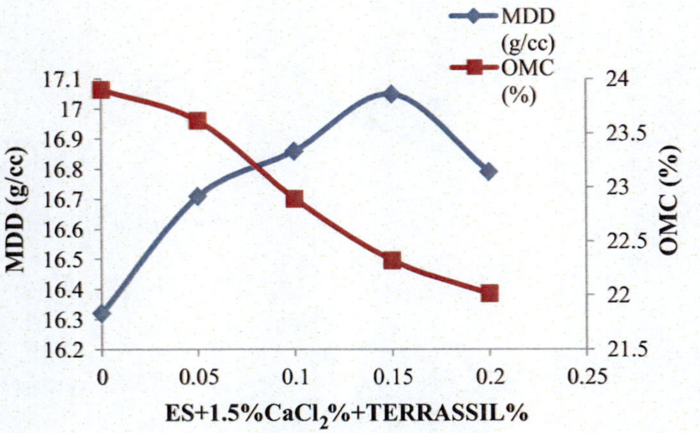

Fig. 33.6 Variations of MDD and OMC for expansive soil treated with optimum

From the Fig. 33.6, the addition of 1.5% calcium chloride with terrassil in increments of 0%, 0.05%, 0.10%, 0.15%, and 0.20% has shown decrement in OMC values from 23.90%, 23.60%, 22.89%, 22.32%, and 22.01%, respectively, and MDD values increased from 16.32%, 16.71%, 16.86%, 17.05%, and 16.79% for the same percentages of terrassil.

Percentage of calcium chloride with different percentages of terrassil.

Effect of Calcium Chloride and Terrassil on California Bearing Ratio (Soaked) Values

The CBR curves for different mix proportions of soil and calcium chloride are 0%, 0.5% 1.0%, 1.5%, 2.0%, respectively, for soaked CBR graphs for all mixes and corresponding values are shown in Figs. 33.6, 33.7 and 33.8. From the Fig. 33.7, addition of calcium chloride has shown increment in CBR values from 1.80%, 2.90%, 4.40%, and 6.70% at with 0%, 0.5%, 1.0%, 1.5% and calcium chloride, respectively.

From the Fig. 33.8, the addition of 1.5% calcium chloride with terrassil in increments of 0.05%, 0.10%, 0.15%, and 0.20% has shown increment in CBR values from 6.70%, 7.60%, 8.20%, and 8.90%, respectively.

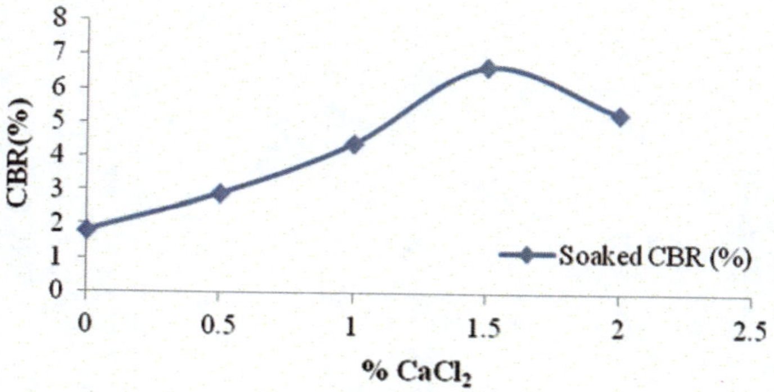

Fig. 33.7 Variation of soaked CBR values for expansive soil with different percentages of calcium chloride

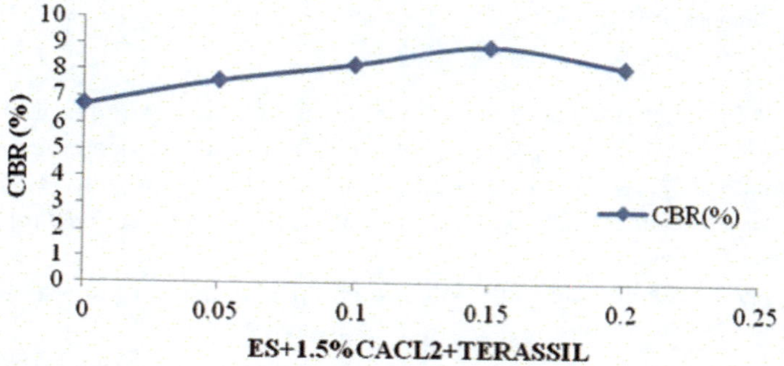

Fig. 33.8 Variation of soaked CBR values for expansive soil treated with optimum percentage of calcium chloride with different percentages of TERRASSIL

Effect of Calcium Chloride and Terrassil on Unconfined Compressive Strength

Unconfined compressive tests were conducted on the samples prepared at MDD and optimum moisture content. From the Fig. 33.9, addition of calcium chloride has shown increment in UCS values from 74, 96, 120, and 155.0 kPa at 0%, 0.5, 1%, 1.50%, and calcium chloride, respectively.

From the Fig. 33.10, the addition of 1.5% calcium chloride with terrassil in increments of 0%, 0.05%, 0.10%, and 0.15% has shown increment in UCS values from 155.0, 181.0, 214.0, and 252.0 kPa, respectively.

From the above results, it can be clearly established that the use of $Cacl_2$ along with terrasil gave a better binding effect in the expansive soil and ultimately enhancement of its strength behavior with the formation of C-S-H gel.

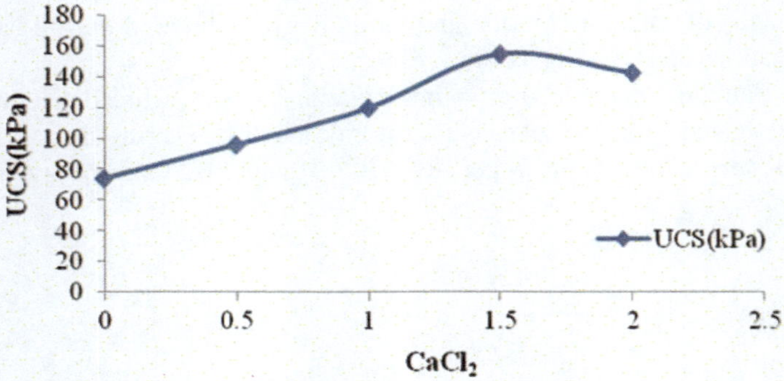

Fig. 33.9 Variation of unconfined compressive strength of expansive soil treated with different percentage of calcium chloride

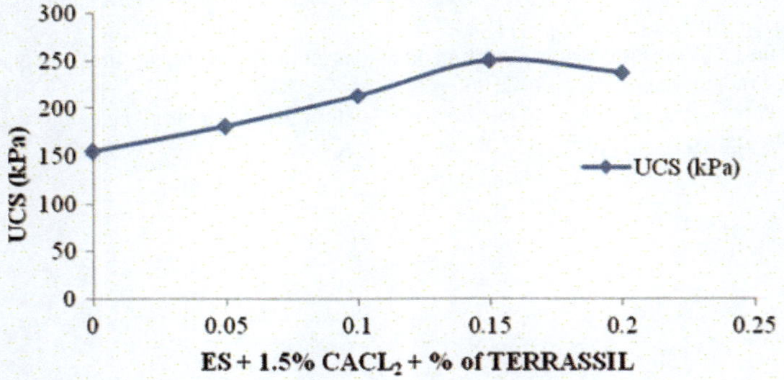

Fig. 33.10 Variation of unconfined compressive strength of expansive soil treated with 1.5% calcium chloride with different percentages of terrassil

Conclusions

Calcium chloride and terrasil treated soil were examined intensively and drawn the following conclusions.

- Addition of calcium chloride and terrassil decreased the DFS value by 75% for the expansive soil. It is evident that liquid limit value decreased by 27.61% and plastic limit value increased by 36.18% and plasticity index value decreased by 67.28% on treating with calcium chloride and terrassil.
- It is clear that OMC value decreased by 25% and MDD value increased by 15% for the expansive soil treated with calcium chloride and terrassil. It is evident that the CBR and UCS value improved by 77.7% and 69%, respectively, for the expansive soil.
- Overall, it can be concluded that calcium chloride and stabilized soil can improve the strength characteristics considerably on expansive clay beds. Optimum

percentages of calcium chloride and terrassil were observed during laboratory investigations are 1.5% and 0.10%, respectively.

- Hence, from the present laboratory investigations, it was included that the expansive soil treated with 1.5% of calcium chloride and 0.10% terrassil as an optimum exhibits satisfactory results as per IRC 37-2001 and 2012 Code of practice.

References

1. Patel NA, Mishra CB, Pancholi VV (2015) Scientifically surveying the usage of Terrasil chemical for soil stabilization. Int J Res Advent Technol 3(6):77–84
2. Johnson R, Rangaswamy K (2015) Improvement of soil properties as a road base material using nano chemical solution. In: 50th IGC, Pune, Maharashtra, India
3. Rathod RG (2017) Efficient way to improve subgrade property of pavement by chemical stabilization. Int J Eng Res Appl 7(1):83–96
4. IS 2720-Part 5 (1985) Methods of test for soils—determination of liquid limit and plastic limit, Bureau of Indian Standards, New Delhi
5. IS 2720-Part 6 (1972) Methods of test for soils—determination of shrinkage factors, Bureau of Indian Standards, New Delhi

Chapter 34
Effect of Sawdust and Sawdust Ash on Expansive Soil

Divyanshu Algotar, Sabbasachi Saha, Rajesh P. Shukla, and Prabir Kumar Basudhar

Introduction

Expansive soils, locally known as black cotton soil, are found in many parts of India [1]. The color black is attributed to the presence of high iron, magnesium minerals, and humus. Expansive soils contain a large quantity of montmorillonite and illite minerals, due to which these soils swell and shrink with alternate wetting and drying in the extreme climatic environment. Due to charge deficiency in the clay minerals, a large amount of water can enter the soil in the rainy season, and the weak bond between soil elements brakes causing very large expansion [2]. In the summer, due to drying, there is shrinkage as the water present in the soil pores is expelled, resulting in the development of cracks that cause distress to roads, buildings, and other civil engineering structures [3].

Due to the rapid infrastructure development and industrialization of various developing countries, lots of waste products arising out of these developmental activities are generated. Agarwal et al. [4] presented a study discussing the scenario of waste produced in India. A similar scenario exists in almost all developing countries. Though the waste management policy is available at the municipal level, there is no national policy or guideline to store and utilize these waste materials for sustainable development. Some waste materials are hazardous in nature, but some waste material is non-hazardous which can be used as a construction material for sustainable development [5]. These materials include fly ash, sawdust, sawdust ash, metal slags, alccofine, ground granulated blast furnace slag, natural and synthetic fibers, marble,

D. Algotar · S. Saha
Adani Institute of Infrastructure Engineering, Ahmedabad 382421, India

R. P. Shukla (✉)
National Institute of Technology Srinagar, Srinagar 190006, India
e-mail: rpshukla.2013@iitkalumni.org

P. K. Basudhar
Indian Institute of Technology Kanpur, Kanpur 208016, India

K. Muthukkumaran et al. (eds.), *Ground Improvement Techniques*, Lecture Notes in Civil Engineering 297, https://doi.org/10.1007/978-981-19-6727-6_34

and other stone dust [6, 7]. The use of waste as resource material in constructional activities may save the cost of waste treatment and save expenditure on construction material.

Some studies explored the application of sawdust/sawdust ash as a potential replacement for various construction materials. Especially in countries and areas where wood is available in abundance and used in constructing wooden houses and making wooden furniture, plenty of such sawdust is available. Such sawdust can be used as ingredients in developing new composite material binding those with glues and reused. But in places where such industries are not present, it can be utilized for the purpose as mentioned. Some studies are available in the literature about the use of sawdust ash in the concrete as an admixture and various other materials [8–10].

Similarly, some studies used sawdust ash also to stabilize soil also [11–13]. The literature shows that very few recent studies studied the effect of sawdust ash in mortar as a filling material and in soil modification. The study is intended to determine the effect of sawdust and sawdust ash, especially on locally available expansive soil in Gujrat, India.

The Material Used and Experimental Testing Methodology

The soil is collected from the Kheda, Gujrat. The Gujrat state is the western state in India. The expansive soils cover almost 30% of the land cover of the state. The sawdust is collected from the waste disposal site of Adani Institute of Infrastructure Engineering, Ahmedabad. The sawdust and sawdust ash as used in the study is shown in Fig. 34.1. The grain size distribution of sawdust, sawdust ash, and soil is shown in Fig. 34.2. These materials were produced during construction work at the institute. The characteristics of used expansive soil are presented in Table 34.1.

All the properties are determined for untreated soil and treated soil following the guidelines given in Indian standards. The liquid limit (LL) of untreated and modified soil was determined by the Casagrande apparatus. The plastic limit (PL) was

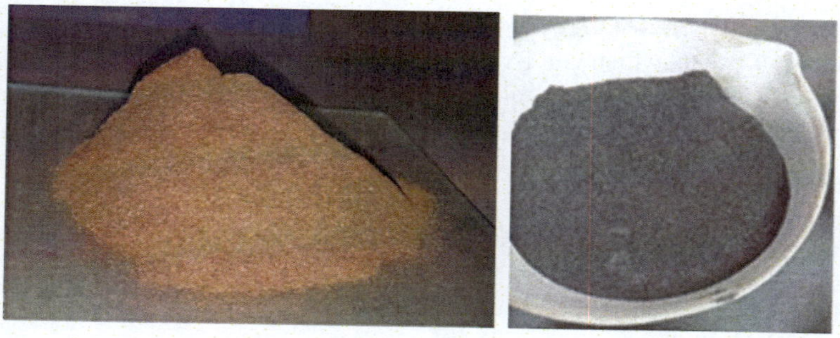

(a) (b)

Fig. 34.1 Materials used in the study **a** sawdust **b** sawdust ash

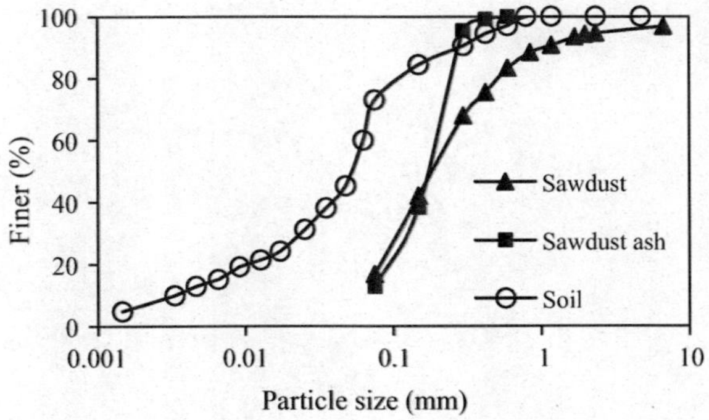

Fig. 34.2 Grain size distribution plot

Table 34.1 Characteristics of expansive soil used in the study

Properties	Description
Liquid limit (LL)	56–62%
Plastic limit (PL)	24–28%
Plasticity index (PI)	30–34%
Shrinkage limit (SL)	12–14%
Optimum moisture content (OMC)	23–25%
Specific gravity (G)	2.48–2.54
Maximum dry density ($\rho_{d\ max}$)	1.53 gm/cm^3

assessed by the most commonly adopted thread rolling method. The comprehensive procedure to determine these index properties is specified in IS 2720: Part-V [14]. The shrinkage limit was evaluated by means of the popular mercury displacement method as specified in IS: 2720: Part-VI [15]. An adequate time-lapse was permitted before testing for uniform mixing of water. As per IS classification laid down by IS 1498 [16], the untreated soil is classified as clayey soil with medium plasticity (CI). The unconfined compressive strength of soil was also determined. The tests were conducted on soil samples (38 mm diameter and 76 mm length) following provisions narrated in IS 2720: Part X [17]. All soil samples were prepared at the optimum moisture content of the soil. The loading was applied at the strain rate of 1.25 mm/min. The soil was tested to check for its susceptibility to swell. The differential free swell of soil was determined for various additives. The tests were performed following the procedure stated in the Indian standard code [18]. The instruments and apparatus used in the study are shown in Fig. 34.3.

Results and Discussion

The liquid limit (LL), plastic limit (PL), and plasticity index (PI) are crucial index properties to characterize and assess engineering properties and to have a general

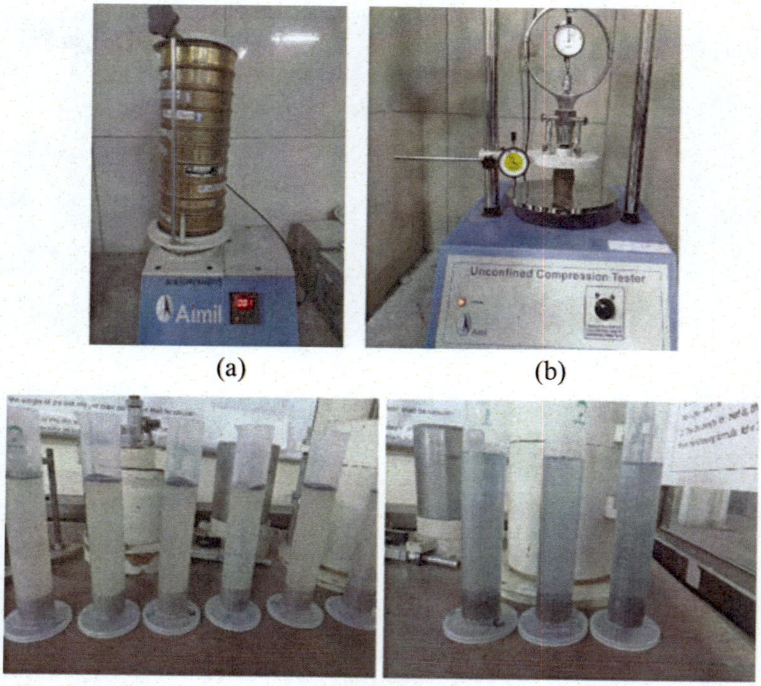

(a) (b)

(c)

Fig. 34.3 **a** Grain size analysis, **b** UCS test, **c** free swelling index test

preliminary idea about the soil behavior. The change in these limits means a change in soil consistency, which is generally reflected in the engineering behavior of soil.

Figure 34.4 depicts the variation in the liquid limit (LL), plastic limit (PL), and plasticity index (PI) of soil addition of sawdust and sawdust ash. It is found that the LL and PL of soil increase with the addition of sawdust and sawdust ash as these ingredients have the capacity to absorb water, which increases the liquid and plastic limits. The increase in plastic limit is more compared to the liquid limit. The effect of sawdust ash is relatively more visible compared to sawdust. Contrary to LL and Pl, change in the PI does not follow any specific trend. The PI reduces with the addition of sawdust and sawdust ash (9% ash content), a favorable change (Fig. 34.4c). However, it should be noted (Fig. .34.4b) that the values of PL do not change when the water content increases from 0 to about 6%, beyond which it increases until the ash content is 9%, after which the same decreases. This is reflected in Fig. 34.4c, which shows that from a 6 to 9% increase in ash content; there is a decrease in PI value and then showing an increase in PI value as a consequence of the decrease in PL values. The decreased values of PL and PI indicate that soil will still be in the plastic state for the reduced moisture content range. Earlier, this range was found to be 32%, which reduces to 22% and 27% with the addition of sawdust and sawdust ash, respectively.

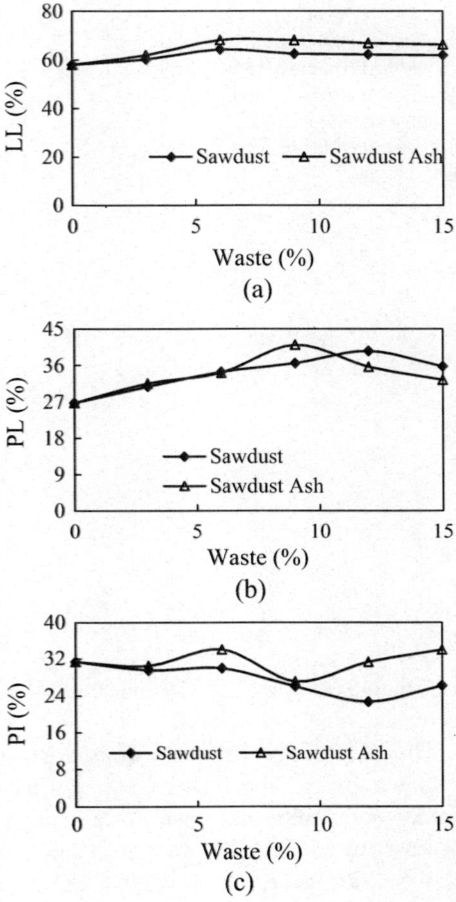

Fig. 34.4 Effect of sawdust and sawdust ash on Atterberg's limits: **a** liquid limit, **b** plastic limit, **c** plasticity index

The addition of coarser particles of additives (Fig. 34.4) reduces the PI due to the complex action of these materials absorbing water, and sawdust ash may be acting as a pozzolanic material reducing the double diffuse layer and probably also acting as a bonding material [1, 7, 19]. However, sawdust and sawdust ash are not very reactive, especially in the short duration reaction between soil, and these additives are not possible. It might be a reason for the lower efficiency of sawdust and sawdust ash. The addition of sawdust and sawdust ash changes soil classification from CH to MH.

Figure 34.5 (a, b) shows the variation in the stress–strain relationship with an increase in sawdust and sawdust ash, respectively. For any sawdust content, the peak stress is found to be more than the untreated soil. However, the strain corresponding to peak stress is found to decreases with the addition of sawdust. The strain corresponding to peak failure reduces from 9 to 6% with the addition of additives. The post-peak behavior changes significantly with the addition of sawdust on the

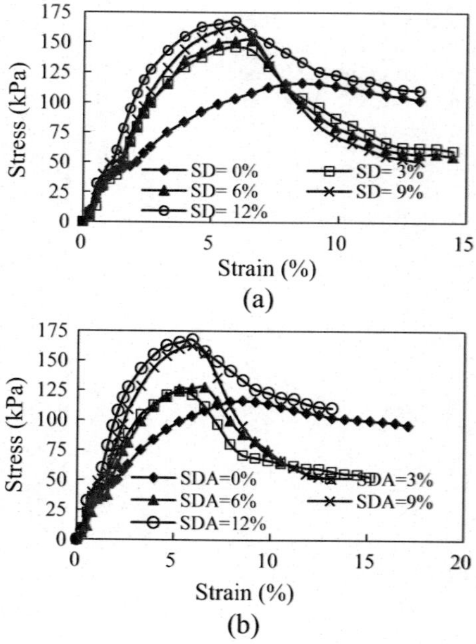

Fig. 34.5 The effect of sawdust and sawdust ash on unconfined compressive behavior: **a** stress–strain in sawdust added to soil, **b** stress–strain with sawdust ash

soil. Figure 34.5(a, b) shows that the soil behavior changes from ductile to brittle, increasing dust and dust ash.

The effect of sawdust and sawdust ash on unconfined compressive strength (UCS) is shown in Fig. 34.6. The UCS of soil increases linearly up to sawdust of 9%. A similar observation has been made in the case of sawdust ash also. The strength increases by 55% for 9% of sawdust and sawdust ash. However, for any content less than 9%, sawdust is relatively more effective. Effect of sawdust as well sawdust ash becomes minimal for any quantity greater than 9%. The increase in the strength is 45%, 50%, 60%, and 65% corresponding to sawdust content of 3%, 6%, 9%, and 12%, respectively. The increase in soil strength may be attributed to the enhanced grain size distribution in modified soil and decreased soil plasticity [20, 21]. The soil-additive matrix improves the index properties with the additive content with a consequent increase in the compressive strength. However, the long-term effect of sawdust and sawdust need to be investigated in the future studies as especially the sawdust being an organic matter may decompose, and there may be deterioration in the engineering behavior of the composite over a period of time.

Figure 34.7 shows the variation in failure of soil samples treated with sawdust (SDA) and sawdust ash (SDA). The failure surface becomes more evident with the increase in sawdust and sawdust ash because the soil behavior changed from ductile to brittle. With increased additives, the samples showed brittle failure, which is consistent with the earlier observation with respect to their stress–strain behavior shown in Fig. 34.5.

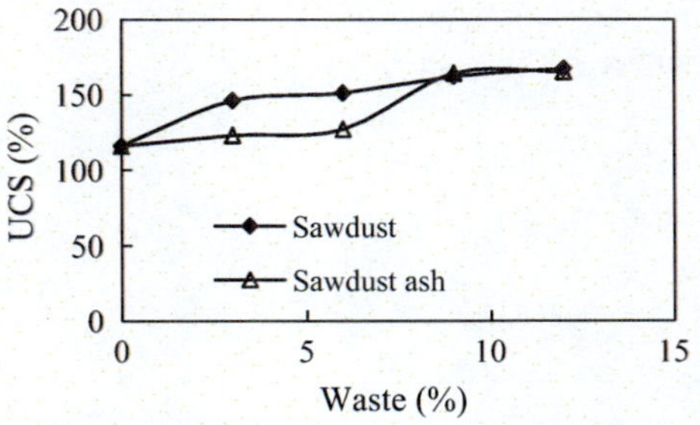

Fig. 34.6 Effect of sawdust and sawdust ash on unconfined compressive strength

Fig. 34.7 Change in failure of soil samples with **a** sawdust (SD), **b** sawdust ash (SDA)

The effect of the additives on the swelling behavior of soil is studied and shown in Fig. 34.8. The free swelling index of soil initially decreases with an increase in sawdust. The free swelling index reduces from 65 to 40% when sawdust content is 9%, but, in contrast, the free swelling index reduces to 23% for sawdust ash content of 15%, which is relatively more. For any number of additives greater than 9%, the change in the free swelling index is almost negligible. The mixing of non-plastic sawdust and sawdust ash reduces the swelling of soil. Similar observations have been made in the previous studies after adding non-plastic additives in the expansive soils [7, 21].

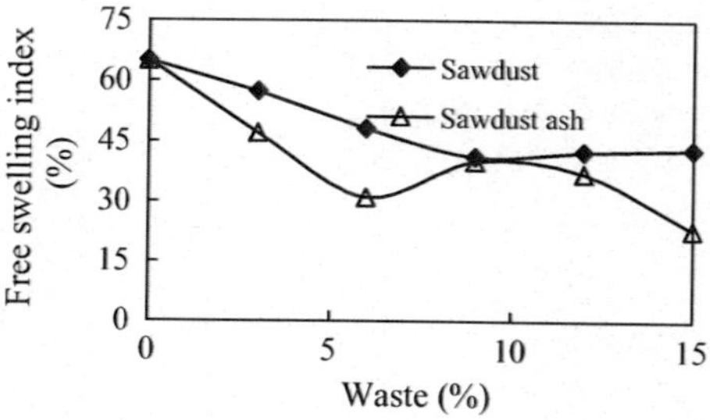

Fig. 34.8 Effect of sawdust on free swelling index

Conclusions

The study explores in a limited way the use of sawdust and sawdust ash for the modification of engineering characteristics (liquid limit, plastic limit, unconfined compressive strength, and swelling potential) of expansive soil (collected from Kheda, Gujrat). Based on the above study, the following conclusions are drawn:

The variation in soil properties is almost identical with the addition of sawdust and sawdust ash. The liquid limit, plastic limit, and unconfined compressive strength increase with sawdust and sawdust ash dust. The increase in soil strength values is about 55% for 9%, corresponding to sawdust and sawdust ash used as additives. However, the plasticity index and free swelling index decrease with the addition of both sawdust and sawdust ash. The optimum content of sawdust and sawdust ash is found to be 9% of soil weight. The addition of sawdust and sawdust ash changes the soil behavior from ductile to brittle. Though the effect of sawdust and sawdust ash is almost identical for a short duration of time, the sawdust is degradable material and, as such, further studies are required to determine the sustainability and degradation characteristics of sawdust and sawdust ash over the long term.

References

1. Shukla RP, Parihar NS (2016) Stabilization of black cotton soil using micro-fine slag. J Inst Eng (India): Ser A 97(3):299–306. https://doi.org/10.1007/s40030-016-0171-1
2. Wong RCK (1998) Swelling and softening behavior of La Biche Shale. Can Geotech J 35(2):206–221. https://doi.org/10.1139/t97-087
3. Shukla RP, Parihar N, Tiwari RP, Agrawal BK (2014) Black cotton soil modification using sea salt. Electron J Geotech Eng 19:8807–8816
4. Agarwal R, Chaudhary M, Singh J (2015) Waste management initiatives in India for human wellbeing. Eur Sci J 105–127
5. Letcher TM, Vallero DA (eds) (2019) Waste: a handbook for management. Academic Press
6. Parihar NS, Shukla RP, Gupta AK (2018) Shear strength of medium plastic expansive soil reinforced with polyester fibers. Slovak J Civ Eng 26(2):1–8. https://doi.org/10.2478/sjce-2018-0007
7. Ikeagwuani CC, Nwonu DC (2019) Emerging trends in expansive soil stabilization; a review. J Rock Mech Geotech Eng 11(2):423–440. https://doi.org/10.1016/j.jrmge.2018.08.013

8. Elinwa AU, Ejeh SP, Akpabio IO (2005) Using metakaoline to improve sawdust-ash concrete. Concr Int 27(11):49–52
9. Elinwa AU, Ejeh SP, Mamuda MA (2008) Assessing of fresh concrete properties of self-compacting concrete containing sawdust ash. Constr Build J 22:1178–1182
10. Mageswari M, Vidivelli B (2009) The use of sawdust ash as fine aggregate replacement in concrete. J Environ Res Dev 3(3):720–726
11. Butt WA, Gupta K, Jha JN (2016) Strength behavior of clayey soil stabilized with saw dust ash. Int J Geo-Engineering 7(1):1–9. https://doi.org/10.1186/s40703-016-0032-9
12. Owamah HI, Atikpo E, Oluwatuyi O, Oluwatomisin AM (2017) Geotechnical properties of clayey soil stabilized with cement-sawdust ash for highway construction. J Appl Sci Environ Manag 21(7):1378–1381
13. Karim H, Al-Recaby M, Nsaif M (2018) Stabilization of soft clayey soils with sawdust ashes. MATEC web of conferences, vol 162. EDP Sciences, p 01006. https://doi.org/10.1051/matecconf/201816201006
14. Indian Standard, IS: 2720 (Part-V) -1985 (2006) Methods of test for soils: Part 5 determination of liquid and plastic limit. Bureau of Indian Standards, India
15. Indian Standard, IS: 2720 (Part-VI)-1985 (2006) Methods of test for soils: Part 6 determination of shrinkage factors. Bureau of Indian Standards, India
16. Indian Standard, IS 1498 (1970) Classification and identification of soils for general engineering purposes. Bureau of Indian Standards, India
17. Standard I (2006) IS 2720 (Part-10)-1973 Methods of test for soils: Part 10—determination of unconfined compressive strength. Bureau of Indian Standards, India
18. Indian Standard, IS: 2720 (Part XL) – 1977 (2002) Methods of test for soils: Part 40 determination of free swell index of soils. Bureau of Indian Standards, India.
19. Sivapullaiah PV, Prashanth JP, Sridharan A (1996) Effect of fly ash on the index properties of black cotton soil. Soils Found 36 (1):97–103. https://doi.org/10.3208/sandf.36.97
20. Shukla RP, Parihar N, Gupta AK (2020) Potassium chloride effect on the efficiency of fine slag used to treat expansive soil. Geo Assets Eng 331(7):87–95. https://doi.org/10.18799/24131830/2020/7/2721 (Bulletin of the Tomsk Polytechnic University)
21. Jain AK, Jha AK (2020) Geotechnical behaviour and micro-analyses of expansive soil amended with marble dust. Soils Found 60(4):737–751. https://doi.org/10.1016/j.sandf.2020.02.013

Chapter 35
Improvisation in the Swelling Behavior of Expansive Soil Using Industrial Waste

P. Devahi, R. Deendayal, and K. Muthukkumaran

Introduction

With the rapid increase in the exploitation of natural resources, there is a huge demand in need of sustainable practices in the construction industry as well as reclamation of contaminated sites [5]. The criteria involved in the effective, sustainable practices are efficient usage of resources, proper measures for pollution reduction, additional incorporation of renewable energy, and enhancement of flora and fauna in the environment [7]. The techniques adopted for implementing such sustainable practices in the construction industry might include prefabricating materials and elements in a controlled environment, construction waste management, improvisation of site environment condition, lean manufacturing concerning less energy consumption, and ultimately choosing the suitable green materials for the conventional construction materials [2].

One of the common issues across the world is soil contamination, which poses a serious threat to human health and the environment [3]. Hence, there is a great need to adopt suitable and sustainable practices in remediating the contaminated sites [8]. In the present work, contaminated soil with heavy metals such as Lead, Zinc, and Iron are studied with respect to their swelling behavior in four different types of expansive soil. Sustainable materials such as industrial waste, i.e., Fly ash and Lime, are mixed for improvising their swelling potential.

P. Devahi (✉) · R. Deendayal · K. Muthukkumaran
National Institute of Technology, Tiruchirappalli, India
e-mail: devahi.civil@gmail.com

K. Muthukkumaran et al. (eds.), *Ground Improvement Techniques*, Lecture Notes in Civil Engineering 297, https://doi.org/10.1007/978-981-19-6727-6_35

Table 35.1 The concentration of heavy metals used

Heavy metal	Industry	Concentration		
		Maximum	Average	Minimum
Pb	Textile	30	15.55	1.1
Zn	Pharmaceutical	50	30.1	10.2
Fe	Paint	120	61.6	3.19

Materials and Methodology

Two categories of soil are selected for this study, namely, expansive soil and virgin soil (i.e., Site Soil), which are designated as ES and VS, respectively. Since the study is to be carried out in extreme conditions, the most swelling type of soil is required, and thus Montmorillonites clay mineral-rich soil Bentonite is chosen, which falls under the category of highly swelling soil. In that commercially available Sodium Bentonite and Calcium Bentonite are chosen and designated as ES1 and ES2, respectively.

Further studies are carried out in two Virgin soils in order to depict the site situation and designated as VS1 and VS2. VS1 is collected from the Sobanapuram located at the outskirts of Trichy city, and VS2 is collected at the National Institute of Technology, Tiruchirappalli.

In order to depict the most contaminated wastewater, industrial wastewater showing the predominant and maximum concentration of heavy metal is chosen. Test analysis is carried out using three concentrations, namely, maximum, average, and minimum. Maximum concentration is fixed from the maximum concentration identified from the untreated industrial wastewater, and similarly, maximum concentration identified from the treated industrial wastewater is fixed as minimum concentration. The average of both maximum and minimum concentration is taken as Average Concentration. The concentration of heavy metal considered for analysis is shown in Table 35.1.

Their three concentrations are taken from the maximum values recorded from Paint, Pharmaceutical, and Textile Industries, respectively. Salts used for making the heavy metals (i.e., Iron, Zinc, and Lead) of required concentration are Iron Sulfate, Zinc Sulfate, and Lead Nitrate, respectively.

Results and Discussion

Expansive Soil Classification

The basic engineering properties of four types of soil are tabulated in Table 35.2. Generally, for clay soils, the specific gravity should lie in the range of 2.4–3.0. Except for VS1, all three soils show the specific gravity of clay. Specific gravity is lesser

Table 35.2 Engineering properties of soil

Property	ES1	ES2	VS1	VS2
Specific gravity	2.63	2.68	2.13	2.43
Liquid limit	260	36	53	35
Plastic limit	71	22	31	24
Plasticity index	189	14	22	11
Optimum moisture content	19	17	21	15
Max dry density	1.40	1.71	1.63	1.82
Free swell index	76.67	0	16.67	10

in VS1 because of the presence of more organic content (peat) in them. Based on the liquid limit and plasticity index, soil expansion classification is done as per IS 1498—1970. ES1 is classified as very high expansive soil, whereas ES2, VS1, and VS2 are classified as low expansive soil.

Analysis with Heavy Metals

Figure 35.1 clearly depicts the variation of the Free Swell Index of four types of soil with all the three Heavy metals. Free Swell Index achieved without any heavy metal is taken as Control Value which was mentioned in the basic engineering Properties (Table 35.1). Overall, ES1 is found to be showing more variation than the Control Value. ES2 showing nil variation, which confirms No Swelling even after Heavy metal Contamination. In VS1, Free Swell Index is found to be increasing only with the maximum concentration of all heavy metals [6]. So, it ensures swelling only with untreated wastewater, and swelling is not induced with treated industrial effluent. VS2 is found to be varying with lead. Thus swelling can be expected to be increasing during Lead contamination.

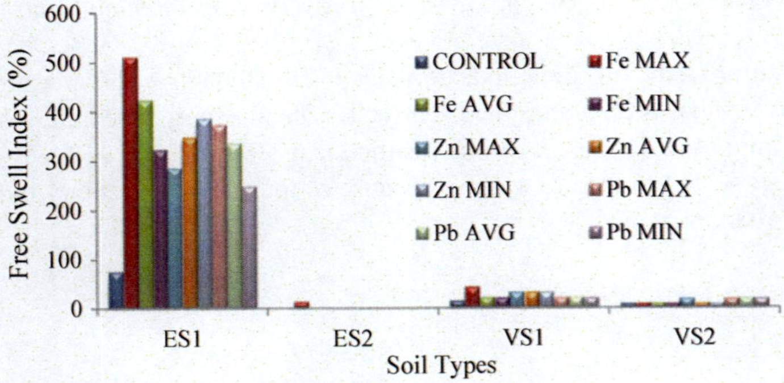

Fig. 35.1 Free swell index with different heavy metal concentration

Table 35.3 Variation of PI for different soil proportion

Soil proportion	LL	PL	PI
5%L + 5% FA	127	85.7	41.3
5%L + 10% FA	115.3	84.9	30.4
5%L + 15% FA	**101.6**	**82.1**	**19.5**
5%L + 20% FA	99.9	75.7	24.2
5%L + 25% FA	95	62.9	32.1

Treatment with Lime and Fly Ash

The susceptibility to swelling is more in ES1 with the presence of heavy metals. Thus further treatment is carried out in that soil alone. The mix proportion adopted for the study is 5% Lime followed by 5, 10, 15, 20, and 25% of Fly ash with respect to their dry weight of soil [4]. The liquid limit and plastic limit are identified in each mixture in expansive soil. Since the plasticity index is directly influencing the swelling behavior, it becomes necessary to study this before identifying their free swell index. Table 35.3 shows the variation of PI for different soil mixtures. Lesser plasticity index at 5% Lime and 15% Fly ash denotes the possibility of minimum swelling than other mixtures.

Swelling Treatment in Lead

After the addition of Lime and Fly ash, FSI with lead is found to be decreasing for the increment of Fly ash till the addition of 15% Fly ash and 5% Lime which is shown in Fig. 35.2. The percentage reduction in the FSI with three concentrations (Max, Avg, and Min) of the lead after adding 5% Lime and 5% Fly ash is 28%, 24%, and 1%, respectively. With the addition of more than 5% Fly ash, percentage reduction is found to be 52%, 45%, and 35%, respectively, with three concentrations. Then, maximum percentage reduction is observed in the 5% Lime and 15% Fly ash Mixture, which is 61%, 61%, and 53%, respectively, with three concentrations of lead.

The variation of the Plasticity index obtained with all the five soil mixtures in the presence of lead has been highlighted. The lesser the plasticity index, the lesser will be the swelling of soil. Thus, it can be justified that 5% Lime and 15% Fly ash soil mixture is subjected to lesser swelling because of their lesser plasticity index than other soil mixtures.

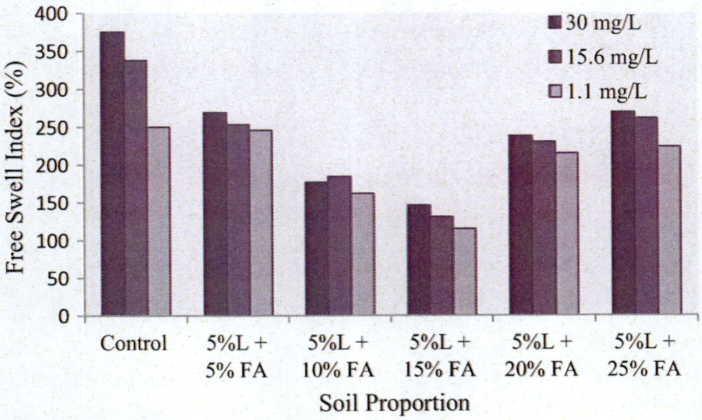

Fig. 35.2 Variation of FSI with three concentrations of the lead after treatment

Swelling Treatment in Zinc

FSI with zinc is found to be decreasing for the increment of Fly ash till the addition of 15% Fly ash and 5% Lime which is shown in Fig. 35.3. The percentage reduction in the FSI with three concentrations (Max, Avg, and Min) of zinc after adding 5% Lime and 5% Fly ash is 6%, 1%, and 2%, respectively. With the addition of more than 5% Fly ash, percentage reduction is found to be 24%, 20%, and 9%, respectively, with three concentrations.

But, the minimum swelling is observed in the 5% Lime and 15% Fly ash Mixture, which is 66%, 64%, and 62%, respectively, with three concentrations of zinc. Table

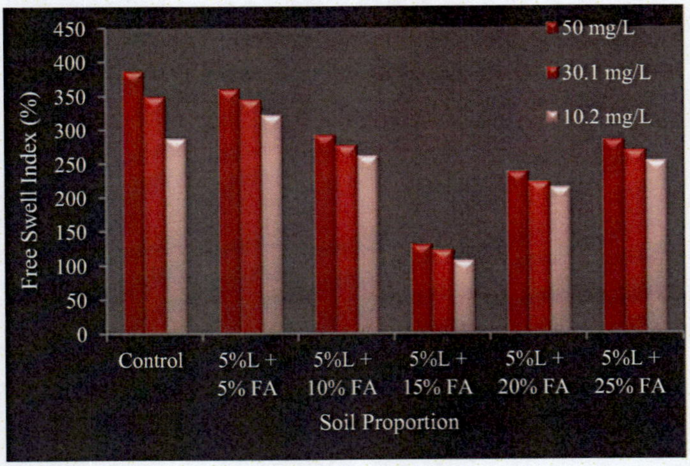

Fig. 35.3 Variation of FSI with three concentrations of zinc after treatment

Table 35.4 Variation of PI for different soil proportion in the presence of zinc

Soil proportion	LL	PL	PI
5%L + 5% FA	122.5	77.9	44.6
5%L + 10% FA	119.4	74.6	44.8
5%L + 15% FA	**108.7**	**73.4**	**35.5**
5%L + 20% FA	112.6	69.8	42.8
5%L + 25% FA	118.7	68.6	50.1

35.4 presents the variation of the Plasticity index obtained with all the five soil mixtures in the presence of zinc.

The lesser the plasticity index, the lesser will be the swelling of soil [1]. Thus, it can be justified that 5% Lime and 15% Fly ash soil mixture is subjected to lesser swelling because of their lesser plasticity index than other soil mixtures.

Swelling Treatment in Iron

After the addition of Lime and Fly ash, FSI with iron is found to be decreasing for the increment of Fly ash till the addition of 15% Fly ash and 5% Lime which is shown in Fig. 35.4. The percentage reduction in the FSI with three concentrations (Max, Avg, and Min) of iron after adding 5% Lime and 5% Fly ash is 25%, 13%, and 8%, respectively. With the addition of more than 5% Fly ash, percentage reduction is found to be 32%, 20%, and 1%, respectively, with three concentrations.

Then, maximum percentage reduction is observed in the 5% Lime and 15% Fly ash mixture, which is 50%, 43%, and 31%, respectively, with three concentrations of iron. Table 35.5 presents the variation of the Plasticity index obtained with all the five soil mixtures in the presence of iron.

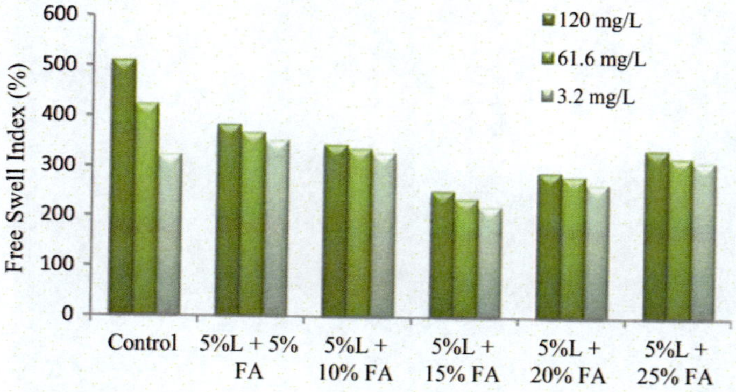

Fig. 35.4 Variation of FSI with three concentrations of iron after treatment

Table 35.5 Variation of PI for different soil proportion in the presence of iron

Soil proportion	LL	PL	PI
5%L + 5% FA	128.3	83.7	44.6
5%L + 10% FA	121.8	78.2	43.6
5%L + 15% FA	**115.4**	**75.4**	**40**
5%L + 20% FA	117.5	74.8	42.7
5%L + 25% FA	119.4	76.4	43.0

The lesser the plasticity index, the lesser will be the swelling of soil. Thus, it can be justified that 5% Lime and 15% Fly ash soil mixture is subjected to lesser swelling because of their lesser plasticity index than other soil mixtures.

Conclusions

The conclusions drawn from the present study are listed below:

1. From the experimental results of the Free Swell Index test, ES1 is found to be highly swelling, and the order of Soil Expansivity is as follows: ES2<VS2<VS1<ES1.
2. By conducting the analysis with the most concentrated heavy metal wastewater till minimum concentrated heavy metal wastewater, changes in their swelling nature are observed. The changes are negligible with minimum concentration and found to be increasing with an average and maximum concentration of heavy metals.
3. In ES1, the increase in the swell index is 5.68, 4.5, and 4% with Cr, Fe, Zn, Cu, and Pb, respectively. In ES2, the increase in the swell index is 14% with Fe and Zn. In VS1, the increase in the swell index is 1 and 0.3% with Zn, and Pb, respectively. In VS2, the increase in the swell index is 1% with Zn, and Pb, respectively.
4. From the experimental results of the Free Swell Index test, it can be concluded that ES1 is susceptible to an increase in swelling unanimously with the three heavy metals, which becomes a negative effect in case of swelling in soil. In other soils, changes in the swell index are minimum or nil.
5. The mixture of Lime and Fly ash is found suitable and sustainable material for the treatment of swelling in expansive soil as well as site soil. The complete swelling was inhibited in the site soil while adding 5% Lime and 5% Fly ash which may be due to the presence of Illite clay mineral.
6. Percentage decrease in the swelling of expansive soil is 50, 66, and 61 with Chromium, Iron, Zinc, and Lead, respectively, while adding 5% Lime and 15% Fly ash which is showing maximum treatment efficiency than other soil mixtures.

7. From the results of the Free Swell Index, it can be suggested that treatment efficiency, i.e., percentage reduction of swelling, is more in the maximum concentration of heavy metals than the average and minimum concentration of Heavy metals. Thus, it can be stated that treatment efficiency increases with an increase in the concentration of heavy metals solutions (i.e., synthetic wastewater).
8. Thus, it can be concluded from the results of the Free Swell Index and Plasticity Index that the optimum percentage of adding Lime and Fly ash for reducing the swelling of expansive soil is identified as 5% and 15%, respectively.

References

1. History P, Balu S, Amman B, Kumar K, Student P (2015) Industrial wastes as additive for stabilization of expansive soils—a review Discovery ANALYSIS The International Daily journal INDUSTRIAL WASTES AS ADDITIVE FOR STABILIZATION OF EXPANSIVE SOILS—A REVIEW
2. Mohanty SK, Pradhan PK, Mohanty CR (2017) Stabilization of expansive soil using industrial wastes. Geomech Eng 12:111–125. https://doi.org/10.12989/GAE.2017.12.1.111
3. Nagaraj HB, Munnas MM, Sridharan A (2010) Swelling behavior of expansive soils. Int J Geotech Eng 4:99–110. https://doi.org/10.3328/IJGE.2010.04.01.99-110
4. R T, Babu S, Naveennayak V, Nirmal R, Lokesh G (2017) A review on stabilization of expansive soil using industrial solid wastes. Engineering 09:1008–1017. https://doi.org/10.4236/eng.2017.912060
5. Soltani A, Deng A, Taheri A, Mirzababaei M, Vanapalli SK (2019) Swell-shrink behavior of rubberized expansive clays during alternate wetting and drying. Minerals 9:1–18. https://doi.org/10.3390/min9040224
6. Soltani A, Deng A, Taheri A, Mirzababaei M, Vanapalli SK (2019) Swell–shrink behavior of rubberized expansive clays during alternate wetting and drying. Miner 9
7. Soundara B, Selvakumar S (2019) Swelling behaviour of expansive soils randomly mixed with recycled geobeads inclusion. SN Appl Sci 1:1–7. https://doi.org/10.1007/s42452-019-1324-4
8. Tiwari N, Satyam N, Sharma M (2021) Micro-mechanical performance evaluation of expansive soil biotreated with indigenous bacteria using MICP method. Sci Rep 11:1–12. https://doi.org/10.1038/s41598-021-89687-2

Chapter 36
Influence of Radial Coefficient of Consolidation on Ground Improvement in Soft Clay with Vertical Drains

C. N. V. Satyanarayana Reddy, G. V. S. S. Sankaranarayana, and R. Sai Chandu

Introduction

Soft clays of marine origin are considered as problematic soils in geotechnical engineering. The problematic behavior of soft clays can be attributed to their affinity to water. Because of higher surface area, soft clays are capable of absorbing high moistures and result in volume changes. The ingress of water into the soil causes further reduction in bearing capacity and gives rise to excessive settlements. Since many years, proper design and construction of structures over soft clays has been an uphill task for civil engineers due to its complex structure resulting from various geological, environmental, and chemical processes [1, 2, 4].

The city of Visakhapatnam, located on the east coast of India, has a long coast line dominated by the presence of soft marine clays extending to depths of 10–20 m below the natural ground surface. Most of the developmental activities and important institutions such as Visakhapatnam Port which is the second largest in the country in terms of the cargos handled are concentrated along the coastline. In order to safeguard these structures and mitigate the detrimental effects, it is often necessary to have a better understanding on the behavior of the soft clays under various environmental conditions [3]. Also, with the scarcity of land and utilization of every bit of available land including sites comprising of soft clays, it is essential to have appreciable knowledge about the ground modification techniques that can be implemented in soft clays so as to make it suitable for constructions.

C. N. V. Satyanarayana Reddy · G. V. S. S. Sankaranarayana
Department of Civil Engineering, Andhra University College of Engineering, Visakhapatnam 530003, India

R. Sai Chandu (✉)
Department of Civil Engineering, ANITS, Visakhapatnam, India
e-mail: ronankisaichandu@gmail.com

K. Muthukkumaran et al. (eds.), *Ground Improvement Techniques*, Lecture Notes in Civil Engineering 297, https://doi.org/10.1007/978-981-19-6727-6_36

For many years, pile foundations have been the solution for constructions over soft clays. However, in recent times, with a lot of research in soft clays and the technological advancements off late, various ground modification techniques were in practice to handle the various problems associated with soft clays. The most commonly used practices include stone columns, preloading with vertical drains such as sand drains, wick drains, and prefabricated vertical drains. However, preloading with PVD's has become the most preferred practice to modify the soft clays due to its ease of installation and increased rate of consolidation [1, 2].

Preloading of soil is merely applying an external loading for a suitable duration to cause desirable changes in the soil. Precompression or surcharging increases the pore water pressure in the soil and initiates consolidation. Upon consolidation, an increase in effective stress takes place in the soil accompanied by surface settlements. However, with preloading alone, the time required for the full consolidation of the soil is higher as only vertical consolidation takes place. This may not be feasible in civil engineering projects with less construction time. The period of consolidation can be greatly reduced to an average time of a very few months, or even a shorter period, by providing vertical drains. Provision of vertical drains in the soil will facilitate radial consolidation along with vertical consolidation of soil. In preloading aided with PVD's, majority of the consolidation is due to the radial consolidation of soil. Hence, coefficient of radial consolidation forms a key design parameter for PVD's. Designers usually adopt the value of coefficient of radial consolidation from the ratio of radial (C_r) to vertical (C_v) coefficients of consolidations. Hence, it is necessary to evaluate the effect of variation in C_r/C_v and spacing of drains on the time rate of consolidation [4].

Literature Review

Barron's theory [9] is widely adopted for the design of vertical drains. One of the input parameter required in Barron's theory is the coefficient of radial consolidation. Several studies were done for determining the coefficient of radial consolidation. Bishop and Henkel [8] were the first to investigate the influence of direction of drainage other than parallel to the axis of the sample, using triaxial cell and filter paper drains. They concluded that the obtained results are not comparable with those obtained from oedometer cells as the filter paper does not form a perfect drain when subjected to the cell pressure.

Rowe and Barden [6] developed a hydraulic consolidation cell known as the Rowe cell at Manchester University. In Rowe cell, the test specimen is loaded hydraulically by water pressure acting on a flexible diaphragm instead of a mechanical lever system. This arrangement allows specimens of larger diameters to be tested and allows for larger settlements. The important features of this cell are its ability to control drainage and to measure the pore water pressure during the course of consolidation tests.

For determining the coefficient of radial consolidation, McKinlay [7] modified the standard oedometer by substituting a porous ring instead of the standard solid ring so that the drainage can occur in horizontal direction toward the periphery direction. However, they reported that the drawback of this method is the friction developed at the interface of soil, and porous ring will affect the coefficient of radial consolidation.

Design of Ground Improvement System with PVDs

The design of ground improvement system with PVDs and preload is done based on the properties of marine clay collected from stacking yard R-10 of Visakhapatnam Port at 3 m depth where the thickness of soft clay layer is 12.5 m below the natural ground surface. The properties of clay determined from the laboratory tests are presented in Table 36.1.

Based on the properties of soft clay, the design of the PVD system is done as per IS:15284 Part-2 [5] considering arrangement of drains in triangular and square pattern. The strip drain of width 100 and 8 mm thickness is considered in design. The equivalent diameter of sand drain for the strip drain is determined as 68.75 mm based on equal circumference. The overall degree of consolidation achieved for three different spacings, i.e., 1.2 m, 1.0 m, and 0.8 m for time periods of 60, 75, and 90 days. The summary of PVD design for strengthening of soft clay for PVDs installed in triangular and square pattern for varying values of C_h/C_v for different spacings at 75 days time period is presented in Tables 36.2 and 36.3.

Selection of Height for stage-wise construction of embankment:

The intensity of preload in each stage shall be limited to the net ultimate bearing capacity of the soil to avoid shear failure in soft clays. The net ultimate bearing capacity of the soft clay is 52 kN/m^2 under local shear failure conditions as per IS 6403. Depending on the target safe bearing capacity of strengthened soft clay, the height of moorum stack (preload) and stages of loading shall be determined. Each

Table 36.1 Engineering properties of marine clay

Property	Value
Specific gravity (G)	2.62
Particle size distribution (a) Gravel (%) (b) Sand (%) (c) Fines (%)	0 15 85
Plasticity characteristics (a) Liquid limit (%) (b) Plastic limit (%) (c) Plasticity index I_p	75 34 41
IS classification	CH
Natural water content (%)	66.8
Vertical coefficient of consolidation (m^2/s)	0.8
Undrained cohesion, C_u (kN/m^2)	18

Table 36.2 Summary of PVD design for triangular pattern for time period $t = 75$ days

S. No.	C_h/C_v	C_v (m^2/s)	C_r (m^2/s)	S (m)	R (m)	r_w	N	F_n	T_r	U_r (%)	T_v	U_v (%)	U (%)
1	1.5	0.8	1.2	1.2	0.63	0.0344	18.31	2.158	0.157	44	0.00425	7.35	48.1
				1.0	0.525		15.11	1.976	0.226	60			63
				0.8	0.42		12.21	1.767	0.353	79			81
2	2.0	0.8	1.6	1.2	0.63	0.0344	18.31	2.158	0.209	53.8	0.00425	7.35	57
				1.0	0.525		15.11	1.976	0.301	70			72
				0.8	0.42		12.21	1.767	0.471	88			89
3	2.5	0.8	2.0	1.2	0.63	0.0344	18.31	2.158	0.262	62	0.00425	7.35	64
				1.0	0.525		15.11	1.976	0.377	78			79.5
				0.8	0.42		12.21	1.767	0.589	93			93.5
4	3.0	0.8	2.4	1.2	0.63	0.0344	18.31	2.158	0.314	68	0.00425	7.35	71
				1.0	0.525		15.11	1.976	0.452	83			85
				0.8	0.42		12.21	1.767	0.707	95.9			96.2
5	3.5	0.8	2.8	1.2	0.63	0.0344	18.31	2.158	0.366	74	0.00425	7.35	76
				1.0	0.525		15.11	1.976	0.528	88			89
				0.8	0.42		12.21	1.767	0.825	97			98
6	4.0	0.8	3.2	1.2	0.63	0.0344	18.31	2.158	0.419	78	0.00425	7.35	80
				1.0	0.525		15.11	1.976	0.6	91			92
				0.8	0.42		12.21	1.767	0.94	98.5			98.6

Table 36.3 Summary of PVD design for square pattern for time period $t = 75$ days

S. No.	C_h/C_v	C_v (m^2/s)	C_r (m^2/s)	S (m)	R (m)	r_w	N	F_n	T_r	U_r (%)	T_v	U_v (%)	U (%)
1	1.5	0.8	1.2	1.2	0.676	0.0344	19.65	2.24	0.13	37	0.00425	7.35	41
				1.0	0.564		16.4	2.05	0.196	53			56
				0.8	0.451		13.11	1.84	0.30	72.7			74.7
2	2.0	0.8	1.6	1.2	0.676	0.0344	19.65	2.24	0.182	48	0.00425	7.35	51.8
				1.0	0.564		16.4	2.05	0.261	64			66.6
				0.8	0.451		13.11	1.84	0.41	83			84
3	2.5	0.8	2.0	1.2	0.676	0.0344	19.65	2.24	0.227	56	0.00425	7.35	59.2
				1.0	0.564		16.4	2.05	0.32	71			73
				0.8	0.451		13.11	1.84	0.51	89			89.8
4	3.0	0.8	2.4	1.2	0.676	0.0344	19.65	2.24	0.27	62	0.00425	7.35	64.5
				1.0	0.564		16.4	2.05	0.39	78			79.6
				0.8	0.451		13.11	1.84	0.61	93			93.5
5	3.5	0.8	2.8	1.2	0.676	0.0344	19.65	2.24	0.318	67.7	0.00425	7.35	70
				1.0	0.564		16.4	2.05	0.45	82.6			83.8
				0.8	0.451		13.11	1.84	0.71	95			95.3
6	4.0	0.8	3.2	1.2	0.676	0.0344	19.65	2.24	0.36	72	0.00425	7.35	74
				1.0	0.564		16.4	2.05	0.52	87			87.7
				0.8	0.451		13.11	1.84	0.81	97			97.2

C_v Coefficient of consolidation for vertical flow, C_r Radial coefficient of consolidation, *S* Spacing adopted between the vertical drains, *R* Radius of cylindrical block of soil, r_w Effective radius of drain, $n = \frac{R}{r_w}$, $F_n = \frac{n^2}{n^2-1}\log_e n - \frac{3n^2-1}{4n^2}$, T_r Time factor for radial flow, U_r Degree of consolidation for radial flow, T_v Time factor for vertical flow, U_v Degree of consolidation for vertical flow, *U* Degree of consolidation for three dimensional flow.

stage of loading shall be maintained till a minimum degree of consolidation of 90% is achieved.

The summary of PVD design at time periods of 60 and 90 days is also performed in a similar manner to 75 days time period, and the results are presented in Figs. 36.1, 36.2, 36.3, 36.4, 36.5 and 36.6.

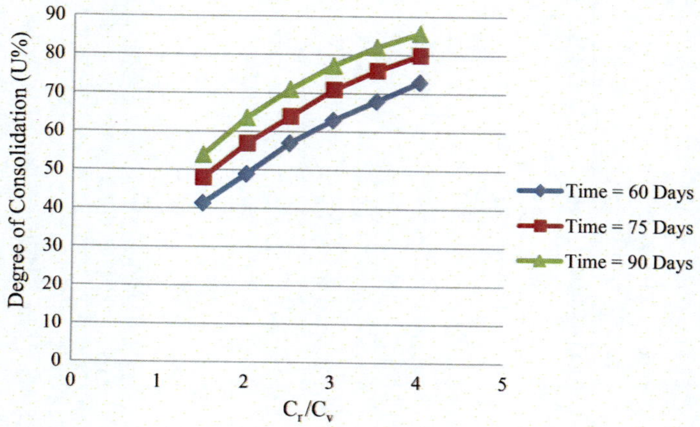

Fig. 36.1 Variation of overall degree of consolidation with ratio of radial to vertical consolidation for triangular pattern of PVDs at spacing of 1.2 m

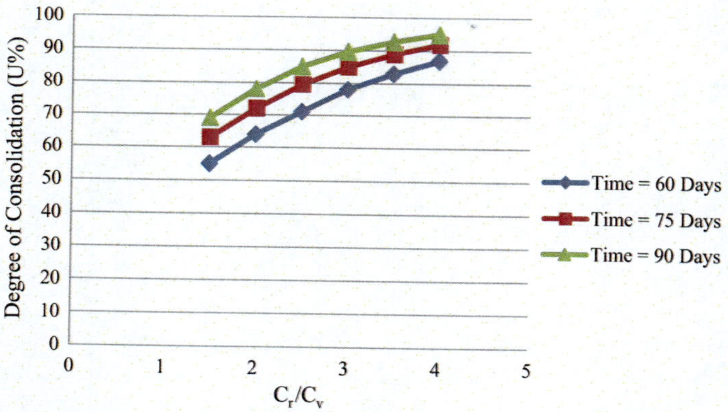

Fig. 36.2 Variation of overall degree of consolidation with ratio of radial to vertical consolidation for triangular pattern of PVDs at spacing of 1.0 m

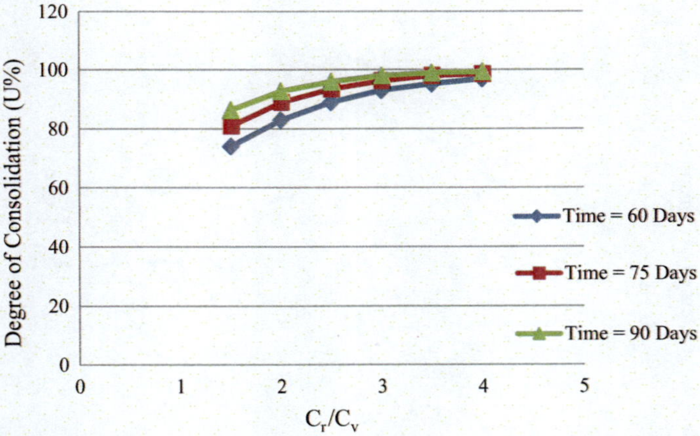

Fig. 36.3 Variation of overall degree of consolidation with ratio of radial to vertical consolidation for triangular pattern of PVDs at spacing of 0.8 m

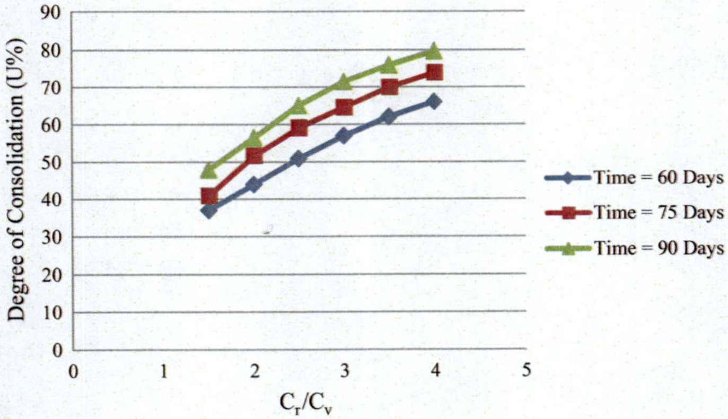

Fig. 36.4 Variation of overall degree of consolidation with ratio of radial to vertical consolidation for square pattern of PVDs at spacing of 1.2 m

Results and Discussion

The overall degree of consolidation of soft clay under study increased with increase in the time of consolidation for a given spacing and C_r/C_v ratio. The percentage increase in the overall degree of consolidation is more pronounced in case of triangular arrangement of drains compared to square pattern. The rate of consolidation is found to be effective at smaller spacings of the PVDs irrespective of the pattern of drains. The rate of vertical consolidation is insignificant on overall consolidation at smaller spacings whereas the rate of vertical consolidation is about 10–20% of overall consolidation at larger spacing (1.2 m).

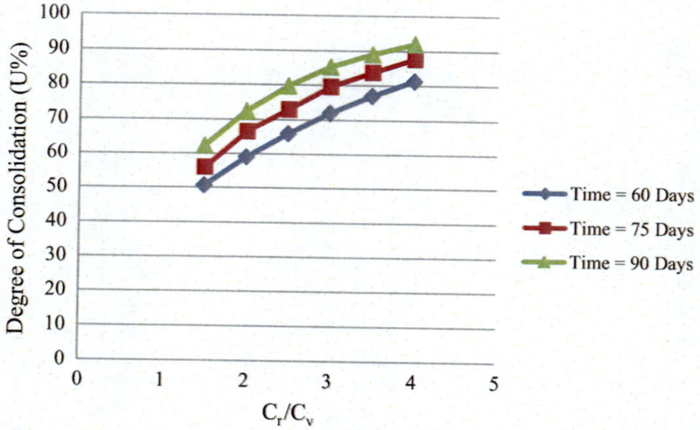

Fig. 36.5 Variation of overall degree of consolidation with ratio of radial to vertical consolidation for square pattern of PVDs at spacing of 1.0 m

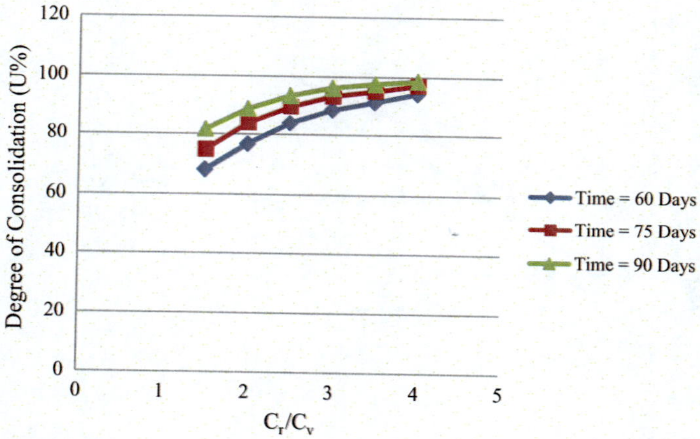

Fig. 36.6 Variation of overall degree of consolidation with ratio of radial to vertical consolidation for square pattern of PVDs at spacing of 0.8 m

Conclusions

From the results of the study, the following conclusions are drawn.

1. The overall degree of consolidation (U) is not much affected for c_r/c_v values of 2.5–4.0 at closer spacing of drains (0.8 m) as the variation in values of U is about 5%. Further, in triangular pattern arrangement of drains, the variation is relatively smaller compared to square pattern arrangement of drains.
2. The value of U_v becomes less significant with decrease in spacing of drains and increase in time of stage consolidation.

3. At larger spacing (1.2 m), the values of U_v are significant as they are about 10–15% of overall consolidation for consolidation periods of 60–90 days.
4. Ground improvement in soft clay with triangular pattern of drains is more effective than square pattern of drains.

References

1. Sarika S, Deepankar C, Salunkhe JS (2015) Ground improvement of marine clay for highway construction at Mumbai, India. IFCEE ASCE 2015
2. Kumar Pichumani N, Aminul I (2021) Reclamation and ground improvement of soft marine clay for development of offshore terminal 4, JNPT, Navi Mumbai. Indian Geotech J-021-00521-y
3. Satyanarayana Reddy CNV, Satyanarayana PVV, Satyanarayana B (2008) Heaving of ground—a case study in marine clay. IGC-2008, Bangalore
4. Chu J, Bo MW, Chang MF, Choa V (2002) Consolidation and permeability properties of Singapore marine clay. J Geotech Geoenvironmental Eng 2002(128):724–732
5. IS:15284-Part 2: Preconsolidation using vertical drains (2004) Guidelines for design of vertical drains
6. Rowe PW, Barden L A new consolidation cell. J Gèotechnique 6(2):162–170
7. McKinlay DG A laboratory study of rates of consolidation in clays with particular reference to conditions of radial pore water drainage. In: Proceedings of 5th international conference on soil mechanics and foundation engineering, Paris, vol 1, pp 225–228
8. Bishop AW, Henkel DJ (1957) The measurement of soil properties in the triaxial test, London: E. Arnold
9. Barron RA (1948) Consolidation of fine grained soils by drain wells. Trans ASCE 113:718–724